Applied Behavior Analysis for Teachers

Second Edition

Paul A. Alberto
Georgia State University

Anne C. Troutman
Memphis State University

Merrill Publishing Company
A Bell & Howell Company
Columbus   Toronto   London   Sydney

Cover Art: © Ben Mahan

Published by Merrill Publishing Company
A Bell & Howell Company
Columbus, Ohio 43216

This book was set in Garamond Light and Helvetica

Administrative Editor: Vicki Knight
Production Editor: Gnomi Schrift Gouldin
Cover Designer: Cathy Watterson

Library of Congress Catalog Card Number: 86-060057
International Standard Book Number: 0-675-20514-X
Printed in the United States of America
1 2 3 4 5 6 7 8 9 — 91 90 89 88 87 86

Contents

Chapter 5
Single-subject Designs

Section 3   *Applying Learning Principles*

Chapter 6
Arranging Consequences that Increase Behavior

Chapter 7
Arranging Consequences that Decrease Behavior

Preface

Why did we prepare an applied behavior analysis text? Because we needed a technically sound, systematically organized, and entertaining book for our own students. We want students to understand concepts of applied behavior analysis, but since this is a skills course we also want you to know how to apply them in the classroom. Applied behavior analysis can make a difference; it can be used to teach Michael to stay in his seat or Liz to work math problems more accurately. In this book we provide examples showing how the principles can be used to teach academic subjects as well as to manage students' social behavior. Applied behavior analysis is an overall management system, not a collection of tricks for keeping students under control.

It is *not* our intent to provide a cookbook. We want you to be able to use the principles to create your own recipes for success. Because we believe so strongly that applied behavior analysis is the most powerful teaching system available, we stress learning to use it appropriately and ethically. In our examples we point out its hazards as well as its strengths. Successful application of the principles requires the full and active participation of a creative teacher. We've included everything we can think of to help you become such a creative teacher—guidelines for the use of procedures like time-out, for example, as well as practical items likes data recording sheets and lots of charts.

Instructors will be interested to know that our book is as technically accurate and as well documented as we could make it. At the same time, we've tried to enliven the text with examples students will enjoy. Our examples describe students from preschool through adolescence functioning at various levels of ability. We describe good teachers and poor teachers; and, after reading his reviews and fan mail, Professor Grundy agreed to an encore in this edition.

In preparing this second edition, we took to heart suggestions from colleagues and thoughtful letters from students using the book. In addition to updating the text overall, we expanded and rewrote Chapters 1, 2, 6, 7, and 8. Hopefully, this effort has added to their clarity as well as content. Chapter 5, "Single-subject Designs," has been reorganized and increased significantly, especially in regard to interpreting data on student performance. We hope these changes will enhance the student's understanding of the principles and techniques of applied behavior analysis and assist the instructor's successful teaching.

Finally, the book remains organized to allow instructors to assign their students a behavior-change project concurrently with lectures and readings. The text progresses from identifying a target behavior to collecting data, selecting an experimental design, arranging consequences, arranging antecedents, and generalizing the behavior change. Chapters 10 and 11 provide suggestions for teaching behavioral procedures to others—parents, paraprofessionals, and even children. We've tried to provide students with the basics of a teaching technology that will serve as a solid foundation for subsequent methods courses.

We welcome your continued personal response to our book. We hope you enjoy reading it as much as we enjoyed writing it.

Acknowledgements

We would like to thank all the people who helped us in the process of producing *Applied Behavior Analysis for Teachers,* and now this second edition, especially the Special Education Editors at Merrill, Marianne Taflinger and Vicki Knight. In spite of her obsessive fealty to deadlines, Vicki managed to keep the proportion of positive reinforcement to punishment fairly high.

We would like to thank again our reviewers of the first edition, especially Sandy Bailey and Colleen Blankenship. We now add to this list the people who extended their time and knowledge to the reviews for the second edition: George Brabner, University of Delaware; Fred Spooner, University of North Carolina, Charlotte; Eric Jones and Jim Krouse, Bowling Green State University; David Gast, University of Kentucky; Marilyn Rouseau, George Peabody/Vanderbilt University; and our special thanks to David Test, University of North Carolina, Charlotte. Their help in reorganizing and expanding this edition was invaluable. We would also like to extend a special acknowledgement to Anita Briggs for relieving us of the task of redoing the instructor's manual.

Nancy Wilder was compelled to type only one version of every chapter for this edition thanks to the efforts of Steven Job. She also edited and verified references, indexed, and was generally indispensable. Finally, special thanks to Bas and Virginia for growing up between editions. Thank you all.

Section One

Behavior Modification in Perspective

1 Roots of Applied Behavior Analysis

Did you know that . . .

- There may be some validity in your mother's claim that "You're just like your father?"
- Chemicals in your brain may affect your behavior?
- Apes could have insight?
- Pretzels preceded M & M's as rewards for good behavior?
- Benjamin Franklin knew about applied behavior analysis?

Why do people behave as they do? Why do some people behave in socially approved ways and others in a manner condemned or despised by society? What can be done to predict what people are likely to do? What can be done to change behavior that is harmful to an individual or destructive to society?

In an effort to answer questions like these, human beings have offered explanations ranging from possession by demons to abnormal quantities of chemicals in the brain. Suggested answers have been debated, written about, attacked, and defended for centuries and continue to be offered today. There are good reasons for continuing to investigate human behavior. Information about the development of certain behaviors in human beings may guide parents and teachers to optimum conditions for child-rearing or for teaching. If we know how people are likely to behave under certain conditions, we can decide whether to provide or avoid such conditions.

Those of us who are teachers are particularly concerned with changing behavior; that is, in fact, our job. We want to teach our students to do some things and to stop doing others.

In order to understand, predict, and change human behavior, we must first understand how human behavior works. In short, we must answer as completely as possible the why questions asked above. Therefore, Alexander Pope's dictum that "The proper study of mankind is man" (perhaps rephrased to "The proper study of humanity is people") needs no other revision; it is as true in the 20th century as it was in the 18th century.

This chapter discusses the requirements for meaningful and useful explanations of human behavior. It then describes several interpretations of human behavior that have influenced large numbers of practitioners, including teachers. The discussion concludes by tracing the historical development of a way of understanding and predicting human behavior called *applied behavior analysis.*

THE USEFULNESS OF EXPLANATIONS

A useful theory has inclusiveness, verifiability, predictive utility, parsimony.

If a way of explaining behavior is to be useful for the practitioner, it must meet several requirements. It should first be *inclusive.* It must account for a substantial quantity of behavior. An explanation has limited use if it fails to account for the bulk of human behavior and thus makes prediction and systematic change of behavior impossible. An explanation must also be *verifiable*; that is, we should be able to test in some way that it does account for behavior. The explanation should have *predictive utility.* It should provide reliable answers about what people are likely to do under certain circumstances, thereby giving the practitioner the opportunity to change behavior by changing conditions. Finally, it should be *parsimonious.* A parsimonious explanation is the simplest one that will account for observed phenomena. Parsimony does not guarantee correctness (Mahoney, 1974)—since the simplest explanation may not always be the correct one —but it prevents our being so imaginative as to lose touch with the reality of observed data. When the bathroom light fails to operate at 3:00 A.M., you should always check the bulb before calling the electric company to report a blackout. There may be a blackout, but the parsimonious explanation is a burned-out bulb.

In examining some of the theories developed to explain human behavior, we shall evaluate each explanation for its inclusiveness, verifiability, predictive utility, and parsimony.

BIOPHYSICAL EXPLANATIONS

Since physicians of ancient Greece first proposed that human behavior was the result of interactions between four bodily fluids or "humors"—blood, phlegm, yellow bile (choler), and black bile (melancholy)—theorists have

Some theorists state that human behavior is controlled by physical influences.

searched for explanations for human behavior within the physical structure of the body. Such theories have included those based on genetic or hereditary factors, those that emphasize biochemical influences, and those that suggest that aberrant behavior is caused by some damage to the brain. The following anecdote offers an explanation for behavior that indicates a belief in hereditary influences on behavior.

Professor Grundy Traces the Cause

Having observed an undergraduate student's behavior for some time, Professor Grundy noticed that the student was consistently late for class (when he came at all), invariably unprepared, and frequently inattentive. Since Grundy was certain that his dynamic, meaningful lectures were not related to this behavior, he decided to investigate the matter. He paid a visit to the high school attended by the student and located his 10th grade English teacher, Ms. Marner. "Yes, DeWayne was just like that in high school," said Ms. Marner. "He just didn't get a good background in junior high."

Professor Grundy then went to visit the junior high school. "You know," said the guidance counselor, "a lot of our kids are like that. They just don't get the foundation in elementary school." At the elementary school, Professor Grundy talked to the principal. "DeWayne was like that from day one. His home situation was far from ideal. If we don't have support from the home, it's hard to make much progress."

Professor Grundy, sure that he would at last find the answer, went to talk to DeWayne's mother. "I'll tell you," said the mother, "he takes after his father's side of the family. They're all *just like that.*"

Genetic and Hereditary Effects

DeWayne's mother explained his inappropriate behavior by referring to hereditary influences. Could she have been right? The effects of heredity on human behavior, both normal and atypical, have been investigated extensively. There is little question that mental retardation (which results in significant deficits in a wide range of behaviors) is sometimes associated with chromosomal abnormalities or with the inheritance of recessive genes (Achenbach, 1974). There is some evidence that other behavioral characteristics have some hereditary basis as well. While few can be as explicitly identified as the syndromes resulting in some mental retardation, we can learn about hereditary influences through other types of investigation.

Researchers studying heredity often compare the characteristics of sets of twins.

Most studies investigating the hereditary components of behavior have been carried out by comparing identical twins with fraternal twins. Identical twins have exactly the same genetic inheritance; fraternal twins are no more alike genetically than any other siblings. Thus, if behavioral characteristics occur more consistently in both of a number of pairs of identical twins than in pairs of fraternal twins, we can assume that the behavioral characteristic results from some inherited factor. Serious behavior disorders, such as that labeled *schizophrenia,* occur more frequently in both members of a pair of identical twins than in both members of a pair of fraternal twins (Kallmann & Roth, 1956). Some less dramatic behavioral characteristics, such as those labeled *social introversion* or *extroversion, depression* (Eysenck, 1956; Vandenburg, 1967), and *reading difficulty* (Bannatyne, 1971; Hallgren, 1950), apparently also have some hereditary component.

Inheritance appears, in addition, to affect some behavioral characteristics that are not necessarily labeled *deviant* or *atypical.* Thomas, Chess, and Birch (1968) conducted a study of 136 children whose development has been closely monitored for a number of years. Each child was observed shortly after birth and then frequently throughout childhood. Interviews were held with parents and teachers. The authors identified nine categories of behavior that they labeled *temperament.* Each characteristic could be reliably observed and was consistent throughout childhood. The categories of temperament included

1. Activity level—the degree to which the child moves around
2. Rhythmicity—the regularity of the child with regard to sleep patterns, eating, and bowel and bladder functions
3. Approach or withdrawal—the reaction of the child to any new stimulus including people, food, places
4. Adaptability—the child's ability to adapt to a new situation even if the initial response was withdrawal
5. Intensity of reaction—the child's tendency to scream or whimper when hungry, shriek with laughter or smile when amused
6. Threshold of responsiveness—the degree of sensitivity to environmental stimuli and to changes in the environment
7. Quality of mood—the child's overall disposition, whether pleasant and friendly or grumpy and unfriendly
8. Distractibility—the ease with which the child can be interrupted in an ongoing activity
9. Attention span and persistence—the length of time a child pursues an activity or persists despite interruptions

Since a number of sets of twins was included in the sample, it was possible to determine by the method described earlier that characteristics 1–7 appear to have some hereditary component. Too little data were available for characteristics 8 and 9. However, only activity level and approach-withdrawal remained more similar for identical twins than for fraternal twins after the first year (Rutter, Korn, & Birch, 1963). The fact that all the characteristics listed show such consistency over time indicates that they have some constitutional, if not genetic, basis.

When DeWayne's mother explained her son's behavior to Professor Grundy, her claim that DeWayne takes after his father's family may have involved a degree of truth. It is possible that certain genetic characteristics may increase the probability of certain behavioral characteristics.

Biochemical Explanations

Some researchers have suggested that certain behaviors may result from excesses or deficiencies of various substances found in the body. These chemical substances are labeled differently from those hypothesized by the ancient Greeks. Some theorists who propose a biochemical basis for behavior hold that brain injury, which we will discuss later, contributes to the biochemical dysfunction (Rimland, 1964).

Some handicapped children have biochemical abnormalities.

Biochemical abnormalities have been found in some children with serious disturbances of behavior labeled *autism* or *childhood psychosis* (Boullin, Coleman, O'Brien, & Rimland, 1971). However, investigation of such factors has established only that biochemical abnormalities exist, not that they cause the disorder.

Other disturbances of behavior characterized as hyperactivity, learning disability, or mental retardation have been linked to biophysical factors such as hypoglycemia (Cott, 1972; Wunderlich, 1977), malnutrition (Cott, 1972; Cravioto & DeLicardie, 1975; Morgane, Stern, & Resnick, 1974; Vitale & Velez, 1968; Wunderlich, 1977), and allergic reactions (Feingold, 1975; Wunderlich, 1977). Feingold's is the best known of such theories. His theory suggests that food containing natural salicylates, artificial flavorings and colorings, and certain antioxidant preservatives contribute to hyperactivity in children. Although Feingold (1975) attributes dramatic effects to removal of such additives from the diet of hyperactive children, most studies that report such effects have been poorly controlled and have relied heavily on such questionable data as parent reports. Some controlled studies using unbiased observers, where neither parents, children, nor teachers were aware of the presence or absence of additive substances, indicate a tentative relationship between hyperactivity and certain food (Conners, 1975; Rose, 1978). A recent review of research (Mattes, 1983) reports that there is a lack of evidence of any dietary effect on hyperactivity. Another biophysical abnormality is used to explain behavior in Professor Grundy's next adventure.

Studies relating hyperactivity to food additives have been inconclusive.

Professor Grundy Learns to Think in Circles

Professor Grundy, as part of his instructional duties, visited student teachers. On his first trip to evaluate Ms. Harper, in a primary resource room, he observed that one student, Ralph, wandered continuously about the room. Curious about such behavior, since the other students remained seated, Professor Grundy inquired, "Why is Ralph wandering around the room? Why doesn't he sit down like the others?" Ms. Harper was aghast at such ignorance on the part of a professor.

"Why, Ralph is hyperactive, Professor Grundy. That's why he never stays in his seat."

"Ah," replied the professor. "That's very interesting. How do you know he's hyperactive?"

With barely concealed disdain, Ms. Harper hissed, "Professor, I know he's hyperactive because *he won't stay in his seat.*"

After observing the class for a few more minutes, during which he noticed Ms. Harper and the supervising teacher whispering and casting glances in his direction, Professor Grundy once again attracted Ms. Harper's attention. "What," he inquired politely, "causes Ralph's hyperactivity?"

The disdain was no longer concealed. "Professor," answered Ms. Harper, "hyperactivity is caused by brain damage."

"Indeed," responded the professor, "and you know he's brain damaged because . . ."

"Of course I know he's brain-damaged, Professor. He's hyperactive, isn't he?"

Brain Damage

The type of circular reasoning illustrated by Ms. Harper is, unfortunately, not uncommon. Many professionals explain a great deal of students' inappropriate behavior in similar ways. The notion that certain kinds of behavior result from brain damage has its roots in the work of Goldstein (1939), who studied soldiers returning from World War I having suffered head injuries. He identified certain behavioral characteristics, including distractibility, perceptual confusion, and hyperactivity. Observing similar characteristics in some retarded children, some professionals concluded that the children must also be brain-injured (Strauss & Werner, 1942; Werner & Strauss, 1940) and that the brain injury was the cause of the behavior. This led to the identification of a hyperkinetic behavior syndrome (Strauss & Lehtinen, 1947), assumed to be the result of brain injury. This syndrome included such characteristics as hyperactivity, distractibility, impulsivity, short attention span, emotional lability (changeability), perceptual problems, and clumsiness. Studying a group of children known to be brain-injured because of cerebral palsy, Cruickshank, Bice, and Wallen (1957) reported the existence of similar behavioral characteristics in children with normal intelligence. Subsequently, the term *minimal brain dysfunction* was used to describe a disorder assumed to exist in children who, although they had no history of brain injury, behaved similarly.

Hyperactivity is not necessarily caused by brain damage.

There is, however, little empirical support for using the theory of brain injury to account for problem behavior in all children who show such behavioral characteristics. Even when brain damage can be unequivocally shown to exist, there is no proof that it causes any particular behavior (Pond, 1961). And, finally, "there is good evidence to show that when psychopathology is associated with brain damage it can take any form, the *least* common of which is the hyperkinetic syndrome" (Werry, 1972, p. 97).

The Usefulness of Biophysical Explanations

The search for explanations of human behavior based on physiological factors has important implications. As a result of such research, the technology for preventing or lessening some serious problems has been developed. The best-known example of such technology is perhaps the routine testing of all infants for phenylketonuria (PKU), a hereditary disorder of metabolism. Placing infants with PKU on special diets can prevent the mental retardation formerly associated with this disorder (Berry, 1969). It is possible that future research may explain a good deal more human behavior on a biological or hereditary basis. Currently, however, only a small part of the vast quantity of human behavior can be explained in this way, so that such biophysical explanations fail to meet our criterion of inclusiveness.

Some biophysical explanations are testable, meeting the second of our four requirements for usefulness. For example, scientists can definitely establish the existence of Down Syndrome by observing chromosomes. Some metabolic or biochemical disorders can also be scientifically verified. However, verification of such presumed causes of behavior as "minimal brain dysfunction" is not dependable (Werry, 1972).

Even with testable evidence of the existence of some physiological disorder, it does not follow that any specific behavior is automatically a result of the disorder. For the teacher, explanations based on presumed physiological disorders have very little predictive utility. To say that Sammy can't walk, talk, or feed himself because he is retarded as a result of a chromosomal disorder tells us nothing about the conditions under which Sammy might learn to perform these behaviors. For Ms. Harper to explain Ralph's failure to sit down on the basis of hyperactivity caused by brain damage does not provide any useful information about what might help Ralph learn to stay in his seat. Even apparently constitutional differences in temperament are so vulnerable to environmental influences (Thomas & Chess, 1977) as to provide only limited information about how a child is apt to behave under given conditions.

The final criterion, parsimony, is also frequently violated when physical causes are postulated for student behaviors. Searching for such causes often distracts teachers from simpler, more immediate factors that may be controlling behaviors in the classroom. Perhaps the greatest danger of such explanations is that some teachers may use them as excuses not to teach: Sammy can't feed himself because he's retarded, not because I haven't taught him. Ralph won't sit down because he's brain-damaged, not because I have poor classroom management skills. Irving can't read because he has dyslexia, not because I haven't figured out a way to teach him. Biophysical explanations may also cause teachers to have low expectations for some students. When this happens, teachers may not even try to teach things students are capable of learning.

The Usefulness of Biophysical Theory	Good	Fair	Poor
Inclusiveness			√
Verifiability		√	
Predictive Utility			√
Parsimony			√

DEVELOPMENTAL EXPLANATIONS

Observation of human beings confirms that many predictable patterns of development occur. Physical growth proceeds in a fairly consistent manner. Most children start walking and talking, along with some social behaviors such as smiling, in fairly predictable sequences and at generally predictable chronological ages (Gesell & Ilg, 1943). Some theorists have attempted to explain many aspects of human behavior—cognitive, social, emotional, and moral—on the basis of fixed, innate developmental sequences. Their proposed explanations are meant to account for normal as well as "deviant" (other than the accepted or usual) human behavior. The following sections review three of the numerous developmental theories and examine their usefulness in terms of inclusiveness, verifiability, predictive utility, and parsimony.

A Freudian by the Garbage Can

Upon returning to the university after observing student teachers, Professor Grundy prepared to return to work on his manuscript, now at least seven months behind schedule. To his horror, his carefully organized notes, drafts, and revisions were no longer "arranged" on the floor of his office. Professor Grundy ran frantically down the hall loudly berating the custodial worker who had taken advantage of his absence to remove what she considered "that trash" from the floor so that she could vacuum.

As Grundy pawed through the outside garbage can, a colleague offered sympathy. "That's what happens when an anal-expulsive personality conflicts

"Well, well, Professor Grundy, did you lose something or are you just doing 're-search' on the things you professors throw away?"

with an anal-retentive." Grundy's regrettably loud and obscene response to this observation drew the additional comment, "Definite signs of regression to the oral-aggressive stage, there, Grundy."

Psychoanalytic Theory

The id says "I wanna"; the superego says "I oughta." The ego is the umpire.

Although many different explanations of human behavior have been described as psychoanalytic, all have their roots in the theories of Sigmund Freud (Fine, 1973), who described human behavior in an essentially developmental manner (Kessler, 1966). Freud's assertion that normal and aberrant human behavior may be understood and explained on the basis of progression through certain crucial stages (Hall, 1954) is perhaps the most commonly accepted and most widely disseminated of his theories.

Freud postulated the existence of certain biologically based drives, specifically a sex drive and an aggressive drive. These drives provide a source of energy referred to by Freud as the *id,* the part of human personality that seeks gratification of desires without reference to any external controls. A second part of personality, the *ego,* develops as an infant learns to relate to its environment. The ego includes such processes as motility, memory, judgment, reasoning, language, and thought (Kessler, 1966). A third component of personality, the *superego,* develops as a function of parental training. The superego includes conscience, morals, ethics, and aspirations. Anxiety-arousing conflict results when the drives of the id encounter the standards of the superego. Freud said that such conflicts are mediated by the ego.

The hypothesized energy of the id centers in different parts of the body during the various stages suggested by Freud. Freud's stage theory was subsequently elaborated by Abraham, whose modifications were incorporated into the theory by Freud (Hall, 1954). The stages are

1. *Oral stage.* In this stage, which lasts from birth through about the second year, gratification is centered around the mouth. The stage is subdivided into oral-dependent and oral-aggressive stages, before and after the acquisition of teeth.

2. *Anal stage.* This stage, from the second to fourth years of life, centers around the elimination of feces. It may be subdivided into an anal-expulsive stage and an anal-retentive stage. The anal-retentive stage, when the child derives gratification from withholding and controlling feces, corresponds to the years when toilet-training is a major parental concern.

3. *Phallic stage.* Gratification in this stage centers around the genitalia. It is during this stage, from ages 4 to 6, that the so-called oedipal complex (named for a mythical Greek who killed his father and married his mother) arises. During the oedipal stage, the child becomes very attached to the parent of the opposite sex and develops hostility to the parent of the same sex.

4. *Latency stage.* This stage represents a sort of rest-stop after the tumultuous development of the first three stages. If the child has mastered the complexes of the prelatency stages, he or she has identified with the parent of the same sex. Children of this stage, from age 6 to puberty, play primarily with other children of the same sex in sex-stereotyped activities.

5. *Genital stage.* At puberty, the child becomes interested in members of the opposite sex.

People who progress through the stages successfully become relatively normal adults. In Freud's view, problems arise when a person fixates at a certain stage because that person's ego cannot resolve conflicts, or when anxiety causes a regression to a previous stage. People who fixate at or regress to the oral-dependent stage may merely be extremely dependent, or they may seek to solve problems by oral means such as overeating or smoking. A person fixated at the oral-aggressive stage may be sarcastic or verbally abusive. Fixation at the anal-expulsive stage results in messiness and disorganization; at the anal-retentive stage, in compulsive orderliness. Freud wrote that failure to resolve oedipal problems results in sexual identification difficulties.

Theories of Cognitive and Moral Development

Psychoanalytic theory is not the only explanation of human behavior that suggests the existence of developmental stages. Other theories have been proposed that describe stages related to cognitive and moral development.

Cognitive Development

When assimilation and accomodation are balanced, a state of equilibration exists.

Jean Piaget was a biologist and psychologist who proposed a stage theory of human development. Piaget's descriptions of the cognitive and moral development of children have had extensive impact among educators. Like Freud, Piaget theorizes that certain forces, biologically determined, contributed to development (Piaget & Inhelder, 1969). The forces suggested by Piaget, however, are those enabling the organism to adapt to the environment—specifically, *assimilation,* the tendency to adapt the environment to enhance personal functioning, and *accommodation,* the tendency to change behavior to adapt to the environment (Achenbach, 1974). The process of maintaining a balance between these two forces is called *equilibration.* Equilibration facilitates growth; other factors that also do so are organic maturation, experience, and social interaction. The cognitive stages suggested by Piaget are

1. *Sensorimotor intelligence* (from birth to 1½ years). During this stage, the infant is preoccupied with differentiating himself from the rest of the world and establishing representations of the objects in his world. He learns how to get to things and how to get things to him. He learns *object constancy,* that things exist even when he can't see them.

2. *Preoperational thought, representational thinking* (from age 1½ years to 4 or 5). This stage includes the development of language. However, Piaget postulates that private symbols and representations precede the development of language.

Some limitations still exist; children at this stage are still unable to take another person's point of view—even in the very concrete sense of "What would this room look like if you were standing over here?" Children's thinking at this stage is typified by their reaction to the familiar conservation experiments: the child who agrees that two glasses hold exactly equal amounts of liquid, will attest that there are unequal amounts when the contents of one glass are poured into a wider or narrower glass while he is watching.

3. *Preoperational thought, intuitional thinking* (from 5 to 7 years). During this stage, the child begins to understand conservation of amount, quantity, number, and weight. She is able to attend to more than one aspect of an object at a time and is less a victim of her immediate perceptions. She begins to understand the reversibility of some operations, but cannot always justify or explain her conclusions. They may be intuitively correct, but the child lacks the ability to systematize them.

4. *Concrete operations* (from 7 to 11 years). During this stage, the child organizes his perceptions and symbols; he becomes able to classify and categorize along several dimensions simultaneously. He can also describe verbally what he is doing, but will not be able to solve problems presented in a purely abstract manner.

5. *Formal operations* (from 12 years through adulthood). Once he or she has reached this stage, the child no longer needs concrete referrents in order to solve problems. The formal-operational thinker can deal with abstractions and with hypothetical situations and can think logically. (Dember & Jenkins, 1970).

Moral Development

A theory of moral development with stages parallel to those of Piaget's cognitive theory has been described by Kohlberg (1964, 1969) and Turiel (1974). The stages of moral development include

1. *Punishment and obedience orientation.* In this stage children behave because of externally imposed rules. Appropriate behavior is performed to avoid punishment.

2. *Naive instrumental hedonism.* This stage emphasizes satisfying personal needs and, to some extent, others' needs. Children begin to view "fairness" as important.

3. *Good-boy morality of maintaining good relations, approval of others.* In this stage, children are attuned to gaining the approval of others. Great emphasis is placed on intentions. "I didn't mean to do it" is a common defense.

4. *Authority-maintaining morality.* In this stage, appropriate behavior is based on respect for authority. Personal values are not an important factor. Such orientation may provide little guidance in situations where authority and fixed rules are absent.

5. *Morality of contract and of democratically accepted law.* At this stage, people show concern for individual rights within the context of societal standards. There is an emphasis on procedures for establishing rules that are fair for the individual as well as good for society.

6. *Morality of individual principles of conscience.* By this stage, the individual has developed a consistent, logical, personal ethical system. Decisions about right and wrong are based on universal principles, not concrete moral rules.

Stage Theories and Intelligence

Piaget and others who propose stage theories of human development have been primarily interested in describing and explaining normal human behavior. Consequently, the chronological age ranges offered in the two previous lists of stages are approximate indicators of when proponents of each theory would expect normal people to develop certain characteristics. It is frequently assumed, however, that cognitive developmental stages are closely tied to mental age rather than to chronological age (Achenbach, 1974), thus leading to the conclusion that mentally retarded people reach each stage later than do people with normal intelligence. Studies concerning moral development (Hoffman, 1970) relate intelligence also to moral development.

The Usefulness of Developmental Theories

Both developmental theories we have discussed are inclusive; they apparently explain a great deal of human behavior, cognitive and emotional, normal and deviant. Verifiability, however, is another matter. Although Piagetian theorists have repeatedly demonstrated the existence of academic and pre-academic behaviors that appear to be age-related (Piaget & Inhelder, 1969), attempts to verify psychoanalytic explanations have not been very successful (Achenbach & Lewis, 1971; Rapaport, 1959). Indeed, considerable resistance to verifying theoretical constructs seems to exist among those who accept the psychoanalytic explanation of human behavior (Schultz, 1969). Although it can be verified that many people act in certain ways at certain ages, this does not prove that the cause of such behavior is an underlying developmental stage or that failure to reach or pass such a stage causes inappropriate or maladaptive behavior. There is little evidence to verify that the order of such stages is invariant, or that reaching or passing through earlier stages is necessary for functioning at higher level stages (Phillips & Kelly, 1975).

Some developmental theories have enough predictive utility to predict what some human beings will do at certain ages; not all of them, however, do even that. Kurtines and Greif (1974) have suggested that, "individuals at

The Usefulness of Developmental Theory			
	Good	Fair	Poor
Inclusiveness	√		
Verifiability			√
Predictive Utility		√	
Parsimony			√

different stages [of Kohlberg's hierarchy] can exhibit the same types of behavior using different types of reasoning, whereas individuals at the same stage can exhibit different behaviors using the same type of reasoning" (p. 459). They provide data that support this concern. Another concern is that, by their nature, these theories offer general information about average persons. However, "a prediction about what the *average* individual will do is of no value in dealing with a particular individual" (Skinner, 1953, p. 19). Marx and Hillix directed a criticism toward psychoanalytic theory: "The unfortunate thing is that the analysts' statements are so general that they can explain whatever behavior occurs. A genuine scientific explanation cannot do this; it must predict one behavior to the exclusion of all other behaviors" (1963, p. 231). Developmental theories do not provide information about what conditions predict an individual's behavior in specific circumstances. The practitioner who wishes to change behavior by changing conditions can expect little help from developmental theories.

Developmental explanations of behavior are equally inadequate when judged by the criterion of parsimony. To say that a child has temper tantrums because he has a weak ego or is fixated at the oral stage of development or to attribute his behavior to his being at the naive instrumental hedonism stage of moral development is seldom the simplest explanation available. Because of their lack of parsimony, developmental explanations may lead the teacher to excuses as unproductive as those prompted by biophysical explanations. Teachers, particularly teachers of handicapped students, may wait forever for a student to become developmentally ready for each learning task. An explanation that encourages teachers to take the student from the student's current level to a subsequent level is clearly more useful than a developmental explanation—at least from a practical point of view. We might expect Professor Grundy's developmental colleagues, for example, to explain Grundy's difficulty with the concept of hyperactivity on the basis of his failure to reach the level of formal operational thinking required to deal with hypothetical constructs. Might there be a more parsimonious and a more useful explanation of his behavior? Professor Grundy continues to collect theories of behavior in the following episode.

Professor Grundy Gains Insight

Having been thoroughly demoralized by his interaction with his student teacher, Professor Grundy decided to pay another surprise visit that afternoon. He was determined to avoid subjecting himself to further ridicule. He did not mention Ralph's hyperactivity but instead concentrated on observing Ms. Harper's teaching. Her lesson plan indicated that she was teaching math, but Professor Grundy was confused by the fact that her group was playing with small wooden blocks of various sizes. Ms. Harper sat at the table with the group but did not interact with the students.

At the conclusion of the lesson, Professor Grundy approached Ms. Harper and asked her why she was not teaching basic addition and subtraction facts as she had planned.

"Professor," stated Ms. Harper, "I conducted my lesson exactly as I had planned. The students were using the blocks to gain insight into the relationship among numbers. Perhaps you are not familiar with the discovery approach, but everyone knows that true insight is vital to the learning process."

Professor Grundy, knowing better but unable to help himself, asked, "Have they discovered yet that $2 + 2 = 4$?"

"Professor," hissed Ms. Harper, "That's not the point. Rote learning is meaningless. I don't care if the children know that $2 + 2 = 4$. I want them to comprehend the meaning of the numerical system."

COGNITIVE EXPLANATION

The educational theory espoused (in a somewhat exaggerated form, to be sure) by Ms. Harper is based on an explanation of human behavior first described in Germany in the early part of this century. The first major proponent of this explanation was Max Wertheimer (Hill, 1963), who was interested in people's perception of reality.

Wertheimer suggested that it was the relationship among things perceived that was important rather than the things themselves. People, he said, tend to perceive things in an organized fashion, so that what is seen or heard is different from merely the parts that compose it. He labeled an organized perception of this type a *gestalt,* using a German word for which there is no exact English equivalent but which may be translated as "form," "pattern," or "configuration." The word *gestalt* has been retained by English-speaking advocates of this view, and we call this explanation *Gestalt psychology.*

The gestaltist seeks to impose structure on chaos to obtain meaning.

Koffka (1935) applied Wertheimer's theories to learning as well as perception. He concluded that learning in human beings is also a process of imposing structure upon perceived information.

The basic principles applied by gestalt psychologists to both perception and learning are referred to as *principles of organization* (Schultz, 1969). These principles are

1. *Proximity.* Parts that are close together in time or space tend to be grouped together. Examples: The *x*'s in Figure 1–1a are seen in three groups. Similarly, when listening to Morse code, the receiver is able to perceive patterns of dots and dashes as letters because they are close together in time.

2. *Similarity.* Parts that are similar tend also to be grouped. Example: The *x*'s in Figure 1–1b are seen in rows rather than columns.

3. *Closure.* Incomplete figures tend to be seen as wholes. Most people would describe what they see in Figure 1–1c as a circle, even though it is not complete.

4. *Pragnanz.* Figures tend to be seen as correct or simple. For example, many people have difficulty proofreading because they "see" words spelled correctly even when they are not.

FIGURE 1–1
Perceptual
Organization

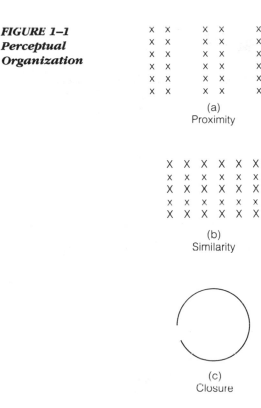

(a)
Proximity

(b)
Similarity

(c)
Closure

Gestalt psychology's emphasis, then, is upon the tendency of human be-ings to impose structure upon their environment and to see relationships rather than individual objects or events as separate entities. This tendency was further described by Kohler (1927) in terms of insight. Kohler was conducting research in the Canary Islands when World War I began and was unable to leave for seven years. He spent these years studying the apes on the island and formulating theories about behavior. His best known ex-periment involved an ape named Sultan.

Sultan was put in a large cage with a variety of objects including some short sticks. He discovered that he could use a stick to rake things toward him when he was feeling lazy. One day Kohler put a banana and a very long stick outside the cage; both objects were too far for Sultan to reach with his arm, but the stick was closer than the banana. Sultan first picked up one of the short sticks in the cage and tried to rake in the banana. The stick wasn't quite long enough, and he threw it down and stomped off to another part of the cage in a fit of pique. As he sat brooding, his eyes suddenly focused on the two sticks and the banana, all arranged in a row. He jumped up, ran over to the small stick, used it to rake in the larger stick, and triumphantly raked in the banana. Kohler showed that the learning experience involved a rearrangement of Sultan's *pattern* of thought. The ape had previously raked things, but the idea of using one stick to rake another and then the banana involved a new application of his prior activity. (Bichler, 1974, p. 238)

According to Kohler, Sultan solved the problem of getting the banana by rearranging in his mind the gestalt of the situation and thus gaining insight.

Wertheimer (1959) extended gestalt theory into human problem-solving. He studied children's and adults' insights into geometric problems and concluded that meaningful solutions depended upon insight, that rote learning—even if it led to correct solutions to problems—was less useful.

Gestalt psychology has had considerable influence on education. The best-known educator to espouse this approach to understanding behavior is Jerome Bruner (1960). What has come to be called the *cognitive theory of education* places an emphasis on rearranging thought patterns and gaining insight as a basis for learning new academic and social behaviors. The resulting teaching practices are called *discovery learning*. Learning is explained on the basis of insight, pattern rearrangement, and intuitive leaps. Teachers do not impart knowledge; they merely arrange the environment to facilitate discovery. Motivation is presumed to occur as a result of innate needs that are met when organization is imposed on objects or events in the arrangement. Motivation is thus intrinsic and need not be provided by the teacher.

Educators who espouse a gestalt theory encourage "discovery learning."

Principles derived from gestalt psychology have also been applied to social behavior, notably in the work of Lewin (1951). His approach has been called *field theory* or *cognitive field theory*. Lewin described human social behavior as based on factors within the person's "life space," the environment as it is perceived by the person and as it affects the person's behavior. He asserts that different people perceive and value environmental objects and events in different ways and that forces exist within people which move them toward or away from these objects or events. Based on a complex procedure for "mapping" or drawing diagrams of people's life spaces, Lewin stated that predictions could be made about what people would do based on the value of the events and the strength of the force. Changing behavior thus depends upon changing people's perceptions of their life space and the relationships among the various events and objects in it.

The Usefulness of Cognitive Theory

Cognitive theory explains a great deal of human behavior. Theorists can account for behavior in both intellectual and social areas. Virtually all behavior can be explained as the result of imposing structure on unstructured environmental events or of perceiving the relative importance of such events. Cognitive theory, then, meets the criterion of inclusiveness.

The theory lacks verifiability, however. Since all of the processes that take place occur internally, there is no way to confirm their existence. Although Kohler, for example, explained Sultan's banana-retrieving behavior on the basis of insight and pattern rearrangement, neither of these processes can be verified. Only the outcome is verifiable—the process is assumed.

The predictive utility of cognitive theory is also extremely limited. In academic areas, the teacher who uses a discovery approach has very little control over what his or her students will discover. Indeed, most advocates of this approach would insist that they do not want to predict outcomes of learning. Unfortunately, this unwillingness to control the outcome of the teaching-learning process has led to rather poor results. In general, educational practices based on a cognitive approach have been less successful than those emphasizing direct instruction (De Cecco, 1968; Engelmann & Carnine, 1982).

The predictive utility of cognitive field theory is somewhat greater than that of cognitive theory. If we know enough about the objects and events in a person's "life space," the value that he assigns them, and his motivation to approach or avoid them, we may be able to predict behavior. Given all this information, of course, we could almost certainly predict behavior without recourse to the theory.

Addressing our final criterion, we must conclude that cognitive theory is not parsimonious. In neither academic nor social areas are the explanations necessary to understanding or predicting behavior.

The Usefulness of Cognitive Theory	Good	Fair	Poor
Inclusiveness	√		
Verifiability			√
Predictive Utility			√
Parsimony			√

Although all of the theories described so far provide information about human behavior, none of them meets all four of our criteria. The explanations we have provided are very general, and our conclusions about their usefulness should not be taken as an indication that they have no value. We simply believe that they provide insufficient practical guidance for classroom teachers. After the following vignette, we shall describe a behavioral explanation of human behavior that we believe most nearly reaches the criteria of inclusiveness, verifiability, predictive utility, and parsimony.

Professor Grundy Takes Action

Professor Grundy had had an absolutely rotten day. A number of the students in his 8 A.M. class—including, of course, DeWayne—had come in late, disrupting his lecture. He had been ridiculed by a student teacher; his precious manuscript had been retrieved from the dumpster in a sadly wrinkled and malodorous condition; his colleague had made repeated references to "anal-expulsive" and "oral-aggressive" tendencies during the day in spite of Grundy's protests.

After arriving at home and pouring himself a large drink for medicinal purposes, Grundy decided that something must be done. He made several detailed plans and retired for the evening, confident that he was on the right

"The librarian sent me over to pick up one of your 48 overdue books . . . if you're . . . uh . . . done with it."

track. The next morning, he arose enthusiastically determined, in spite of a slight headache, to put his plans into action.

His first step was to arrive at his 8 A.M. class 5 minutes early—somewhat of a novelty since he usually arrived several minutes after the bell. He spent the extra 5 minutes chatting affably with students and clarifying points from the previous day's lecture when asked to do so. When the bell rang, he presented each of the five students present with an "on-time slip" worth two points on the next exam.

After the morning lecture Professor Grundy proceeded to his office, where he affixed to the door a large sign reading "PLEASE DO NOT CLEAN THIS OF-FICE TODAY." He then opened the window, wondering just what the biology department had deposited in the dumpster to cause so strong a smell. He spent an hour reorganizing his notes.

Next Grundy once again visited Ms. Harper, this time suggesting that she would receive a failing grade for student teaching unless she learned to control Ralph's behavior and to teach basic math facts. Her habitual expression of disdain changed to one of rapt attention.

Professor Grundy had observed that Ralph, since he was too "hyperactive" to remain in his seat, spent the time while other students worked wandering from toy to toy in the free-time area of the classroom. He suggested that Ms. Harper allow Ralph to play with the toys only after remaining in his seat for a specified length of time: very short periods at first, gradually increasing in length.

Grundy further suggested that the student teacher make flash cards of basic addition and subtraction facts and allow the students to play with the colored blocks after they had learned several combinations.

Returning happily to his office, the professor encountered his psychoanalytically oriented colleague, who once again jocularly repeated his insights into

Grundy's character. Ignoring the comments, the Professor began an animated conversation with his secretary, praising the rapidity with which she was retyping his manuscript. She assured him that it had first priority, since she couldn't wait to dispose of the stinking pages. She did suggest that if he had written his original draft on the available word processor, it would be safely stored on a disk.

Within a short time, Professor Grundy felt that he had things under control. Most of the students enrolled in the 8 A.M. class were present and on time every morning, even though Grundy had begun to give "on-time slips" only every few days. Ms. Harper had stopped sneering and had started teaching. Ralph's wandering had decreased dramatically and the math group had learned to add and subtract. Grundy continued to ignore his colleague's comments, which gradually ceased when no response was forthcoming, and the fresh copy of his manuscript was typed in record speed. The only negative outcome was a sharp note from campus security to the effect that the condition of his office constituted a fire hazard and that it must be cleaned immediately.

BEHAVIORAL EXPLANATION

In the preceding vignette, Professor Grundy emerged as the behaviorist he is. To solve some of his problems, he used techniques derived from yet another explanation of human behavior. The behavioral explanation states that human behavior, both adaptive and maladaptive, is learned. Learning occurs as a result of the consequences of behavior. To put it very simply, behavior that is followed by pleasant consequences tends to be repeated and thus learned. Behavior that is followed by unpleasant consequences tends not to be repeated and thus not learned. By assuming that his students, including DeWayne, came to class late, that the custodian cleaned, that the student teacher ridiculed, that Ralph wandered, and that the psychoanalytic colleague teased because they had learned to do so, Professor Grundy was able to teach them to do other things instead. In doing so, he applied several learning principles (Homme, C'deBaca, Cottingham, & Homme, 1970) underlying the behaviorists' view of human behavior. The following sections introduce these principles, each of which will be discussed in detail in later chapters.

Positive Reinforcement

Chapter 6 discusses reinforcement in detail.

Positive reinforcement* describes a functional relationship between two environmental events: a **behavior** (any observable action) and a **consequence** (a result of that act). When a behavior is followed by a consequence that increases the behavior's rate of occurrence, positive reinforcement has been demonstrated.

Many human behaviors are learned as a result of positive reinforcement. Parents who praise their children for putting away toys may teach the

---

*Words printed in **boldface** in the text are defined in the Glossary at the end of the book.

children to be neat; parents who give their children candy to make them stop screaming in the grocery store may teach the children to scream. The cleaning behavior of Professor Grundy's custodian undoubtedly was learned and maintained through positive reinforcement, as was the wit of Grundy's psychoanalytic colleague. Grundy used positive reinforcement (on-time tickets, conversation, and time with toys) to increase his students' rate of coming to class on time and the amount of time Ralph stayed in his seat.

Negative Reinforcement

Negative reinforcement describes a relationship between events in which the rate of a behavior's occurrence increases when some (usually aversive or unpleasant) environmental condition is removed or reduced in intensity. Human beings learn many behaviors when acting in a certain way results in the termination of unpleasantness. Professor Grundy, for example, had learned that opening windows results in the reduction of unpleasant odors in closed rooms. Similarly, his secretary typed rapidly, because, when she finished, she could throw away the smelly papers.

Punishment

Punishment also describes a relationship: a behavior is followed by a consequence that decreases the behavior's future rate of occurrence. An event is described as a **punisher** only if the rate of occurrence of the preceding behavior decreases. Behaviorists use the word *punishment* as a technical term to describe a specific relationship; confusion may arise because the same word is used in a nontechnical sense to describe unpleasant things done to people in an effort to change their behavior. To the behaviorist, punishment occurs only when the preceding behavior decreases. In the technical sense of the term, something is not necessarily punishment merely because the consequent event is perceived by someone as unpleasant. A behaviorist can never say, "I punished him, but it didn't change his behavior," as do many parents and teachers. It's punishment *only* if the functional relationship can be established. We could say that Professor Grundy's verbal threat to Ms. Harper, for example, was apparently a punisher: her ridiculing comments to him stopped. We wish, of course, that he had used a more positive approach.

Extinction

Chapter 7 discusses punishment and extinction in detail.

When a previously reinforced behavior is no longer reinforced, its rate of occurrence decreases. This relationship is described as **extinction.** Recall from our vignette that when Grundy no longer reacted to his colleague's ridicule, the behavior stopped. For a behaviorist, all learning principles are defined on the basis of what actually happens, not what we think is happening. Grundy may have thought he was punishing his colleague by yelling or otherwise expressing his annoyance. In reality, the rate of the behavior increased when Grundy reacted in this way; the real relationship was that of positive reinforcement. The behavior stopped when the positive reinforcer was withdrawn.

Stimulus Control

Stimulus control is the focus of Chapter 8.

Stimuli, such as environmental conditions or events, become linked to particular behavior over time. Any behavior is more likely to occur in the presence of those stimuli that have accompanied the behavior when it was reinforced in the past. This is the principle of **stimulus control,** which describes a relationship between behavior and **antecedent stimuli** (events or conditions occurring before the behavior is performed) rather than behavior and its consequences. The consequences must have been present during the development of the relationship, but the antecedent condition or event now serves as a sort of signal or cue for the behavior. In our vignette, the custodian's adherence to posted notices had apparently been reinforced in the past, so that Professor Grundy's sign was effective even in the absence of a positive reinforcer on this occasion.

Other Learning Principles

In addition to these major learning principles, Professor Grundy illustrated the use of several other influences on human behavior described by behaviorists. These influences include **modeling** and **shaping.** *Modeling* is the demonstration of behavior. The professor had been modeling inappropriate behavior—coming to class late—and his students had apparently been imitating that behavior. Many behaviors, both appropriate and inappropriate, are learned by imitation of a model. Infants learn to talk by imitating their parents; adults can learn to operate complex machinery by watching a demonstration.

Shaping uses the reinforcement of successive approximations to a desired behavior to teach new behavior. Grundy suggested that Ms. Harper use shaping to teach Ralph to stay in his seat. She initially reinforced sitting behavior when it occurred for short periods of time and gradually increased the sitting time required for Ralph to earn the reinforcer. Many behaviors are taught by shaping. Parents may praise a young child effusively the first time she dresses herself, even if her blouse is on inside out and her shorts are on backwards. Later, she may earn a compliment only if her outfit is perfectly coordinated and her shoes are polished.

The Task of the Behaviorist

If you can see it, hear it, feel it, or smell it, it's observable. If you can count it or measure it, it's quantifiable.

Behaviorists explain the development of both normal and atypical human behavior in terms of several principles just described. An important aspect of this approach is its emphasis on behavior (Ullmann & Krasner, 1965). To qualify as a behavior something must be *observable* and *quantifiable* (Baer, Wolf, & Risley, 1968). We must be able to see (or sometimes hear, feel, or even smell) the behavior. In order to make such direct observation meaningful, some way of measuring the behavior in quantitative terms (how much? how long? how often?) must be established. Behaviorists cannot reliably state that any of the relationships described as learning principles exist unless these criteria are met.

Skinner (1953) has suggested that behaviorists are less concerned with explaining behavior than with describing it. The emphasis, he says, is on which environmental factors increase, decrease, or maintain the rate of occurrence of specific behaviors. Although it may be an interesting exercise to speculate on what environmental consequences have led to the development of certain behaviors, it is much more important to determine what factors in the immediate environment have a functional relationship to the behaviors in question. The reinforcers or punishers that led to the acquisition or suppression of behaviors in the past cannot be observed in the present and may not be the events which currently maintain it (Ullmann & Krasner, 1965). The child who learned to scream in the supermarket because his parent gave him candy may well scream in school when he is asked to do arithmetic. There may be no candy around in the classroom, but careful observation of the environment will probably help his teacher locate some environmental condition reinforcing this behavior. We cannot go back into the past and remove the candy or other past reinforcers; we must be concerned with present rather than past events.

It is important to note that behaviorists do not deny the existence of physiological problems which may contribute to some behavioral problems. Nor do most behaviorists deny the effects of heredity (Mahoney, 1974) or even developmental stages (Ferster, Culbertson, & Boren, 1975). Their primary emphasis, however, is on present environmental conditions maintaining behavior and on establishing and verifying functional relationships between such conditions and behavior.

The Usefulness of the Behavioral Explanation

One of the most common criticisms of the behavioral approach is that it leaves much of human behavior unexplained (Schultz, 1969). Emphasis on observable behavior has led many to assume that behavioral principles cannot account for any but simple motor responses. However, Skinner (1953, 1957, 1971) applied basic learning principles to explain a wide variety of complex human behavior, including verbal behavior and sociological, economic, political, and religious beliefs.

The fact that behavioral principles have not accounted for all aspects of human behavior should not lead to the assumption that they cannot. Skinner had an interesting reaction to assumptions about the limitations of behavioral principles:

> Patience with respect to unexplored parts of a field is particularly important in a science of behavior because, as part of our own subject matter, we may be overwhelmed by the facts which remain to be explained. Subtle illusions, tricks of memory, the flashes which solve problems—these are fascinating phenomena, but it may be that genuine explanations, within the framework of a science of behavior, as distinguished from verbal principles or "laws" or neurological hypotheses, are out of reach at the present time. To insist that a science of behavior give a rigorous account of such phenomena in its present state of knowledge is like asking the Gilbert of 1600 to explain a magnetic amplifier or the Faraday of 1840 to explain superconductivity. Early physical scientists enjoyed a natural simplification of their subject matters. Many of the most subtle phenomena were to come into existence only through technical advances in the sciences themselves. . . . The behavioral scientist enjoys no such natural protection. He is faced with the full range of the phenomena he studies. He must

therefore more explicitly resolve to put first things first, moving on to more difficult things only when the power of his analysis permits. (1966, p. 218)

Many advances have been made since Skinner's 1966 statement. Many phenomena have yet to be explained. Because behaviorists refuse to theorize about what they have not observed, explanation must await verification. Behaviorists are ready temporarily to sacrifice some degree of inclusiveness for verifiability.

Verifiability is the essence of the behavioral explanation. Other theorists posit a theory and attempt to verify it through experimental investigation. Behaviorists, on the other hand, investigate before formulating what may be described as generalizations rather than theories. That adult attention serves as a positive reinforcer for most children (Baer & Wolf, 1968; Harris, Johnston, Kelley, & Wolf, 1964) is an example of such a generalization. This statement was made only after repeated observations established a functional relationship between children's behavior and adult attention.

The Usefulness of Behavioral Theory			
	Good	Fair	Poor
Inclusiveness		√	
Verifiability	√		
Predictive Utility	√		
Parsimony	√		

The focus of the behavioral approach is changing behavior (Ullmann & Krasner, 1965). Predictive utility is an essential part of any behavioral explanation. Functional relationships are established and generalizations are made precisely so that they can be used to change maladaptive or inappropriate behavior and increase appropriate behavior. Behaviorists are reinforced by changing behavior, not by discussing it. Unless it is possible to use generalizations to predict what people will do under certain conditions, behaviorists see little point in making the statements. An enormous body of evidence exists representing the application of learning principles to human behavior. The *Journal of Applied Behavior Analysis,* from 1968 to the present, attests to this wealth of evidence. Such data makes possible the prediction of behavior under a wide variety of conditions.

An "explanatory fiction" explains nothing. Behaviorists explain behavior on the basis of observation, not imagination.

Behavioral explanations are parsimonious, satisfying our fourth criterion for usefulness. Describing behavior solely in terms of observable, verifiable, functional relationships avoids the use of "explanatory fictions" (Ullmann & Krasner, 1969). Such fictions are defined only in terms of their effects, resulting in the kind of circular reasoning we discussed earlier. Rather than invoking "hyperactivity"—an example of an explanatory fiction—to explain Ralph's out-of-seat behavior, Professor Grundy chose a behavioral approach, to look at what happened before and after Ralph left his seat. In this way, behaviorism avoids explanations distant from observed behavior and its relationship to the environment. It is unacceptable to explain out-of-seat behavior by labeling the cause as hyperactivity or to explain messiness on the basis of fixation at or regression to the anal-expulsive stage of behavior. Neither explanation adds anything useful to our information about the problem.

Haughton and Ayllon (1965) offer one example of the fluency with which many professionals are willing to invoke unparsimonious explanations of behavior. The authors were working with a hospitalized mental patient whose behavior for many years had been limited to sitting and smoking cigarettes. After a period during which smoking was limited, the patient was given cigarettes only when standing up and holding a broom. The patient began carrying the broom most of the time. Two psychiatrists were asked to observe and evaluate the patient's behavior. The report of the first psychiatrist read:

> The broom represents to this patient some essential perceptual element in her field of consciousness. How it should have become so is uncertain; on Freudian grounds it could be interpreted symbolically, on behavioral grounds it could perhaps be interpreted as a habit which has become essential to her peace of mind. Whatever may be the case, it is certainly a stereotyped form of behavior such as is commonly seen in rather regressed schizophrenics and is rather analogous to the way small children or infants refuse to be parted from some favorite toy, piece of rag, etc. (Haughton & Ayllon, 1965, p. 97)

The second psychiatrist theorized:

> Her constant and compulsive pacing holding a broom in the manner she does could be seen as a ritualistic procedure, a magical action. When regression conquers the association process, primitive and archaic forms of thinking control the behavior. Symbolism is a predominant mode of expression of deep-seated unfulfilled desires and instinctual impulses. By magic, she controls others, cosmic powers are at her disposal and inanimate objects become living creatures.

Her broom could be then

> 1. a child that gives her love and who she gives in return her devotion;
>
> 2. a phallic symbol;
>
> 3. the sceptre of an omnipotent queen.

> Her rhythmic and prearranged pacing in a certain space are not similar to the compulsions of a neurotic, but because this is a far more irrational, far more controlled behavior from a primitive thinking, this is a magical procedure in which the patient carries out her wishes, expressed in a way that is far beyond our solid, rational and conventional way of thinking and acting. (Haughton & Ayllon, 1965, pp. 97–98)

When staff members stopped giving cigarettes to the woman while she was carrying the broom, she stopped carrying the broom. Although we stated earlier in this chapter that the parsimonious explanation may not always be correct, in this case it was. Even when the development of unusual behavior is not as easy to trace as in this example, the assumption that such behaviors are being maintained by current environmental conditions and that the behavior may be changed by changing the environment is not merely parsimonious, it is supremely optimistic. The teacher who concentrates on searching out and changing the environmental conditions maintaining students' inappropriate or maladaptive behavior does not give up on them because they are retarded, brain-damaged, emotionally disturbed, hyperactive, or developmentally unready to learn; she or he teaches them. If students' behavior is described in terms of behavioral excesses (too much moving around) or deficits (too little reading) as suggested by Gelfand and Hartmann (1975) and Hersen and Bellack (1977) rather than in terms of explanatory fictions, the teacher can go about the business of teaching—decreasing behavioral excesses and remediating behavioral deficits.

HISTORICAL DEVELOPMENT OF BEHAVIORISM

Behaviorism as a science has roots in philosophical and psychological traditions dating from the 19th century. However, the learning principles described earlier in this chapter certainly existed before being formally defined. People's behavior has been influenced since people have existed. The following section examines several historical descriptions of how people have used the relationship between behavior and its consequences. After that, we will trace the development of behaviorism as a formal way of explaining, predicting, and changing human behavior.

Historical Precedents

The arrangement of environmental conditions in order to influence behavior is by no means a recent invention. It is said that the ancient Romans put eels in the bottom of wine cups in order to decrease excessive drinking. Not all examples are so negative. Birnbaum (1962) provides the following more positive example:

> A parable comes down to us across the centuries, which are but a moment in God's eye. It tells us that when a teacher wishes a young boy to study the Torah, a boy too young to understand how meaningful a thing this is to do, the teacher says to him: "Read, and I shall give you nuts, and figs, and honey." And the boy makes an effort, not because of the sweetness of reading, but because of the sweetness of the eating. As the boy grows older, no longer tempted by sweets, his teacher says to him: "Read, and I shall buy you fine shoes and garments." Again the boy reads, not for the fine text, but for the fine clothing. As the boy reaches young manhood, and the new clothes become less important to him, his teacher now tells him: "Learn this paragraph, and I shall give you a dinar, or perhaps even two dinars." The young man studies now to attain not the learning, but the money. And still later, as his studies continue into his adult life, when even a bit of money comes to mean less to him, his teacher will now say to him: "Learn, so that you may become an elder and a judge, that the people honor you and rise before you, as they do before this one and that." Even at this stage of life, then, this student learns, not in order to exalt the Lord, but so that he himself will be exalted by other men. (p. 32)

Crossman (1975) provides another example of the use of positive reinforcement:

> There is a fascinating history behind the pretzel. About 610 A.D. an imaginative Alpine monk formed the ends of dough, left over from the baking of bread, into baked strips folded into a looped twist so as to represent the folded arms of children in prayer. The tasty treat was offered to the children as they learned their prayers and thereby came to be called "Pretiola"—Latin for "little reward." [From the back of a Country Club Foods pretzel bag, Salt Lake City.] (p. 348)

Educators used behavioral principles long before the principles were formally identified.

Several innovative educators developed elaborate programs of rewards and punishment to manage their students' behavior. In the early 19th century, Lancaster (Kaestle, 1973) instituted a system in Great Britain that was later also used in the United States. Students earned tickets that could be exchanged for prizes or money. Tickets were lost when students misbehaved. A similar program, the Excelsior School system (Ulman & Klem, 1975), was marketed and used in several states.

Benjamin Franklin demonstrated that adults' behavior could also be changed, using a rather different positive reinforcer:

> "We had for our chaplain a zealous Presbyterian minister, Mr. Beatty, who complained to me that the men did not generally attend his prayers and exhortations. When they enlisted, they were promised, besides pay and provisions, a gill of rum a day, which was punctually serv'd out to them, half in the morning, and the other half in the evening; and I observ'd they were as punctual in attending to receive it; upon which I said to Mr. Beatty: 'It is, perhaps, below the dignity of your profession to act as steward of the rum, but if you were to deal it out and only just after prayers, you would have them all about you.' He liked the tho't, undertook the office, and, with the help of a few hands to measure out the liquor, executed it to satisfaction, and never were prayers more generally and more punctually attended; so that I thought this method preferable to the punishment inflicted by some military laws for non-attendance on divine service." From: Franklin, Benjamin, American Philosophical Society. (Reprinted by Skinner, 1969, p. 247)

Parents and teachers have likewise applied the principles of learning in their efforts to teach children. "Clean up your plate and then you can have dessert," says the parent hoping for positive reinforcement. "When you finish your arithmetic, you may play a game," promises the teacher. Parents and teachers, whether they are aware of it or not, also use punishment: the child who runs into the street is spanked; the student who finishes his assignment quickly is given more work to do. All of us have heard 'Just ignore him and he'll stop—he's only doing it for attention." If he does stop, we have an example of extinction.

Behavioral principles operate whether or not anyone is consciously using them.

Of course, many parents and teachers extinguish appropriate behavior as well, paying no attention to children who are behaving nicely. Negative reinforcement is demonstrated in many homes every day: "You don't play outside until that room is clean." Teachers also use negative reinforcement when they require students, for example, to finish assignments before going to lunch or to recess. The kindergarten teacher who asks his charges to use their "inside voices" is trying to establish stimulus control. Whenever teachers show their students how to do something, they are modeling.

It becomes apparent that a person does not need to know the names of the relationships involved to use them. Indeed, applying behavioral learning principles sounds a lot like common sense. If it is so simple, why must students take courses and read books? Why have such quantities of material been written, so much research conducted?

The answer is that it is inefficient to fail to arrange environmental conditions so that functional relationships are established, or to allow such relationships to be randomly established, or to assume that such relationships have been established on the basis of common sense alone. This inefficiency has resulted in high levels of maladaptive behavior in schools as well as sometimes frighteningly low levels of academic and pre-academic learning. It is our aim in writing this book to help teachers become applied behavior analysts. The derivation and definition of the term *applied behavior analysis* will be discussed in the remaining sections of this chapter.

Philosophical and Psychological Antecedents

The roots of the behavioral viewpoint are firmly planted in a 19th century philosophical movement known as *positivism*. Its earliest proponent, Auguste Comte (Schultz, 1969), emphasized that the only valid knowledge was that which was objectively observable. Comte apparently arrived at such a standard as a result of his attempt to make a systematic survey of all knowledge. In order to limit his task, he decided to accept only facts or knowledge which resulted from direct observation.

A second important contribution came from *animal psychology*, influenced by the work of Charles Darwin (Boring, 1950), which emphasized the continuity between animal and human behavior and thus suggested that something about human beings could be learned through the careful observation of lower animals. Animal psychology focused on the adaptation of physical structures in the body to the environment. This focus led to consideration of mental processes in the same light and to a psychological movement known as *functionalism* (Schultz, 1969).

Functionalism was a third important influence on the development of a behavioral approach to explaining human behavior. William James, whose work was a precursor of behaviorism (Boring, 1950), emphasized that when people think and act, they do so for a purpose, or function. John Dewey and James Angell were also influential in turning the emphasis in American psychology from an introspective, theorizing model to one emphasizing a practical, observational approach.

Respondent Conditioning

A famous physiologist who had won a Nobel Prize for his studies of digestion in dogs noticed a curious thing in his laboratory. If the attendant who ordinarily fed the dogs came into the room when he was studying a dog's stomach, he observed that the dog began to secrete gastric juices, just as if it had been given some food. The physiologist called this *psychic secretion* and noted that it affected the whole digestive system, starting the flow of saliva in the mouth as well as the secretion of acids in the stomach.

The advice of his friends and colleagues in physiology was to forget the whole mysterious business. He was assured that he had nothing to gain and everything to lose by getting associated with something so unscientific, so insubstantial, and so unphysiological as "psychic secretions." From the point of view of mechanistic physiology, there was every reason to avoid these intangible sources that supposedly produced tangible secretions. The "mind," the "life force," and the emergent spirit had all been swept out of physiology, and they were not to be allowed to sneak back via psychic secretions. His friends thought that his observations were probably in error in some way. And even if there were such secretions, the whole matter would be cleared up when the nervous system was completely explored.

After a good deal of soul-searching, the physiologist made the painful decision to desert the work that had brought him fame and turn to the pursuit of the nature and origins of psychic secretions. He embarked on a career of studies that was to last 30 years and provide support for theories of learning unformulated at that time. Thus, Ivan Pavlov committed himself to his classic research on the conditioned reflex. (Dember & Jenkins, 1970, p. 3)

Pavlov's work has been extremely influential in the development of contemporary psychology and education. His precise observation and measurement has served as a model for experimental research to this day (Schultz, 1969). Pavlov is best known for the model of behavior derived from his observations of the salivation of dogs, a study inspired by the chance observation described in the preceding quotation. Pavlov ultimately did come to discuss his observations in terms of observed behavior rather than "psychic secretions." His classic experiment involved pairing food powder (which elicits salivation—an automatic reflex) with a tone that would normally have no effect on dogs' salivation. The presentation of the tone preceded the presentation of the food powder, and after repeated pairings, salivation occurred when only the tone was presented. The food powder was labeled the *unconditioned stimulus* (UCS); the tone, the *conditioned stimulus* (CS). Salivation is an unconditioned response to food powder and a conditioned response to the tone. The relationship may be represented as shown in the accompanying diagram showing two stages.

Stage 1:   Food (UCS)
         Tone . . . .→Salivation
Stage 2:
         Tone ——————————————————————→Salivation

The term *conditioning,* often used to describe any form of learning, came to be attached to this process of *respondent,* or *classical, condition-*

ing by a quirk of translation. Pavlov had used the terms *conditional* and *unconditional* stimuli. The words were then translated from Russian as "conditioned" and "unconditioned," and thus *conditioning* was applied as a label for this learning process and, ultimately, for others as well (Hill, 1970).

The process of pairing stimuli so that an unconditioned stimulus elicits a response is known as *Pavlovian, classical,* or *respondent conditioning.* Such conditioning is the basis of a method of behavior change known as *behavior therapy.* Behavior therapists concentrate on breaking up maladaptive conditioned reflexes and building more adaptive responses. These therapists often work with people who have problems such as irrational fears or phobias. They also help those who want to change habits like smoking, overeating, or excessive alcohol consumption. A detailed discussion of behavior therapy is beyond the scope of this text. For the student who is interested in learning more about it, we recommend texts by Wolpe (1958, 1969), Lazarus (1971, 1976) and Wolpe and Lazarus (1966).

Associationism

Another influential experimenter whose research paralleled that of Pavlov was Edward Thorndike. Thorndike studied cats rather than dogs, and his primary interest was discovering associations between situations and responses (Thorndike, 1931). He formulated two laws that profoundly influenced the subsequent development of behavioral science. The Law of Effect (Thorndike, 1905) states that "any act which in a given situation produces satisfaction becomes associated with that situation, so that when the situation recurs the act is more likely than before to recur also" (p. 203). Second is the Law of Exercise, which states that a response made in a particular situation becomes associated with the situation. The relationship of the Law of Effect with the principle of positive reinforcement is obvious. The Law of Exercise is similarly related to the stimulus control principle discussed earlier.

Behaviorism

If we were all Watsonians, we couldn't say: "She hurt my feelings," "My mind wandered," "Use your imagination."

The use of the term *behaviorism* was originated by John Watson (1914, 1919, 1925). Watson advocated the complete abolition of any datum in psychology which did not result from direct observation. He considered such concepts as mind, instinct, thought, and emotion both useless and superfluous. He denied the existence of instinct in human beings and reduced thought to subvocal speech, emotion to bodily responses. A Watsonian behaviorist of our acquaintance once responded to a question by saying "I've changed my mind (you should excuse the expression)." The true Watsonian does not acknowledge the existence of any such entity as "mind."

Watson and Raynor (1920) conditioned a startle response in a baby, Albert, by pairing a white rat (CS) with a loud noise (UCS). Watson contended that all "emotional" responses such as fear were conditioned in similar ways. In an interestingly related procedure, Jones (1924) desensitized a 3-year-old child who showed a fear response to white rabbits and other white furry objects by pairing the child's favorite foods with the rabbit. This procedure was unfortunately not carried out with Albert, who moved away before his conditioned fear could be eliminated. Albert may still be scared of white rats, which may have created a number of problems in his life, including preventing his employment as a behavioral psychologist.

Operant Conditioning

The learning principles described at the beginning of this section are those suggested by proponents of an operant conditioning model for explaining, predicting, and changing human behavior. The best-known operant conditioner is B.F. Skinner (1938, 1953), who first distinguished operant from respondent conditioning.

Operant behaviors are emitted voluntarily; respondent behaviors are elicited by stimuli.

Respondent conditioning, you will remember, deals with behaviors elicited by stimuli that precede them. Most such behaviors are reflexive, that is, they are not under voluntary control. *Operant conditioning* (sometimes called *instrumental conditioning*), on the other hand, deals with behaviors usually thought of as voluntary rather than reflexive. Operant conditioners are concerned primarily with the consequences of behavior and the establishment of functional relationships between behavior and consequences. The behavioral view described earlier in this chapter is that of operant conditioning, which will be the emphasis of the entire text.

Skinner's early work was with animals, primarily white rats. In this, he followed in the tradition of earlier behaviorists, to whom this particular animal was so important that one researcher (Tolman, 1932) dedicated a major book to *Mus norvegious albinius,* a strain of white rats. Bertrand Russell, the philosopher, is said to have suggested facetiously that the different emphases in European (primarily Gestalt, introspective, and theorizing) and American (primarily behavioral, active, observational) studies may have resulted from differences in the breeds of rats available. Whereas European rats sat around quietly waiting for insight, American rats were active go-getters, scurrying around their cages and providing lots of behaviors for psychologists to observe.

Skinner also worked with pigeons. He explained (1963) that, while in the military, he was assigned to a building whose window sills were frequented by these birds. Because there was very little to do, he and his colleagues began to train the pigeons to perform various behaviors.

Early application of operant conditioning techniques to human beings was directed toward establishing that the principles that govern animal behavior also govern human behavior. The use of these principles to change human behavior—usually called *behavior modification*—did not really emerge in nonlaboratory settings until the 1960s. One of the authors remembers being told in an experimental psychology course in 1961 that there was some indication that operant conditioning could be applied to simple human behavior. As an example, the instructor laughingly described college students' conditioning their professor to lecture from one side of the room simply by looking interested only when he stood on that side. The instructor insisted that it would not be possible with him, because he was aware of the technique. He was wrong; he ended up backed into one corner of the room by the end of the next lecture.

At that time, however, in spite of Skinner's (1953) theoretical application of operant conditioning techniques to complex human behavior, and pioneer studies like those of Ayllon and Michael (1959) and Birnbrauer, Bijou, Wolf, and Kidder (1965), few people anticipated the enormous impact that the use of such principles would have on American psychology and education as well as on other disciplines, including economics (Kagel & Winkler, 1972). The application of behavior modification in real-life settings had become so prevalent by 1968 that a new journal, *The Journal of Applied Behavior Analysis,* was founded to publish the results of research. In Volume 1, Number 1, of the journal, Baer, Wolf, and Risley (1968) defined *applied behavior analysis* as the "process of applying sometimes tentative principles of behavior to the improvement of specific behaviors, and simultaneously evaluating whether or not any changes noted are indeed attributable to the process of application" (p. 91).

Applied behavior analysis must deal with socially important behavior. Observable behaviors must be chosen for change, and relationships between the behavior and the intervention must be verified.

Baer et al. (1968) suggest that in order for research to qualify as applied behavior analysis, it must change socially important behavior, chosen because it needs change, not because its study is convenient to the researcher. It must deal with observable and quantifiable behavior, objectively defined or defined in terms of examples, and there must be clear evidence of the existence of a functional relationship between the behavior to be changed and the experimenter's intervention. Applied behavior analysis is more rigorously defined than behavior modification. In our earlier vignette, Professor Grundy did succeed in modifying behavior, but he failed to meet the criterion of analysis—he had no way of knowing for sure whether his techniques changed behavior or whether the change was mere coincidence. This book is designed to help teachers become applied behavior analysts—effective modifiers of behavior and efficient analyzers of the principles of learning involved in all aspects of their students' performance.

SUMMARY

In this chapter, we described a number of approaches to explaining human behavior. We evaluated these approaches in terms of their inclusiveness, verifiability, predictive utility, and parsimony. We also described an explanation of human behavior that appears to us to be the most useful—the behavioral explanation.

In tracing the history of the behavioral approach to human behavior, we emphasized the development of a science of applied behavior analysis. We discussed the necessity for concentrating on socially useful studies of human behavior as well as careful observation and the establishment of functional relationships. In the next chapter, we will describe the difficulty of defining the term *socially useful* and the related ethical and professional concerns about applied behavior analysis. We will also address other issues concerning the responsible use of this technology.

REFERENCES

ACHENBACH, T.H., & LEWIS, M. 1971. A proposed model for clinical research and its application to encopresis and enuresis. *Journal of the American Academy of Child Psychiatry, 10,* 535–554.

ACHENBACH, T.M. 1974. *Developmental psychopathology.* New York: Ronald Press.

AYLLON, T.A., & MICHAEL, J. 1959. The psychiatric nurse as a behavior engineer. *Journal of the Experimental Analysis of Behavior, 2,* 323–334.

BAER, D.M., & WOLF, M.M. 1968. The reinforcement contingency in preschool and remedial education. In R.D. Hess & R.M. Bear (Eds.), *Early education: Current theory, research, and action.* Chicago: Aldine.

BAER, D.M., WOLF, M.M., & RISLEY, T.R. 1968. Some current dimensions of applied behavior analysis. *Journal of Applied Behavior Analysis, 1,* 91–97.

BANNATYNE, A. 1971. *Language, reading and learning disabilities.* Springfield, Ill.: Charles C Thomas.

BERRY, H.K. 1969. Phenylketonuria: Diagnosis, treatment and long-term management. In G. Farrell (Ed.), *Congenital mental retardation.* Austin: University of Texas Press.

BIEHLER, R.F. 1974. *Psychology applied to teaching.* Boston: Houghton Mifflin.

BIRNBAUM, P. (Ed.). 1962. *A treasury of Judaism.* New York: Hebrew Publishing.

BIRNBRAUER, J.S., BIJOU, S.W., WOLF, M.M., & KIDDER, J.D. 1965. Programmed instruction in the classroom. In L.P. Ullmann & L. Krasner (Eds.), *Case studies in behavior modification* New York: Holt, Rinehart & Winston.

BORING, E.G. 1950. *A history of experimental psychology.* New York: Appleton-Century-Crofts.

BOULLIN, D.J., COLEMAN, M., O'BRIEN, R.A., & RIMLAND, B. 1971. Laboratory predictions of infantile autism based on 5-hydroxtryptamine efflux from blood platelets and their correlation with the Rimland E-2 score. *Journal of Autism and Childhood Schizophrenia, 1,* 63–71.

BRUNER, J.S. 1960. *The process of education.* Cambridge, Mass.: Harvard University Press.

CONNERS, C.K. 1975. *Food additives and hyperkinesis: A controlled double-blind experiment.* Pittsburgh: University of Pittsburgh. (ERIC Document Reproduction Service No. ED 117 877.)

COTT, A. 1972. Megavitamins: The orthomolecular approach to behavior and learning disabilities. *Therapy Quarterly,* 7(3), 245–258.

CRAVIOTO, J., & DELICARDIE, E. 1975. Environmental and nutritional deprivation in children with learning disabilities. In W.M. Cruickshank and D.P. Hallahad (Eds.), *Psychoeducational practices: Perceptual and learning disabilities in children* (vol. 1). Syracuse, N.Y.: Syracuse University Press.

CROSSMAN, E. 1975. Communication. *Journal of Applied Behavior Analysis, 8,* 348.

CRUICKSHANK, W.M., BICE, H.V., & WALLEN, N.E. 1957. *Perception and cerebral palsy.* Syracuse, N.Y.: Syracuse University Press.

DE CECCO, J.P. 1968. *The psychology of learning and instruction: Educational psychology.* Englewood Cliffs, N.J.: Prentice-Hall.

DEMBER, W., & JENKINS, J. 1970. *General psychology: Modeling behavior and experience.* Englewood Cliffs, N. J.: Prentice-Hall.

ENGELMANN, S., & CARNINE, D. 1982. *Theory of instruction: Principles and applications.* New York: Irvington Publishers, Inc.

EYSENCK, H.H. 1956. The inheritance of extroversion-introversion. *Acta Psychologica, 12*, 95–110.

FEINGOLD, B.F. 1975. *Why your child is hyperactive.* New York: Random House.

FERSTER, C.B., CULBERTSON, S., & BOREN, M.C.P. 1975. *Behavior principles* (2nd ed.). Englewood Cliffs, N.J.: Prentice-Hall, Inc.

FINE, S. 1973. Family therapy and a behavioral approach to childhood obsessive-compulsive neurosis. *Archives of General Psychiatry, 28*, 695–697.

GELFAND, D.L., & HARTMANN, D.P. 1975. *Child behavior analysis and therapy.* New York: Pergamon Press.

GESELL, A., & ILG, F.L. 1943. *Infant and child in the culture of today..* New York: Harper.

GOLDSTEIN, K. 1939. *The organism.* New York: American Book.

HALL, C.S. 1954. *A primer of Freudian psychology.* Cleveland: World Publishing.

HALLGREN, B. 1950. Specific dyslexia (congenital word blindness): A clinical and genetic study. *Acta Psychiatriea et Neurologica, 65*, 1–287.

HARRIS, F.R., JOHNSTON, M.K., KELLEY, C.S., & WOLF, M.M. 1964. Effects of social reinforcement on repressed crawling of a nursery school child. *Journal of Educational Psychology, 55*, 34–41.

HAUGHTON, E., & AYLLON, T. 1965. Production and elimination of symptomatic behavior. In L.P. Ullmann & L. Krasner (Eds.), *Case studies in behavior modification* (pp. 94–98). New York: Holt, Rinehart and Winston, Inc.

HERSEN, M., & BARLOW, D.H. 1976. *Single-case experimental designs: Strategies for studying behavior change.* New York: Pergamon Press.

HERSEN, M., & BELLACK, A.S. 1977. Assessment of social skills. In A.R. Ciminero, K.S. Calhoun, & H.E. Adams (Eds.), *Handbook for behavioral assessment.* New York: Wiley.

HILL, W.F. 1963. *Learning: A survey of psychological interpretations.* San Francisco: Chandler.

HILL, W.F. 1970. *Psychology: Principles and problems.* Philadelphia: J.B. Lippincott.

HOFFMAN, M.L. 1970. Moral development. In P.H. Mussen (Ed.), *Carmichael's manual of child psychology* (3rd ed., vol. 2, pp. 261–359). New York: John Wiley & Sons, Inc.

HOMME, L., C'DE BACA, P., COTTINGHAM, L., & HOMME, A. 1970. What behavioral engineering is. In R. Ulrich, T. Stacknik, & J. Mabry (Eds.), *Control of human behavior* (vol. 2, pp. 17–23). Glenview, Ill.: Scott, Foresman & Co.

JONES, M.C. 1924. A laboratory study of fear: The case of Peter. *The Pedagogical Seminary and Journal of Genetic Psychology, 31*, 308–315.

KAESTLE, C.F. (Ed.). 1973. *Joseph Lancaster and the monitorial school movement: A documentary history.* New York: Teachers College Press.

KAGEL, J.H., & WINKLER, R.C. 1972. Behavioral economics: Areas of cooperative research between economics and applied behavior analysis. *Journal of Applied Behavior Analysis, 5*, 335–342.

KALLMANN, F.J., & ROTII, B. 1956. Genetic aspects of preadolescent schizophrenia. *American Journal of Psychiatry, 112*, 599–606.

KESSLER, J.W. 1966. *Psychopathology of childhood.* Englewood Cliffs, N. J.: Prentice-Hall.

KOFFKA, K. 1935. *Principles of Gestalt psychology.* New York: Harcourt, Brace & World.

KOHLBERG, L. 1964. Development of moral character and moral ideology. In M. Hoffman & L. W. Hoffman (Eds.), *Review of child development research* (vol. 1). New York: Russell Sage.

KOHLBERG, L. 1969. *Stage and sequence: The developmental approach to morality.* New York: Holt, Rinehart & Winston.

KOHLER, W. 1927. *The mentality of apes* (E. Winter, trans.). New York: Humanities Press.

KURTINES, W., & GREIF, E.B. 1974. The development of moral thought: Review and evaluation of Kohlberg's approach. *Psychological Bulletin, 81*, 453–470.

LAZARUS, A.A. 1971. *Behavior therapy and beyond.* New York: McGraw-Hill.

LAZARUS, A.A. 1976. *Multimodal behavior therapy.* New York: Springer.

LEWIN, K. 1951. *Field theory in social science.* New York: Harper and Row.

MAHONEY, M.J. 1974. *Cognition and behavior modification*. Cambridge, Mass.: Ballinger Publishing Company.

MARX, M.H., & HILLIX, W.A. 1963. *Systems and theories in psychology*. New York: McGraw-Hill.

MORGANE, P.J., STERN, W., & RESNICK, O. 1974. Early protein lack: Malevolent effects. *Science News, 106*, 229.

MATTES, J.A. 1983. The Feingold diet: A current reappraisal. In G.M. Senf & J.K. Torgensen (Eds.), *Annual review of learning disabilities* (vol. 1). 127–131.

PHILLIPS, D.C., & KELLY, M.E. 1975. Hierarchical theories of development in education and psychology. *Harvard Educational Review, 45*, 351–375.

PIAGET, J., & INHELDER, B. 1969. *The psychology of the child*. New York: Basic Books.

POND, D. 1961. Psychiatric aspects of epileptic and brain-damaged children. *British Medical Journal, 2*, 1377–1382, 1454–1459.

RAPAPORT, D. 1959. The structure of psychoanalytic theory: A systematizing attempt. In J.S. Koch (Ed.), *Psychology: A study of science*. New York: McGraw-Hill.

RIMLAND, B. 1964. *Infantile autism*. New York: Appleton-Century-Crofts.

ROSE, T.L. 1978. The functional relationship between artificial food colors and hyperactivity. *Journal of Applied Behavior Analysis, 11*, 439–446.

RUTTER, M., KORN, S., & BIRCH, H.G. 1963. Genetic and environmental factors in the development of primary reaction patterns. *British Journal of Social and Clinical Psychology, 2*, 161–173.

SCHULTZ, D. P. 1969. *A history of modern psychology*. New York: Academic Press.

SKINNER, B.F. 1938. *The behavior of organisms*. New York:Appleton-Century-Crofts.

SKINNER, B.F. 1953. *Science and human behavior*. New York: Macmillan.

SKINNER, B.F. 1957. *Verbal behavior*. New York: Appleton-Century-Crofts.

SKINNER, B.F. 1963. Operant behavior. *American Psychologist, 18*, 503–515.

SKINNER, B.F. 1966. What is the experimental analysis of behavior? *Journal of the Experimental Analysis of Behavior, 9*, 213–218.

SKINNER, B.F. 1969. Communication. *Journal of Applied Behavior Analysis, 2*, 247.

SKINNER, B.F. 1971. *Beyond freedom and dignity*. New York: Knopf.

STRAUSS, A.A., & LEHTINEN, L.E. 1947. *Psychopathology and education of the brain-injured child*. New York: Grune & Stratton.

STRAUSS, A.A., & WERNER, H. 1942. Disorders of conceptual thinking in the brain-injured child. *Journal of Nervous and Mental Disease, 96*, 153–172.

THOMAS, A., & CHESS, S. 1977. *Temperament and development*. New York: Brunner/Mazel.

THOMAS, A., CHESS, S., & BIRCH, H.G. 1968. *Temperament and behavior disorders in children*. New York: New York University Press.

THORNDIKE, E.L. 1905. *The elements of psychology*. New York: Seiler.

THORNDIKE, E.L. 1931. *Human learning*. New York: Appleton-Century-Crofts.

TOLMAN, E.C. 1932. *Purposive behavior in animals and men*. New York: Appleton-Century-Crofts.

TURIEL, E. 1974. Conflict and transition in adolescent moral development. *Child Development, 45*, 14–29.

ULLMANN, L.P., & KRASNER, L. (Eds.). 1965. *Case studies in behavior modification*. New York: Holt, Rinehart and Winston, Inc.

ULLMANN, L.P., & KRASNER, L. 1969. *A psychological approach to abnormal behavior*. Englewood Cliffs, N.J.: Prentice-Hall.

ULMAN, J.D., & KLEM, J.L. 1975. Communication. *Journal of Applied Behavior Analysis, 8*, 210.

VANDENBURG, S.G. 1967. Hereditary factors in normal personality traits as measured by inventories. In J. Wortis (Ed.), *Recent advances in biological psychiatry* (vol. 9). New York: Plenum Press.

VITALE, J.J., & VELEZ, H. 1968. In R.A. McCance & E.M Widdowson (Eds.), *Calorie deficiencies and protein deficiencies*. London: Churchill.

WATSON, J.B. 1914. *Behavior: An introduction to comparative psychology*. New York: Holt, Rinehart & Winston.

WATSON, J.B. 1919. *Psychology from the standpoint of a behaviorist*. Philadelphia: Lippincott.

WATSON, J.B. 1925. *Behaviorism*. New York: Norton.

WATSON, J.B., & Raynor, R. 1920. Conditioned emotional reactions. *Journal of Experimental Psychology, 3*, 1–4.

WERNER, H., & STRAUSS, A.A. 1940. Causal factors in low performance. *American Journal of Mental Deficiency, 45*, 213–218.

WERRY, J.S. 1972. Organic factors in childhood psychopathology. In H.G. Quay & J.S. Werry (Eds.), *Psychopathological disorders of childhood*. New York: Wiley.

WERTHEIMER, M. 1959. *Productive thinking*. New York: Harper & Row.

WOLPE, J. 1958. *Psychotherapy by reciprocal inhibition*. Palo Alto, Cal.: Stanford University Press.

WOLPE, J. 1969. *The practice of behavior therapy*. New York: Pergamon Press.

WOLPE, J., & LAZARUS, A.A. 1966. *Behavior therapy technique*. New York: Pergamon Press.

WUNDERLICH, R.C. 1977. The hyperactivity complex. *Journal of Optometric Vision Development, 8*(1), 8–45.

2 Responsible Use of Applied Behavior Analysis Procedures

Did you know that . . .

- Some people think that behavior modification and brain surgery are the same thing?
- It's possible to be both a humanist and a behaviorist?
- Effective procedures scare some people?
- Applied behavior analysis can make students more creative?
- Some behavior change programs may violate students' constitutional rights?

The use of applied behavior analysis in classrooms and other environments has engendered considerable controversy. The debate has not been limited to professionals: arguments about the use of operant principles to change behavior have appeared in popular magazines and newspapers as well as in scholarly journals. Discussion occurs on social occasions as well as at conventions and symposia. Psychiatrists, psychologists, social workers, and teachers weigh the merits of behavioral procedures, as do judges, lawyers, legislators, parents, and students themselves. Much of the discussion, however, has come from applied behavior analysts (cf. Wood, 1975). In addition to defending Applied Behavior Analysis, ABA professionals have felt it necessary to clarify ethical issues and suggest guidelines for the responsible use of the technology.

This chapter addresses many of the issues raised by the current popular interest. First, we'll consider some of the causes of controversy. Then we'll examine and respond to some of the specific criticisms of behavioral procedures, particularly as these methods are used in educational settings. The final sections provide teachers with guidelines for avoiding common pitfalls in the use of applied behavior analysis procedures, and Professor Grundy answers some common questions asked by uninformed people.

CONCERNS ABOUT APPLIED BEHAVIOR ANALYSIS

Resistance to the use of operant procedures to change behavior has come from several sources. The term *behavior modification,* which is most commonly used to describe such techniques, has caused some confusion. Because *modification* is synonymous with *change*, the term has often been misused to refer to any procedure that has the potential to change behavior. This contamination of the term is one reason that this book uses the term *applied behavior analysis.*

Some humanists state that changing behavior invariably infringes upon personal freedom.

Other objections to operant procedures have come from those who feel that any systematic effort to change behavior is coercive and, thus, inhumane. Those who take this position often describe themselves as "humanists." Their objections are based on a rejection of a deterministic viewpoint and advocacy of free will and personal freedom. The intuitive appeal of these humanistic values makes humanists' rejection of behavioral procedures a formidable objection, although as we shall see, such objections frequently rely on a rather shaky logical foundation.

The very effectiveness of applied behavior analysis procedures is one source of much concern about this approach. It is ironic that many people are very comfortable with ineffective techniques ineffective or techniques whose effectiveness at least lacks verification. The same people often reject other procedures—such as those based on applied behavior analysis—because their use results in predictable, consistent behavior change.

Confusion with Other Procedures

Applied behavior analysis is not hypnosis, prefrontal lobotomy, brain implants, drug therapy, or shock treatment.

Much of the public outcry against what is popularly called *behavior modification* results from the use of this term to describe procedures that are totally unrelated to applied behavior analysis. Popular journalists (Holden, 1973; Mason, 1974; Wicker, 1974) and even behavior modification professionals (McConnell, 1970) have done incalculable harm to the image of applied behavior analysis by including unrelated treatment procedures under the heading of behavior modification. Hypnosis, psychosurgery, brain implants, drug therapy and electroconvulsive shock treatment have all been lumped under this label. Such procedures undoubtedly change behavior, but they are not related to the systematic changing of behavior by application of behavioral principles. It would be equally logical and equally erroneous to list under the title of behavior modification the entire array of therapeutic interventions including "psychoanalysis, Gestalt therapy, primal screams, lectures, books, jobs and religion" (Goldiamond, 1975, p. 26).

Used correctly, the term *behavior modification* refers only to procedures derived from the experimental analysis of human behavior. Unfortunately, the term has become so unalterably linked with other procedures in the minds of the public that its use is best avoided. The teacher who tells parents that he or she uses behavior modification must be prepared for negative reactions. By no means do we advocate abandoning the technology of behavior modification: we simply suggest that teachers avoid using the term with uninformed or misinformed people. In many cases, other professionals, including administrative staff and fellow teachers, may be as confused as parents and school board members. It may be as necessary to educate these fellow professionals as it is to teach children.

Applied behavior analysis certainly does not include such treatments as electroconvulsive therapy or brain surgery; neither does it involve the use of drugs. Indeed, the effective application of appropriate behavioral procedures often reduces the need for such drastic interventions. This has been strongly demonstrated in studies using positive reinforcement as an alternative to medication for children labeled *hyperactive* (Ayllon, Layman, & Kandel, 1975). It is possible that behavior modification, in the proper sense of the term, will ultimately diminish the use of surgery, drugs, and other such behavior change techniques. It is, therefore, particularly unfortunate that the improper use of the term has caused so much public hostility to a technology that is so potentially benign.

Reaction to Controversial Procedures

Not all of the misunderstanding or hostility has resulted from those outside the field. Procedures derived from the experimental analysis of behavior are frequently rejected by professionals as well as the public. Such procedures include, among others, the use of contingent electric shock (Lovaas & Simmons, 1969; Risley, 1968) for suppressing certain extremely maladaptive behaviors and the use of exclusion (Bostow & Bailey, 1969; Wolf, Risley, & Mees, 1964). Although these procedures are only a few of the tools of the applied behavior analyst, their use has received a disproportionate share of attention from the press, the public, and the judiciary (Stolz, 1977). It is sufficient to note that aversive or exclusionary procedures may create problems in two ways:

Guidelines for the use of aversive and exclusionary procedures are provided in Chapter 7.

1. Their misuse is common and often described by the users as behavior modification.

2. Their use, even when appropriate, causes more concern than other behavioral procedures.

Unfortunately, there have been many abuses. Risley (1975) describes a "behavior modification" program in a Florida institution where bizarre punishments were used. The mildest of these was washing residents' mouths out with detergent when they spoke, in the staff members' opinions, inappropriately. One of us was recently asked to work with a student who, while a resident in a "behavior modification unit" in a private facility, had been in what was called *time-out* for several weeks. Time-out, used properly, is measured in minutes. Isolation for days or weeks is more properly called *solitary confinement* and is certainly not an ABA procedure. It is fully understandable that procedures causing pain or discomfort to children or young people, particularly those who are handicapped, cause concern. Many, but not all, professionals, however, believe that certain controversial procedures are sometimes justified, just as many, but not all, parents believe that spanking children is sometimes appropriate. Later we shall discuss attempts to define the circumstances under which these procedures may be considered appropriate.

Concerns About Coercion

The notion that applied behavior analysis is inhumane rests on the assumption that each human being should be free to choose a personal course of behavior. It follows, for those who criticize behavioral procedures, that any systematic attempt to alter the behavior of another human being is coercive and thus inhumane.

This criticism of behavioral techniques is based on the philosophic concept of free will. Advocates of the assumption of free will tend to attribute human behavior to forces arising from within the individual and, thus, not subject to prediction or control. In other words, people just do what they do because they decide to do it. A deterministic position, on the other hand, holds that behavior is "lawful" (subject to lawful prediction) and its causes can be identified in environmental events. A determinist recognizes systematic relationships among such events (Mahoney, Kazdin, & Lesswing, 1974) and considers human behavior as part of the system. This contrasting view concludes, then, that human behavior is subject to lawful prediction. People do things, or decide to do things, because of past events and present circumstances. It is important to distinguish between the use of the term *lawful*, in the sense of an orderly relationship between events, and any implication of authoritarian control (Becker, Engelmann, & Thomas, 1975). Lawful, in the sense used here, refers simply to relationships among events that occur naturally, not to attempts to legislate human behavior.

A belief that behavior is lawful does not imply that human beings are not free to choose what they will do.

Applied behavior analysts, by definition, are also determinists. Their position is predicated upon solid evidence that "the assumption of determinism is both justified and essential in dealing with human behavior" (Craighead, Kazdin, & Mahoney, 1976, p. 172). This confirmation has come from a large body of psychological research, some of it conducted by those who call themselves applied behavior analysts.

Behaviorists define freedom in terms of a person's ability to make choices and to exercise options.

The assumption of lawful relationships among events and behavior does not imply a rejection of human freedom. For the applied behavior analyst, "freedom is defined in terms of the number of options available to people and the right to exercise them" (Bandura, 1975, p. 865). The goal of the behavior analyst is to increase such options not decrease them (Bandura, 1975) and thus to increase the freedom of the individual. The high school student who repeatedly fails English is not free to attend college. The child who is afraid to interact with peers is not free to make friends. People who have severe behavioral deficits may have no options at all; they cannot move around, take care of their basic needs, or control their environment in any way.

The last example points out a crucial concept in understanding the deterministic position. The relationship between behavior and the environment is reciprocal (Bandura, 1969; Craighead et al., 1976). Environmental events control behavior, but behavior also inevitably alters the environment, too. This reciprocal relationship exists between people. The behavior modifier's behavior is changed by the actions of the subject of the modification. Thus everyone influences and controls others' behavior. It is impossible to abandon control; we inevitably influence the behavior of other people (Bandura, 1975; Rogers & Skinner, 1956). For example, if Ms. Johnson decides to reinforce her student Max's smiling behavior, Max will probably smile more often. A smiling student is more reinforcing to most teachers, so Ms. Johnson will probably interact more often with Max when he smiles frequently. This increased interaction provides more opportunities for her to reinforce Max's smiling, and so on.

Seen in this context, behavioral technology is neither dehumanizing nor inhumane. When goals are humane, we must offer the most effective means available to reach them. In many cases, the proven effectiveness of applied behavior analysis procedures makes them the most humane choice.

The following anecdote illustrates a clash between humanists and behaviorists. We urge you to consider the values in practice on each side of this particular example.

A colleague at a different university showed us a deeply moving film. The heroine was an institutionalized primary-grade girl. She was a head-banger, so a padded football helmet was put on her head. Because she could take it off, her hands were tied down in her crib. She kept tossing her neck and tore out her hair at every opportunity. She accordingly had a perpetually bruised face on a hairless head, with a neck almost as thick as that of a horse. She was nonverbal.

My colleague and his staff carefully planned a program for her, using all kinds of reinforcers. She was remanded to their program, but persisted in her typical behavior. In desperation, the ultimate weapon was unwrapped. When she tossed her head, my colleague yelled "Don't!", simultaneously delivering a sharp slap to her cheek. She subsided for a brief period, tossed again, and the punishment was delivered. My colleague reports that less than a dozen slaps were ever delivered and the word "Don't!" yelled even from across the room was effective. Its use was shortly down to once a week and was discontinued in a few weeks. In the meantime, the football helmet was removed and the girl began to eat at the table. She slept in a regular bed. Her hair grew out, and she turned out to be a very pretty little blond girl with delicate features and a delicate neck. In less than a year, she started to move toward joining a group of older girls whose behavior, it was hoped, she would model. She smiled often.

The initial institution and her parents discovered that she had been slapped. They immediately withdrew her from the custody of my colleague's staff. The last part of the film shows her back at the institution. She is strapped down in her crib. Her hands are tied to a side. She is wearing a football helmet. Her hair is torn out, her face is a mass of bruises and her neck is almost as thick as that of a horse. (Goldiamond, 1975, pp. 62–63)

ETHICAL USE OF ABA PROCEDURES

Awareness of potential criticism may help avert interference from uninformed persons.

All teachers—whether or not they are also applied behavior analysts—are concerned with ethics. This concern has at least two sources. A primary motivation is the desire to do what is best for students. Another strong motivator is that teachers must also be cognizant of other people's worries about teaching techniques. Previous sections have acknowledged that people are especially apt to worry about behavioral techniques. Unless teachers take particular care to act ethically and to assure others that they do, they may find noneducators seeking and acquiring more and more control over what may and may not be done in classrooms. Problems may arise because of the way programs are described even when the programs themselves are appropriate. Risley (1975) describes a time-out procedure that was disallowed primarily because staff members referred to the free standing structures built for short-term exclusion as "boxes" and to the procedure as "putting him (the resident) in the box." That the "boxes" were large, adequately lighted structures made no difference. The use of the wrong words resulted in withdrawal of approval for the program. Those of us who tend toward flippant labels would be especially wise to guard our tongues when discussing procedures with people who might misunderstand.

A number of factors must be considered when attempting to determine whether a proposed intervention is ethical. These include "community standards, laws, prevailing philosophies, individual freedom and responsibility of the clients through informed consent as well as the clients attitudes and feelings" (Sulzer-Azaroff, Thaw, & Thomas, 1975). In the case of schoolchildren or residents of an institution, it is important to seek the opinions of the parents or guardians of the students. It may seem strange for behaviorists to concern themselves with such subjective criteria as attitudes and feelings, but Wolf (1978) makes a strong case for considering these factors. If participants do not like a program, he suggests, "they may avoid it, or run away, or complain loudly" (p. 206). Wolf suggests that social validity should be established for goals, procedures, and outcomes.

Stainback and Stainback (1984) suggest that increased attention be given to qualitative research methods that provide "more attention to the social and educational relevance of research efforts" (p. 406). Simpson and Eaves (1985) urge that attempts be made to quantify such subjective measurements. It is clear that teachers using ABA procedures must concern themselves with factors in addition to those within their classrooms. Goals, procedures, and outcomes must be acceptable to the consumers of education—students, parents, and the community.

Misunderstanding of behavioral procedures, as well as concern about their misuse, has resulted in increasing pressure for mandatory guidelines for the use of applied behavior analysis procedures. Such mandatory guidelines and restrictions are becoming increasingly common, especially in institutional settings (Martin, 1975). Particularly likely to be restricted are aversive or exclusionary procedures and items or events that may be used as positive reinforcers. It is not generally permissible, for example, to require residents of institutions to earn the right to food, privacy, or basic activities (*Wyatt v. Stickney*, 1972). Teachers who want to use applied behavior analysis procedures in an ethical and responsible manner must take these factors into account:

1. Teacher and staff competence
2. Appropriate goals and procedures
3. Voluntary participation
4. Accountability

The teacher who thoughtfully considers each of these factors not only increases the effectiveness of his or her program but also decreases potential criticism from other staff, parents, and members of the community.

Competence

Implementing ABA procedures is not so easy as it sounds.

Because many applied behavior analysis procedures seem so simple, they are often misused by persons who do not adequately understand them. A common example is the teacher who attends a short workshop on ABA techniques, buys a bag of candy, and proceeds to hand out "reinforcement" indiscriminately. A common result is that the teacher in question concludes, as have many others, that behavior modification doesn't work. An unfortunate side effect is that the children treated in this manner may become more disruptive than ever because they understand neither why they got candy nor why they stopped getting it. Moreover, parents become upset because their children's teeth are rotting and their appetites spoiled; the principal expresses annoyance because she receives numerous irate phone calls from those parents; other teachers become enraged because their students demand candy too; and the escutcheon of applied behavior analysis receives another blot.

It is not possible to learn, in a few days, enough about applied behavior analysis to implement ethical, effective programs (Franks & Wilson, 1976). One of the authors attended a seminar several years ago during which she was asked to develop a packet for other faculty members to read that would enable them, after a few hours' reading, to include behavior management techniques in their methods courses. When she retorted that she had taken 8 courses in applied behavior analysis, had a background in animal research, had been practicing the procedures for 17 years, and was still learning, the reaction was the usual: "But behavior mod is so simple!"

The principles of applied behavior analysis are, indeed, easy to understand. Their effective implementation is not so easy, however. In addition to a thorough understanding of the principles, acquired from qualified instructors, supervised practice is desirable. This is particularly important in the case of difficult procedures, such as shaping, or procedures subject to abuse, such as aversive or exclusionary techniques.

Shaping is described in Chapter 8. Aversive and exclusionary procedures are discussed in Chapter 7.

Some professionals within the field who are concerned about adequate training have suggested that some sort of licensing procedures be implemented. Most applied behavior analysts, however, reject a plan for licensing practitioners of applied behavior analysis unless the same standards are imposed on practitioners of other types of intervention procedures. Licensing only applied behavior analysts implies that this group is less responsible than other practitioners (Kazdin, 1978).

In addition to ensuring personal competence, a responsible teacher makes certain that others who interact with students understand the principles and techniques of applied behavior analysis. Chapter 11 describes programs for training parents, paraprofessionals, and classroom volunteers to use these principles efficiently and responsibly.

Good supervision includes training, observation, and evaluation.

Initial training must always be followed by adequate supervision. Martin (1975) suggests that such supervision include inservice training, both formal and informal, and regular evaluation. Ongoing supervision will continuously upgrade the performance of competent staff. It is equally important to ensure that incompetent staff members do not implement procedures that may be detrimental to students or that they fail to follow through with procedures that are in the students' best interests.

Selection of Appropriate Goals

Ethical use of behavioral procedures starts with selection of appropriate goals. Since Baer, Wolf, and Risley stated in 1968 that applied behavior analysis must address socially important behavior, the issue of goal selection has concerned professionals. Wolf (1978) suggests that the way to accomplish this social validation, that is, to ensure that socially important behaviors are being changed, is to consult the consumers of our services. What do the parents or guardians of our students want them to learn? What behaviors are acceptable or unacceptable in a particular community? A clear demonstration of the pitfalls of failing to ascertain the answers to these questions is demonstrated by the reaction in some communities to the introduction of sex education in public schools.

A number of considerations affect goal selection for behavior change procedures. In the case of handicapped students, recent legislation requires that such goals be formulated by a group of concerned professionals in conjunction with the students' parents (or advocates) and, if possible, with the student as well. The goals are then recorded on the student's Individualized Education Program (IEP). Several specific factors (Martin,

1975) should be considered by those responsible for choosing target behavior (goals) for any student:

1. The necessity for stating goals in objective, measurable terms.

2. The assurance that achieving the goal will benefit the student more than the school or institution.

3. A reasonable degree of certainty that achieving the goal is realistic for the student.

4. The assurance that the program will help the student develop appropriate behaviors, not just suppress inappropriate ones.

5. The certainty that a behavior to be changed is not one which is protected by students' constitutional rights.

Stating Goals Objectively

Chapter 3 examines the whole process of writing behavioral objectives and offers extensive guidelines for ensuring that such goal statements are written in both observable and measurable terms. It is crucial that any goal statement clearly specifies what behaviors will be changed and what criteria will be used to measure the change.

The objective statement of goals leads to improved communication among professionals and others involved in the student's education. It also facilitates agreement as to whether goals have been reached. If a teacher states as a goal the vague objective of improving a student's attitude or making the student more independent, teacher, parents, aides, and the student are bound to disagree about whether or not the goal is reached. Each person involved will be looking at different behaviors as indicative of a "good attitude" or "independence." Thus, one of the primary characteristics of responsible goal setting is describing the specific behaviors that are to be changed.

Benefitting the Student

Applied behavior analysis procedures may be abused if students' rights and best interests are not considered.

It may seem obvious that the behaviors targeted for change should be those whose change will benefit the student. Nevertheless, accusations have been made that both residential institutions (*Wyatt v. Stickney*, 1972) and schools (Winett & Winkler, 1972) use behavior-change programs primarily to reduce behaviors that disrupt the smooth functioning of the institution or school but are not detrimental to residents or students. Winett and Winkler examined articles detailing behavior-change programs in the *Journal of Applied Behavior Analysis* from 1968 through 1970. They stated that the majority of the articles concerned the attempted suppression of talking, moving around, and such disruptive behaviors as whistling, laughing, and singing. Winett and Winkler concluded that the technology of applied behavior analysis was being used merely to establish "law and order" (p. 499) rather than to serve the best interests of students. They further mentioned that similar goals were set in residential institutions.

Although O'Leary (1972) agreed with Winett and Winkler that careful examination of goals is important, he disagreed with their conclusions. He cited numerous studies that demonstrated the researchers' concern with such behaviors as academic response rates (Lovitt & Curtiss, 1969), talking (Reynolds & Risley, 1968), prosocial interactions (O'Connor, 1969), and language and reading skills (Bereiter & Engelmann, 1966). O'Leary did agree with Winett and Winkler's call for "extensive community dialogues concerning those behaviors and values we wish to develop in our children" (p. 511).

Applied behavior analysis procedures can help students improve academic skills, become more creative, and make friends.

Applied behavior analysis procedures may be used to decrease or increase any behavior. Even such behaviors as creativity (defined in terms of observable behavior, of course) have been increased using this technology (Goetz & Baer, 1973; Goetz & Salmonson, 1972; Maloney & Hopkins, 1973; Reese & Parnes, 1970). There is no justification for asserting that the procedures, in and of themselves, contribute to maintenance of behaviors for the convenience of those other than students. Appropriate goal selection should prevent the use of such institution-oriented criteria for change as "Be still, be quiet, be docile" (Winett & Winkler, 1972, p. 499).

Setting Achievable Goals

When setting goals for students, the professionals involved must consider not only the appropriateness of the goals but also the abilities of the student. Setting unrealistically high goals is a disservice to students and to their parents, as well as a source of frustration to teachers and other staff. Applied behavior analysis procedures are so effective that it is easy to lose sight of their limitations. Even the most efficient teaching cannot magically make severely handicapped children exactly like their nonhandicapped peers.

It is equally important that goals be sufficiently ambitious. Some teachers, fearing that they will be criticized if students fail to achieve stated goals, may set goals that are too low. It is difficult but necessary to estimate students' potential progress fairly. Any Individualized Education Program should be revised if goals are found to be either unrealistically high or unnecessarily low.

Information for goal-setting comes from many sources: tests, records, parents, teachers, observation, and the students themselves.

Goal selection is a very difficult process. Participation in such goal selection is one responsibility that sets teachers apart from nonprofessional staff. Examination of all pertinent information, including test results, previous records, and personal observation, should precede goal selection, as should consultation with parents and other professionals who may provide additional information about the student's abilities and potential. The process of goal selection is discussed in Chapter 3, along with preparation of behavioral objectives.

Developing Appropriate Behaviors

It is sometimes necessary to eliminate or reduce the rate of some student behaviors. A child who bites himself must be stopped from doing so. A student who hurts others cannot be allowed to continue. Students who are so disruptive that they cannot be maintained in a classroom must learn to stop running, screaming, or destroying property. However, merely eliminating such behavior is indefensible in the absence of a plan to develop constructive behavior. A student who just sits quietly is not much better off than she was before intervention. Teachers must pay attention to developing behaviors in the student that will lead to improved learning or social interaction.

In some cases, inappropriate behavior may be decreased by reinforcing constructive behavior rather than by directly attempting to decrease destructive behavior. For example, decreasing disruptive behavior does not automatically result in improved academic performance (Ferritor, Buckholdt, Hamblin, & Smith, 1972), whereas increasing academic output may result in decreased inappropriate behavior (Ayllon & Roberts, 1974; Kirby & Shields, 1972). In general, for students who display any appropriate behavior at all, the teacher should try reinforcing such behavior and monitoring the effects of this procedure on the inappropriate behavior. Some students' repertoires of appropriate behavior are so limited and their performance of inappropriate behavior so continuous that there is little or no opportunity for a positive-reinforcement approach. In such cases, the teacher may have to undertake elimination of the maladaptive behavior as a first step. This should only be a first step, however. As soon as possible, the student must be taught to substitute constructive behaviors.

Protecting Students' Rights

Some student behaviors are protected by law and may not be chosen for change without risk of litigation. Choosing to change such behaviors is more than unethical; it is illegal. Martin (1975) lists speech, dress, assembly, and publication as examples of constitutional rights that may be protected by law. Students who criticize school policy, either verbally or in writing, are often labeled disruptive or insubordinate and referred for help to behavioral specialists. Teachers who provide such help may be violating students' constitutional rights. Some schools may have arbitrary rules about students' dress or grooming. The teacher who makes reinforcers contingent upon adherence to such rules may also risk violation of students' rights.

There is, of course, a difference between interfering with students' constitutional rights to dress as they choose and teaching appropriate grooming. For many handicapped students, personal hygiene and clothing selection are an important part of the curriculum. Similarly, teaching students to use good manners when addressing an adult is unlikely to interfere with their rights. However, teachers should keep in mind ethical constraints in choosing behaviors to change, selecting those that will benefit students and that will lead to increased options and greater personal freedom. A clean, well-groomed, polite person has more options than does a filthy, unkempt, rude one.

Programs that may be inappropriate are those that seek to change students' attitudes, values, or their religious or political beliefs (Martin, 1975). A bill introduced in the United States Senate in 1974 (PL 93–380, Section 438) included prohibition of such programs. The prohibition was deleted before the bill was passed, partly because of its vague language, which would probably have made very difficult the job of changing students' social behaviors. Nevertheless, the ethical teacher must consider whether the social behaviors she or he wishes to change are really maladaptive for the students or merely antithetical to the teacher's or employer's value systems. For example, even a program designed to teach children to maintain eye contact with adults who are speaking to them may cause problems. In certain subcultures, such behavior is considered bold, and parents may object.

Is it ethical to teach students respect for authority? How about ambition? Or, tolerance for diversity of religious beliefs?

Teachers of handicapped students must deal with yet another aspect of legality. PL 94–142 (The Education of All Handicapped Children Act) requires that students receive educational services in the "least restrictive environment." For many mildly handicapped students, this means education in the regular classroom with their nonhandicapped peers for most of the school day. In order to make such placement possible, special education teachers may find themselves in a position where they must select as targets for change behaviors that seem relatively trivial. Such behaviors as arguing with teachers or failing to say "Yes, ma'am" or "Yes, sir" when addressing adults may make a child unacceptable in some regular classrooms. The special education teacher may have to make difficult decisions in some cases as to which course is most ethical and most protective of students' rights.

Selection of Appropriate Procedures

Writing an Individualized Education Program also involves selecting an intervention procedure for changing a behavior. Two primary considerations guide professionals and parents in writing this program: the proven effectiveness of a technique in changing similar behaviors in similar students and the characteristics of the student in question.

The Criterion of Effectiveness

In general, the most ethical and responsible procedure to use in changing behavior is one that has been established as most effective. Throughout this text, we shall review literature related to changing specific behaviors and provide suggestions about effective procedures. Teachers who plan behavioral programs should also continually review current professional journals in order to keep abreast of new developments. The *Journal of Applied Behavior Analysis* is an excellent source of information for teachers as well as professionals in other fields. Many other journals also provide information on behavior-change procedures for use with students who have specific handicapping conditions and with students in regular classes who display deficits or excesses in certain areas.

Aversive or Seclusionary Procedures   In some cases, it may not be possible or ethical to use a procedure that has been proven effective. The use of aversive procedures—specifically including contingent electric shock—though proven effective in eliminating some very maladaptive behaviors such as self-mutilation (Bucher & Lovaas, 1968; Lovaas, Freitag, Gold, & Kassorla, 1965; Wolf, Risley, & Mees, 1964), has been restricted (*Wyatt v. Stickney*, 1972). Such procedures may now be used only in cases where consent has been obtained from the subject (or parents or an advocate) and where a supervisory committee has confirmed that less drastic procedures have not eliminated or sufficiently decelerated the maladaptive behavior (Martin, 1975). Limitations have also been placed on the use of **time-out** from positive reinforcement, a technique that may involve seclusion (confining a student in an isolated area) or exclusion (removing a student from a potentially reinforcing situation). Seclusionary time-out is particularly subject to regulation (*Morales v. Turman*, 1974). Before using any aversive or seclusionary procedure, teachers should examine their employers' guidelines or regulations that pertain to such procedures, as rules may vary considerably. The unauthorized use of even short-term seclusion, a relatively mild but effective technique (Birnbrauer, Wolf, Kidder, & Tague, 1965; Madsen & Madsen, 1970; Wolf et al., 1964; Zeilberger, Sampen, & Sloane, 1968), may result in criticism or misunderstanding.

Time-out procedures are discussed in Chapter 7.

The use of aversive or seclusionary interventions should, in any event, be reserved for severely maladaptive behaviors that have not been successfully modified using positive means. Many behaviors that are targeted for deceleration may be eliminated using reinforcement of incompatible behavior, differential reinforcement of other behavior (DRO), or extinction.

Techniques for decreasing behavior will be discussed in Chapter 7.

Simplest Approach   When the goal is to increase the rate of a targeted behavior instead of decreasing an undesirable behavior, the teacher must also review techniques reported in the literature. A general rule to follow is, Don't kill flies with an elephant gun. In other words, the least elaborate, least expensive, least time-consuming procedure possible should be tried before more complicated efforts are undertaken. Teachers should also be well acquainted with the policy of their employers toward the use of primary reinforcers (such as candy or other food) or other tangibles (such as toys) before instituting a program including such reinforcers.

Students' Needs

The second major consideration when choosing a procedure is the particular student whose behavior is to be changed. Parents and former teachers may be excellent sources of information about kinds of techniques that have been effective in the past, items or events that are likely to be reinforcers for the student, and potential difficulties that may arise. The student may also provide helpful information about effective procedures. Some evidence (Glynn, 1970; Lovitt & Curtiss, 1969) exists that involving students in choosing aspects of the behavior change procedure may, in fact, increase its effectiveness. Such involvement should be even more extensive when self-management procedures are to be used. Parents and students are also an excellent source of information about what the student needs to learn. It may seem obvious that all students need to learn self-care skills and that schools are places where students learn to read. Many things that teachers teach, however, do not so clearly meet generally accepted student needs. It is not appropriate for a teacher arbitrarily to decide what a given student's needs are. The teacher must consider input from the student, her parents, and the community in determining these needs.

Self-management techniques are discussed in Chapter 10.

In addition to considering the effectiveness of the procedure and its appropriateness for a particular student, teachers will want to pay attention to the likelihood that the new behavior can be maintained in the student's normal environment by the people concerned with managing the student's behavior on a daily basis. A strategy, for example, that requires the full-time effort of an adult is unrealistic for a child who must function in a classroom. If such an intervention is required in order to bring about the original change (and it may be in some cases), the program should include a mechanism for maintaining the behavior change in the natural environment and transferring control of the behavior to the people responsible for the student's day-to-day educational progress.

Procedures for maintenance and transfer are described in Chapter 9.

Voluntary Participation

Federal legislation requires that parents consent to programs planned for their handicapped children. If parents are not available, an advocate must be named to determine that any proposed program is in the best interests of the child. Such a requirement is intended to ensure that participation in programs is voluntary. It is not necessary to acquire parental consent for all aspects of a teaching program, however. Martin (1975) suggests that widely accepted strategies for overall classroom management and student motivation do not require anyone's consent, even if the teacher decides to change from one strategy to another. Consent is required for procedures that are not yet widely accepted and for procedures applied only to individual students.

More provisions of PL 94–142.

The consent that ensures voluntary participation in behavior-change programs must be both informed and voluntary. Informed consent is based on full understanding of the planned program. Informed consent does not occur unless parents or other advocates demonstrate that they comprehend all aspects of the program, including possible risks. If necessary, information must be provided in the native language of those involved.

Voluntary consent may be obtained only if neither threats nor rewards are used to acquire such consent (Martin, 1975). It is not acceptable, for example, to tell parents that unless a particular procedure is used, their child will have to be institutionalized. Nor is it ethical to assure parents that if a procedure is used, their child will no longer require placement in a special class.

Sulzer-Azaroff and Mayer (1977) suggest that the voluntariness of students' participation in behavior-change programs should also be considered. Voluntary participation is facilitated by avoiding threats and incentives that are too powerful and by involving the subjects of the programs in selecting as many aspects of the program as possible. Such involvement leads naturally to eventual self-management—the ultimate goal for most students.

Accountability

The final consideration in the ethical and responsible use of applied behavior analysis is accountability. Accountability implies publication of goals, procedures, and results so that they may be evaluated. Applied behavior analysis lends itself easily to such accountability. Goals are stated behaviorally, procedures described clearly, and results defined in terms of direct, functional relationships between interventions and behaviors. It is, indeed, impossible to conduct applied behavior analysis as described by Baer, Wolf, and Risley (1968) without being accountable. The entire process is visible, understandable, and open to evaluation. The result of such accountability is that parents, teachers, administrators, and the public can judge for themselves whether an approach is working or whether a change is needed.

Accountability is a major benefit of applied behavior analysis.

Teachers should not view the requirement of accountability as negative or threatening. It is to a teacher's advantage to verify the effectiveness of his or her teaching. This approach enables teachers to monitor their own competence and to demonstrate this competence to others. It is much more impresssive to face a supervisor at a yearly evaluation conference armed with charts and graphs showing increases in reading ability and decreases in disruptive behavior than it is to walk in with only vague statements about a pretty good year.

To whom are teachers accountable? In terms of ethical behavior, we're afraid the answer is everyone. Teachers are accountable to their profession, to the community, to their administrative superiors, to the parents of their students, to those students, and to themselves.

The teacher who follows the suggestions provided in this chapter should avoid many problems associated with the use of applied behavior analysis procedures in the classroom. Table 2–1 provides a summary of these suggestions. No amount of prevention can forestall all criticism; nor

TABLE 2–1
Suggestions for ethical use of applied behavior analysis

Assure competence of all staff members.
Choose appropriate goals.
Ensure voluntary participation.
Be accountable.

can a teacher avoid making mistakes. However, systematic attention to ethical standards in using ABA procedures can minimize criticism and enable the teacher to learn from mistakes rather than become discouraged by them.

In the final section of this chapter, let's listen in on Professor Grundy, whose workshop discussion may address concerns that you have. All of the questions the professor answers here are inevitably addressed to anyone who undertakes a career as an applied behavior analyst.

Professor Grundy Conducts a Workshop

The superintendent of schools in a large metropolitan area near the university asked Professor Grundy to conduct a one-day workshop on behavior modification for all elementary and secondary teachers. Although aware of the limitations of such short-term workshops (Franks & Wilson, 1976), Grundy concluded that if he confined himself to a description of basic learning principles, no harm would be done. On the appointed day, Grundy, dressed in his best tweed coat with leather elbow patches, stood before 700 teachers, wondering how he got himself into this mess.

After a slow start, during which several teachers fell asleep and numerous others openly graded papers, Grundy hit his stride. He delivered a succinct, snappy talk full of humorous anecdotes and sprinkled with just enough first-name references to his friends, all ''biggies'' in applied behavior analysis and all totally unfamiliar to the teachers. As Grundy reached the conclusion of his presentation, glowing with satisfaction, he noticed to his horror that he was coming up about 45 minutes short of the amount of time agreed to in his contract. Over the thunderous applause (resulting at least partially from the fact that the teachers thought they were going to be released early), Grundy called faintly for questions. There was considerable rumbling and shifting about, but when the superintendent mounted the stage and glared fixedly at the audience, the hands began to go up. The nature of the questions made Grundy vow never to be caught short again, but he did his best to answer each.

Question: Isn't what you're suggesting bribery?

Answer: I'm glad you asked that question. *Webster's New World Dictionary of the American Language* (1956) defines a bribe as something given to induce a person to do something illegal or wrong. In that sense, the use of the principles I have described is certainly not bribery. Teachers use the principles of learning to motivate their students to do things that will benefit them—things like reading, math, and social skills.

A second definition is that a bribe is anything promised or given to induce a person to do something against his or her wishes. Some people might say that's exactly what I'm advocating. As a behaviorist, I have some difficulty with the word *wishes,* since I cannot see wishes but only actions. It appears to me that students have a free choice as to whether they will perform a behavior for which they know they will receive a reinforcer. My interpretation is that if Joanie, for example, chooses to perform the behavior, she has demonstrated her ''wishes.''

The word *bribery* definitely implies something underhanded. I prefer to think of applied behavior analysis procedures as open, honest attempts to change students' behavior in a positive direction. Any other questions? If not . . .

Question: But shouldn't children be intrinsically motivated? Surely they don't have to be rewarded for learning. They should want to learn.

Answer: Madam, why are you here today? I'm sure that given the choice of spending the day at the beach or coming to an inservice session, your intrinsic motivation for learning might have wavered just a little. All of us here are being paid to be here; most adults, even those who enjoy their work enormously, would not continue to perform it in the absence of some very concrete application of the law of positive reinforcement. Why should we expect children to perform difficult tasks for less than we expect of ourselves?

Question: But won't our students expect rewards for everything they do?

Answer: Certainly. And why not? However, as your students become more successful, they will begin to respond to the reinforcers available in the natural environment—the same reinforcers that maintain the appropriate behavior of students who are already successful. Good students do not work without reinforcers; their behavior is reinforced by good grades, by parental approval and, yes, by the love of learning. When doing good work has been consistently reinforced, it does eventually become a secondary, or conditioned, reinforcer. We cannot, however, expect this to happen overnight with students who have had very little experience with success in learning tasks. Does that answer your questions? Thank . . .

Question: Doesn't this kind of behavior management just suppress the symptoms of serious emotional problems without getting at the root cause?

Answer: Oh, my. That's a very complicated question. Behaviorists don't accept the concept of emotional problems caused by some underlying root cause. We have found that if we deal with the problem behaviors, the roots just seem to die out. Human beings are not like weeds whose roots lurk under the surface of the ground waiting to send up shoots as soon as it rains.

Question: Yes, but everyone knows that if you suppress one symptom, a worse one will take its place. Doesn't that prove that there are underlying problems?

Answer: No, sir, everyone does *not* know that. Human beings are no more like piston engines than like weeds. Just because one symptom "goes down" another one will not necessarily "pop up." My colleagues (Baer, 1971; Bandura, 1969; Rachman, 1963; Yates, 1970) have reported extensive research indicating that removal of so-called "symptoms" does not result in the development of new ones. As a matter of fact, when children's inappropriate behaviors are eliminated, they sometimes learn new, appropriate behaviors without being taught (Chadwick & Day, 1971; Morrow & Gochros, 1970; O'Leary, Poulos, & Devine, 1972). Even if new maladaptive behaviors do occur—and they occasionally do (Balson, 1973)—there is no evidence to show that they are alternative symptoms of underlying deviance. Such behaviors may be easily eliminated using applied behavior analysis procedures. Now if . . .

Question: Isn't what you're talking about based on the behavior of animals like rats and monkeys? Are you saying we should treat our kids like animals?

"With automotive 'behavior,' prof, you gotta get right to the cause. Now, after you had the oil changed in 1953 . . ."

Answer: Early research studying the laws of behavior was conducted with animals. This doesn't mean that we control human beings as if they were *nothing but* white rats or pigeons. Such animal research—also called *analogue research* (Davison & Stuart, 1975)—provides only a basic foundation for studying human behavior. The applied behavior analyst uses procedures tested with human beings in real-world conditions—not in laboratories. These procedures take into account the complexity of human behavior and the undeniable freedom of human beings to choose their course of action.

Question: This stuff may work on those special education kids, but my students are smart. Won't they catch on?

Answer: Good heavens, of course they'll catch on. The laws of behavior operate for all of us. We can change behavior in very severely handicapped youngsters, but it's a very complex process. With your students, you can shorten and simplify the procedure. You just tell them what the contingencies are. You don't have to wait for the students to learn from experience. Applied behavior analysis procedures work on everyone; even professors. Take punishment, for example. If I ever agree to do another workshop, it'll be a cold day in. . . . Pardon me. Any more questions?

Question: But how can behavior mod work with my kids? I don't care how many candies you gave them, they still couldn't read.

Answer: Perhaps, madam, you missed some of my presentation during your nap. Applied behavior analysis is not just giving students candy. If your students do not respond verbally to the written word, then you must bring their responses under stimulus control. That's applied behavior analysis. If they have no vocal language, you shape it; that's applied behavior analysis. If they just sit there and do nothing, you get their attention. AND THAT'S APPLIED BEHAVIOR ANALYSIS! ARE THERE ANY MORE QUESTIONS?

Question: I think the whole thing sounds like too much trouble. It seems awfully tedious and time-consuming. Is it really worth the trouble?

Answer: If .. it's .. not .. worth .. the .. trouble .. don't .. do .. it. Behaviors that are serious enough to warrant more complicated procedures take up an enormous amount of your time. You don't use a complicated procedure to solve a simple problem. Try timing yourself with a stopwatch. How much time is this problem taking up the way you're handling it now? (or not handling it now?) Try applying systematic contingencies and keeping records. Then compare the amount of time you've spent. You might be surprised! Now, I really must . . .

Question: I have only one student with really serious problems. If I use some systematic procedure with him, won't the others complain? What do I say to them?

Answer: The problem will not occur as often as you think. Most of the students know that a student who is not performing well needs extra help and are neither surprised nor disturbed when he gets it. Few students will even ask why the problem student is treated differently. If they do, I suggest that you say to them, "In this class everyone gets what he needs. Harold needs a little extra help remembering to stay in his seat." If you consistently reinforce appropriate behavior for all of your students, they will not resent it when a more systematic procedure is implemented for a student with special problems. If that's all, I . . .

Question: Most of my students with problems can't learn much because they come from very bad home situations. There's just nothing you can do in such cases, is there?

Answer: Madam, pigeons can learn to discriminate between environments and to perform the behaviors that will be reinforced in each. Are you implying that your students are less capable than birds? Such an assumption is inhumane. Blaming poor learning or inappropriate behavior on factors beyond your control is a cop-out. You may have very little influence on your students' environment outside your classroom, but you have an enormous influence on that classroom environment. It is your job to arrange it so that your students learn as much as possible, both academically and socially. What do you think teaching is, anyway? *Teach* is a transitive verb. You're not teaching unless you're teaching somebody something.

Question: Have you ever taught school?

At this point Professor Grundy became incoherent and had to be helped from the podium by the superintendent. As he drove home, he realized that he had made a number of mistakes, the first of which was agreeing to do the workshop. He had assumed that teachers expecting to receive concrete help with classroom management problems would be interested in a theoretical discussion of learning principles. He had also assumed that the teachers would immediately see the relationships between these principles and the behaviors of their students. Grundy realized that it was unreasonable of him to expect this. He did decide, however, that he needed to include more practical application in his courses on applied behavior analysis.

THEORY OR RECIPES

Professor Grundy was undoubtedly correct in his belief that the effective use of applied behavior analysis requires a knowledge of the basic principles. Teachers often reject theory and seek immediate practical solutions to specific problems. Such a cookbook approach, however, has serious limitations. While students supplied with cookbook methods may acquire competencies more quickly, the students who are required to spend more time on basic principles tend to show more competence in the long run (White, 1977).

To this point we have spent considerable time discussing definition, historical background, basic principles, and ethical concerns of applied behavior analysis. It is now finally time to go on to what teachers call "the good stuff." In succeeding chapters, we shall discuss practical procedures for changing behavior in classroom settings and provide a more thorough discussion of basic learning principles. By striking a balance, we hope to offer not only a review of tried-and-true solutions to classroom problems, but also the knowledge to enable teachers to develop their own solutions.

SUMMARY

This chapter described several objections to the use of applied behavior analysis techniques. These techniques have been criticized on the basis that they interfere with personal freedom and that they are inhumane. We have described our reasons for disagreeing with these objections. Properly used, applied behavior analysis procedures enhance personal freedom by increasing options. Applied behavior analysis procedures are humane because they are an effective tool for increasing options and teaching appropriate skills.

Applied behavior analysis procedures will be used ethically if the program includes competent practitioners, appropriate goals, voluntary participation in behavior-change programs, and a commitment to accountability. Teachers who choose applied behavior analysis procedures and consider these factors know that they are acting in the best interests of their students.

REFERENCES

AYLLON, T., LAYMAN, D., & KANDEL, H.J. 1975. A behavioral-educational alternative to drug control of hyperactive children. *Journal of Applied Behavior Analysis, 8*, 137–146.

AYLLON T., & ROBERTS, M.D. 1974. Eliminating discipline problems by strengthening academic performance. *Journal of Applied Behavior Analysis, 7*, 71–76.

BAER, D.M. 1971. Behavior modification: You shouldn't. In E.A. Ramp and B.L. Hopkins (Eds.), *A new direction for education: Behavior analysis* (vol. 1). Lawrence: University of Kansas, Support and Development Center for Follow Through.

BAER, D.M., WOLF, M.M., & RISLEY, T.R. 1968. Some current dimensions of applied behavior analysis. *Journal of Applied Behavior Analysis, 1*, 91–97.

BALSON, P.M. 1973. Case study: Encopresis: A case with symptom substitution. *Behavior Therapy, 4,* 134–136.

BANDURA, A. 1969. *Principles of behavior modification*. New York: Holt, Rinehart and Winston, Inc.

BANDURA, A. 1975. The ethics and social purposes of behavior modification. In C.M. Franks and G.T. Wilson (Eds.), *Annual review of behavior therapy, theory & practice* (Vol. 3, pp. 13–20). New York: Brunner/Mazel.

BECKER, W.C., ENGELMANN, S., & THOMAS, D.R. 1975. *Teaching 1: Classroom management*. Chicago: Science Research Associates, Inc.

BEREITER, C., & ENGELMANN, S. 1966. *Teaching disadvantaged children in the preschool*. Englewood Cliffs, N.J.: Prentice-Hall.

BIRNBRAUER, J.S., BIJOU, S.W., WOLF, M.M., & KIDDER, J.D. 1965. Programmed instruction in the classroom. In L.D. Ulman & L. Krasner (Eds.), *Case studies in behavior modification*. New York: Holt, Rinehart & Winston.

BIRNBRAUER, J.S., WOLF, M. M., KIDDER, J.D., & TAGUE, C.E. 1965. Classroom behavior of retarded pupils with token reinforcement. *Journal of Experimental Child Psychology, 2,* 219–235.

BOSTOW, D.E., & BAILEY, J. 1969. Modification of severe disruptive and aggressive behavior using brief timeout and reinforcement procedures. *Journal of Applied Behavior Analysis, 2,* 31–37.

BUCHER, B., & LOVAAS, O.I. 1968. Use of aversive stimulation in behavior modification. In M. R. Jones (Ed.), *Miami Symposium on the Prediction of Behavior, 1967: Aversive Stimulation*. Coral Gables, Fla.: University of Miami Press.

CHADWICK, B.A., & DAY, R.C. 1971. Systematic reinforcement: Academic performance of underachieving students. *Journal of Applied Behavior Analysis, 4,* 311–319.

CRAIGHEAD, W.E., KAZDIN, A.E., MAHONEY, M.J. 1976. *Behavior modification: Principles, issues, and applications*. Boston: Houghton Mifflin Company.

DAVISON, G.C., & STUART, R.B. 1975. Behavior therapy and civil liberties. *American Psychologist, 30*(7), 755–763.

FERRITOR, D.E., BUCKHOLDT, D., HAMBLIN, R.L., & SMITH, L. 1972. The noneffects of contingent reinforcement for attending behavior on work accomplished. *Journal of Applied Behavior Analysis, 5,* 7–17.

FRANKS, C.M., & WILSON, G.T. (Eds.). 1976. *Annual review of behavior therapy, theory & practice*. New York: Bruner/Mazel.

GLYNN, E.L. 1970. Classroom applications of self-determined reinforcement. *Journal of Applied Behavior Analysis, 3,* 123–132.

GOETZ, E.M., & BAER, D.M. 1973. Social control of form diversity and the emergence of new forms in children's blockbuilding. *Journal of Applied Behavior Analysis, 6,* 209–217.

GOETZ, E.M., & SALMONSON, M.M. 1972. The effect and general and descriptive reinforcement on "creativity" in easel painting. In G. Semb (Ed.), *Behavior analysis and education—1972*. Lawrence: Kansas University Department of Human Development.

GOLDIAMOND, I. 1974. Toward a constructional approach to social problems. *Behaviorism, 2,* 54–78.

GOLDIAMOND, I. 1975. Toward a constructional approach to social problems: Ethical and constitutional issues raised by applied behavior analysis. In C.M. Franks & G.T. Wilson (Eds.), *Annual review of behavior therapy, theory & practice* (Vol. 3, pp. 21–63). New York: Brunner/Mazel Publishers.

HOLDEN, C. 1973. Psychosurgery: Legitimate therapy or laundered lobotomy? *Science, 173,* 1104–1112.

KAZDIN, A.E. 1978. *History of behavior modification*. Baltimore, Md.: University Park Press.

KIRBY, F.D., & SHIELDS, F. 1972. Modification of arithmetic response rate and attending behavior in a seventh-grade student. *Journal of Applied Behavior Analysis, 5,* 79–84.

LOVAAS, O.I., FREITAG, G., GOLD, V.J., & KASSORLA, I.C. 1965. Experimental studies in childhood schizophrenia: Analysis of self-destructive behavior. *Journal of Experimental Child Psychology, 2,* 67–84.

LOVAAS, O.I., & SIMMONS, J.Q. 1969. Manipulation of self-destruction in three retarded children. *Journal of Applied Behavior Analysis, 2,* 143–157.

LOVITT, T.C., & CURTISS K.A. 1969. Academic response rate as a function of teacher- and self-imposed contingencies. *Journal of Applied Behavior Analysis, 2,* 49–53.

MADSEN, C.K., & MADSEN, C.H. 1970. *Parents-children-discipline: A positive approach*. Boston: Allyn & Bacon.

MAHONEY, M.J., KAZDIN, A.E., & LESSWING, N.J. 1974. Behavior modification: Delusion or deliverance? In C.M. Franks & G.T. Wilson (Eds.), *Annual review of behavior therapy, theory & practice* (vol. 2, pp. 11–40). New York: Brunner/Mazel Publishers.

MALONEY, K.B., & HOPKINS, B.L. 1973. The modification of sentence structure and its relationship to subjective judgements of creativity in writing. *Journal of Applied Behavior Analysis, 6*, 425–433.

MARTIN, R. 1975. *Legal challenges to behavior modification: Trends in schools, corrections, and mental health*. Champaign, Ill.: Research Press.

MASON, B. 1974. Brain surgery to control behavior. *Ebony, 28*(4), 46.

MCCONNELL, J.V. 1970. Stimulus/response: Criminals can be brain-washed now. *Psychology Today, 3*, 14–18, 74.

Morales v. Turman, 383 F. Supp 53 (E.D. Tex. 1974).

MORROW, W.R., & GOCHROS, H.L. 1970. Misconceptions regarding behavior modification. *The Social Service Review, 44*, 293–307.

O'CONNOR, R.D. 1969. Modification of social withdrawal through symbolic modeling. *Journal of Applied Behavior Analysis, 2*, 15–22.

O'LEARY, K.D. 1972. The assessment of psychopathology in children. In H.C. Quay & J.S. Werry (Eds.), *Psychopathological disorders of childhood* (pp. 234–272). New York: John Wiley & Sons, Inc.

O'LEARY, K.D., POULOS, R.W., & DEVINE, V.T. 1972. Tangible reinforcers: Bonuses or bribes? *Journal of Consulting and Clinical Psychology, 38*, 1–8.

RACHMAN, S. 1963. Spontaneous remission and latent learning. *Behavior Research and Therapy, 1*, 3–15.

REESE, H.W., & PARNES, S.J. 1970. Programming creative behavior. *Child Development, 41*, 413–423.

REYNOLDS, N.J., & RISLEY, T.R. 1968. The role of social and material reinforcers in increasing talking of a disadvantaged preschool child. *Journal of Applied Behavior Analysis, 1*, 253–262.

RISLEY, T.R. 1968. The effects and side effects of punishing the autistic behaviors of a deviant child. *Journal of Applied Behavior Analysis, 1*, 21–34.

RISLEY, T.R. 1975. Certified procedures not people. In W.S. Wood (Ed.), *Issues in evaluating behavior modification,* pp. 159–181. Champaign, Ill.: Research Press.

ROGERS, C.R., & SKINNER, B.F. 1956. Some issues concerning the control of human behavior: A symposium. *Science, 124*, 1057–1066.

SIMPSON, R.G., & EAVES, R.C. 1985. Do we need more qualitative research or more good research? A reaction to Stainback and Stainback. *Exceptional Children, 51*, 325–329.

STAINBACK, S., & STAINBACK W. 1984. Broadening the research perspective in special education. *Exceptional Children, 50*, 400–408.

STOLZ, S.B. 1977. Why no guidelines for behavior modification? *Journal of Applied Behavior Analysis, 10*, 541–547.

SULZER-AZAROFF, B., & MAYER, G.R. 1977. *Applying behavior-analysis procedures with children and youth*. New York: Holt, Rinehart and Winston.

SULZER-AZAROFF, B., THAW, J., & THOMAS, C. 1975. Behavioral competencies for the evaluation of behavior modifiers. In W.S. Wood (Ed.), *Issues in evaluating behavior modification,* pp. 47–98. Champaign, Ill.: Research Press.

Webster's new world dictionary of the American language. 1956. Cleveland: World Publishing.

WHITE, O.R. 1977. Behaviorism in special education: An arena for debate. In R.D. Kneedler & S. G. Tarber (Eds.), *Changing perspectives in special education*. Columbus, Ohio: Charles E. Merrill.

WICKER, T. February 8, 1974. A bad idea persists. *The New York Times*, p. 31.

WINETT, R.A., & WINKLER, R.C. 1972. Current behavior modification in the classroom: Be still, be quiet, be docile. *Journal of Applied Behavior Analysis, 5*, 499–504.

WOLF, M.M. 1978. Social validity: The case for subjective measurement or how applied behavior analysis is finding its heart. *Journal of Applied Behavior Analysis, 11*. 203–214.

WOLF, M.M., RISLEY, T.R., & MEES, H.L. 1964. Application of operant conditioning procedures to the behavior problems of an autistic child. *Behavior Research and Therapy, 1*, 305–312.

WOOD, W.S. (Ed.). 1975. *Issues in evaluating behavior modification.* Champaign, Ill.: Research Press.

Wyatt v. Stickney, 344F. Supp. 373, 344F. Supp. 387 (M.D. Ala. 1972) affirmed sub nom. *Wyatt v. Aderholt,* 503 F. 2nd 1305 (5th Cir. 1974).

YATES, A.J. 1970. *Behavior therapy.* New York: Wiley.

ZEILGERGER, J., SAMPEN, S.E., & SLOANE, H.N., JR. 1968. Modification of a child's problem behaviors in the home with the mother as therapist. *Journal of Applied Behavior Analysis, 1,* 47–53.

Section Two

Technology of Behavior-Change Programs

3 *Preparing Behavioral Objectives*

This chapter discusses the first step in carrying out a program for behavior change—defining the target behavior; that is, the behavior to be changed. As Chapter 1 stated, a target behavior may be selected because it represents either a behavioral deficit (such as too little reading) or a behavioral excess (such as too much screaming). After the behavior to be changed has been identified, teachers formulate a written **behavioral objective.**

Definition

A behavioral objective is a statement that communicates a proposed change in behavior. It describes a level of performance and serves as a basis for evaluation.

A behavioral objective for a student who demonstrates a deficit in reading behavior would describe the level of reading performance the student should reach. A behavioral objective for a student who screams excessively would describe an acceptable level of screaming. Anyone reading a behavioral objective should be able to understand exactly what a student is working to accomplish. Because behavioral objectives are such an integral part of planning for student behavior change, they are required as part of the IEP for handicapped students. The relationship between objectives and the IEP will be discussed later in the chapter.

In this chapter, you'll also meet some teachers who are learning to use a behavioral approach in their teaching. Through them, you'll encounter some of the difficulties of putting behavioral programs into effect. Consider the plight of Ms. Samuels, the resource teacher, in the following vignette.

Are We Both Talking About the Same Thing?

Ms. Wilberforce, the third-grade teacher, was in a snit.

"That resource teacher," she complained to her friend, Ms. Folden, "is absolutely useless. I asked her two months ago to work on vowels with Martin and he still doesn't know the short sounds."

"You're absolutely right," agreed Ms. Folden, "I told her last September that Melissa Sue had a bad attitude. The longer Melissa Sue goes to see the resource teacher, the worse it gets. All she does is giggle when I correct her. It seems to me that we were better off without resource teachers."

Meanwhile, Ms. Samuels, the resource teacher, was complaining bitterly to her supervisor.

"Those regular classroom teachers are so uncooperative. They don't seem to appreciate what I'm doing at all. Just look at what I've done with Martin. He can name all the vowels when I ask him, and he even knows a little song about them. And Melissa Sue, who used to pout all the time, smiles and laughs so much now. I've done exactly what the teachers asked—why don't they appreciate it?"

DEFINITION AND PURPOSE

Behavioral objectives improve communication.

The preceding vignette illustrates one of the most important reasons for writing behavioral objectives: to clarify the goals of a behavior-change program and thus to facilitate communication among people involved in the program. Because it is a written statement targeting a specific change in behavior, the objective serves as an agreement among school personnel and parents about the academic or social learning goals for which school personnel are taking responsibility. The statement may also serve as a vehicle for informing the student of the goals he is expected to reach.

A second important reason for writing behavioral objectives is that a clearly stated target for instruction facilitates effective programming by the teacher and ancillary personnel. A clearly stated instructional target provides a basis for selecting appropriate materials and instructional strategies. Mager (1962) pointed out that "a machinist does not select a tool until he knows what operation he intends to perform; a composer does not orchestrate a score until he knows what effects he wishes to achieve; a builder does not select his materials or specify a schedule until he has his blueprints" (p. 3). Clearly written behavioral objectives should prevent the classroom teacher from using materials simply because they are available or strategies simply because they are familiar. The selection of materials and teaching strategies is more likely to be appropriate if goals are clearly defined.

There is yet another excellent reason for writing behavioral objectives. Consider the following account.

A Matter of Opinion

Mr. Henderson, the teacher of a self-contained class for severely retarded students, hurried to the principal's office in a state of complete panic. The parents of Alvin, one of his students, had just threatened to remove the boy from school. They insisted that Mr. Henderson was not teaching Alvin anything. Mr. Henderson had agreed in August to work on toilet training with Alvin and felt that the boy had made excellent progress. Alvin's parents, however, were upset because Alvin still had several accidents a week. they insisted that Mr. Henderson had not reached his stated goal.

"I have toilet trained Alvin," howled Mr. Henderson. "Wouldn't you consider only two or three accidents all right?"

Behavioral objectives help evaluate progress.

Mr. Henderson's problem could have been prevented if a clearly written objective statement had come out of the August meeting. If a definition of *toilet-trained* had been stated at the beginning of the year, there would have been no question as to whether the goal had been achieved. Behavioral objectives provide for precise evaluation of instruction. When a teacher identifies a deficit or an excess in a student's behavioral repertoire, he or she had identified a discrepancy between real and expected levels of functioning. If the teacher states a performance criterion (the ultimate goal) and records ongoing progress toward this goal, both formative (ongoing) and summative (final) evaluation of intervention procedures becomes possible for program alteration and future planning. Such ongoing evaluation and measurement enables the teacher, the student, or a third party to monitor the student's progress continuously and to determine when goals have been reached. Continuous monitoring ensures that individual interpretations or prejudices in judgment are minimized in the evaluation of an instructional procedure or a student's performance.

Pinpointing Behavior

Before initiating a behavior-change program for a student, it is necessary to describe the target behavior clearly. Referral information may often be vague and imprecise. In order to write effective objectives for guiding educational programs, the applied behavior analyst must refine broad generalizations into specific, observable, measurable behaviors. This process is frequently referred to as *pinpointing behavior.*

Pinpointing may be accomplished by asking a series of questions, usually including, "Could you please tell me what he *does*?" or "What exactly do you want her to *do*?" For example, teachers often refer students to a behavior analyst because of "hyperactivity." The referring teacher and the applied behavior analyst must define this hyperactive behavior by describing exactly what is occurring. Is it that the student, like Ralph in Ms. Harper's class, wanders around the room? Does he tap his pencils on the desk? Does he weave back and forth in his chair?

There are many categories of behavior that may result in referrals and require pinpointing. Here are a few examples, with some questions that may help to refine the definition:

"Jean can't do math": Is the problem that he doesn't have basic arithmetic computation skills or that he can't finish his problems within the time limit set?

"Virginia is always off-task": Is the problem that she stares out the window, or is it that she talks to her neighbor, or is it that she scribbles in her book instead of looking at the chalkboard?

"Arthur is always disturbing others": Is he grabbing objects from someone, or talking to others during lessons, or hitting his neighbor, or knocking neighbors' books off the desk, or pulling someone's hair?

"Lauren throws tantrums": Is she crying and sobbing, or is she throwing herself on the floor, or is she throwing objects around the room?

A teacher may ask a similar series of questions in describing more complex or abstract categories of behaviors. If the referring teacher said "Carol doesn't use the critical thinking skills in reading," the applied behavior analyst would want to know if Carol

1. distinguishes between facts and opinions
2. distinguishes between facts and inferences
3. identifies cause-effect relationships
4. identifies errors in reasoning
5. distinguishes between relevant and irrelevant arguments
6. distinguishes between warranted and unwarranted generalizations
7. formulates valid conclusions from written material
8. specifies assumptions needed to make conclusions true. (Gronlund, 1978, p. 14)

With certain behaviors, questioning or redefining may need to go a step further. If a student is often out of his chair, the teacher's concern may be either the number of times he gets out of his chair or the length of time he stays out. Each is a very different behavior and will, therefore, require different intervention strategies and data-collection techniques. In the case of broad behavior categories such as temper tantrums, for which a teacher must pinpoint multiple behaviors occurring simultaneously it will be necessary to list the behaviors in some order of priority. For instance, they might be listed in order from least-to-most interference to the child or to the environment. After referral information has been refined so that target behaviors can be clearly described, educational goals and eventually behavior objectives can be written.

EDUCATIONAL GOALS

Goals precede objectives.

Behavioral objectives should not be created on the spur of the moment. Objectives should be derived from a set of **educational goals** that provide the framework for the academic year. These goals should evolve from an accumulation of evaluation information and should be correlated with curriculum planning. Goals define the anticipated academic and social development for which the school is taking responsibility. During goal selection, educators estimate what proportion of the student's educational potential is to be developed within the next academic year. Thus, educational goals (long-term objectives) are statements of program intent whereas behavior objectives (short-term or instructional objectives) and statements of actual classroom instructional intent.

Establishing Goals

A "staffing committee" or "multidisciplinary team" usually is responsible for setting goals for students who need special services and have been formally referred. When gathering data upon which to base a student's educational program, the staffing committee (or an equivalent group of professionals from various disciplines) will review the results of various evaluations. These evaluations represent information that has been generated (often by educational specialists and consultants) to determine the student's current levels of functioning. These data may include screening reports from

Formal sources of information for goal-setting.

1. School psychology: for example, scores on the Wechsler Intelligence Scale for Children (Wechsler, 1974), Bayley Scales of Infant Development (Bayley, 1969), or Behavior Problem Checklist (Quay & Peterson, 1975).

2. Education: for instance, scores from the Peabody Individual Achievement Test (Dunn & Markwardt, 1970) or the Developmental Activities Screening Inventory (DuBose & Langley, 1977).

3. Medicine: for example, neurological, pediatric, vision, hearing screenings.

4. Therapeutic services: such as results of physical therapy, occupational therapy, speech pathology, recreation therapy.

5. Adaptive behavior: for instance, scores on the AAMD Adaptive Behavior Scales (Nihira, Foster, Shellhaas, & Leland, 1975) or the System of Multicultural Pluralistic Assessment (Mercer, 1979).

This information pool may also include diagnostic information from instruments such as the Brigance Inventory of Basic Skills (Brigance, 1977), the Uniform Performance Assessment System (Benderesky, 1978), the Behavioral Characteristics Progression (Santa Cruz County, 1973), or from diagnostic reading or mathematics tests.

Informal sources of information for goal-setting.

In addition to these more formal sources, the goal-setting group should also consider the concerns of the parent who is automatically included in such meetings. Recommendations from previous teachers will be considered as well. The environmental demands of the present classroom, the home, the projected educational placement, or a projected worksite should also be examined.

These data reveal the student's current functioning in the various areas, certain deficits defined by chronological-age or mental-age expectancies, or specific behavior excesses, any of which may become likely targets for a behavior-change program. Based upon this accumulated information, the committee proposes a set of educational goals for the student. An estimate of progress is then included in the long-term objectives prepared for the student's Individualized Education Program (IEP).

In many regular classrooms, the establishment of educational goals may not involve such an extensive accumulation of information. Assessment may be limited to group achievement tests supplemented by informal teacher-made assessments. Goal setting may also be constrained by the adopted grade-level curriculum guide. For example, the school board and district administrators or curriculum coordinators usually provide a standard set of educational goals for each class of fourth graders. Such goals assume equal background and achievement. Under a standard curriculum, for example, all students at a certain grade level are to be instructed in natural resources of Peru, the excretory system of the earthworm, multiplication of fractions, and reading comprehension. The teacher's task, then, becomes translating these goals into reasonable objectives for each member of a particular class. The teacher may write behavioral objectives for the class as a whole, giving consideration to the general characteristics of the group. In addition, if the teacher is to help a particular student who is having problems or to deal with the "tortoise" reading group, that teacher may need to write additional behavioral objectives to prescribe a course of instruction that will facilitate learning.

Educational goals for individual students must be developed on the basis of evaluation data but should also consider other factors:

1. The student's past and projected rate of development compared with long-range plans for his or her future.

2. The presenting physical and communicative capabilities of the student.

3. Inappropriate behaviors that must be brought under control because they interfere with learning.

4. Skills the student lacks for appropriate functioning in the home, school, and social environments.

5. The amount of instructional time available to the student within the school day and within the total school experience.

6. The prerequisites necessary for acquiring new skills.

7. The functional utility of the skills (what additional skills may be built upon these?)

8. The availability of specialized materials, equipment, or resource personnel (speech pathologist, remedial reading teacher, and so on).

Write goals in observable and quantifiable terms.

Because educational goals are projected over long periods of time, they are written in broad terms. For practical application, however, they need to write in terms that are observable and quantifiable. As you learned in Chapter 1, applied behavior analysts deal only with observable behaviors.

For students who are not handicapped or who have only mild handicaps, goals are needed only for each curriculum area. For severely handicapped or very young students, goals should be written in a number of *domains* of learning:

1. Cognitive
2. Language
3. Motor
4. Social
5. Self help
6. Maladaptive behavior

Hypothetical long-term goals for Bill, a mildly handicapped student with learning problems in the area of math, and for a severely handicapped student, Thom, are listed below.

Long-term educational goals for a mildly handicapped student.

Bill will

Arithmetic	Master basic computation facts at the first grade level.
Social studies	Demonstrate knowledge of the functions of the three branches of the federal government.
Reading	Be able to identify relevant parts of a story he has read.
Science	Demonstrate knowledge of the structure of the solar system.
Language arts	Increase the creative expression of his oral language.
Physical education	Increase his skills in team sports.

Bill's regular classroom teacher would be responsible for setting all goals except the one in arithmetic. Bill will probably go to a part-time special education class (resource room) for arithmetic. Compare Bill's goals with Thom's.

*Long-term educational
goals for a severely
handicapped student.*

Thom will

Cognitive	Be able to categorize objects according to their function.
Language	Demonstrate increased receptive understanding of functional labels.
Motor	Develop gross motor capability of his upper extremities.
Social	Learn to participate appropriately in group activities.
Maladaptive behavior	Decrease out-of-seat behavior.
Self-help	Demonstrate the ability to dress himself independently.

The teacher must convert these broad goals into statements of instructional intent (behavioral objectives) for actual classroom use. A behavioral objective is not a restatement of a goal; instead it breaks a goal into instructional components. Because of the complexity of certain goals, a given goal may have more than one objective. A goal that states that a student will learn to play cooperatively with other children, for example, may require individual objectives identifying the need to share with others, to take turns, and to follow the rules of the game.

Components of a Behavioral Objective

In order to communicate all of the intended information and provide a basis for evaluation, a complete behavioral objective should

1. Identify the learner
2. Identify the target behavior
3. Identify the conditions under which the behavior is to be displayed
4. Identify criteria for acceptable performance

Identify the Learner

Use the student's name.

Behavioral objectives were initially designed to promote individualization of instruction (Gagne, 1970). As part of the continuing effort to emphasize the process of individualization, in each objective statement the teacher must re-identify the specific student or students for whom the objective has been developed. Restatement reinforces the teacher's focus upon the learner in question and communicates this focus to others who may become involved in the intervention program. Thus we include in a behavioral objective statements such as

John will . . .

The fourth-graders will . . .

The participants in the training program will . . .

Identify the Target Behavior

After pinpointing and goal setting have resulted in agreement upon the deficient or excessive target behavior, the teacher must state what the stu-

State what the student will do.

dent will be doing when the desired change has been achieved. This statement must identify a precise response that is representative of the target behavior.

There are three basic purposes for including this component in the behavioral objective:

1. It ensures that the teacher is consistently observing the same behavior. The observation and recording of the occurrence or nonoccurrence of exactly the same behavior allows for an accurate and consistent reflection of the behavior in the data to be collected.

2. The statement of the target behavior allows for confirmation by a third party that the change observed by the teacher has actually occurred.

3. The precise definition of the target behavior facilitates continuity of instruction when people other than the teacher are involved.

In order to achieve these three purposes, the target behavior must be described so that its occurrence is verifiable. Precise description will minimize differing interpretations of the same behavior. A student's performance of a given behavior can best be verified when the teacher is able to see or hear the behavior or to see or hear a direct product of the behavior. To attain this precision and clarity in an objective, the verb used to delineate the behavioral response should describe a behavior that is directly *observable, measurable,* and *repeatable.*

Though teachers of the gifted would like students to "discover" and art teachers would like students to "appreciate," objectives described in this manner are open to numerous interpretations. For example, it would be difficult for a third party to decide whether a student were performing the following behaviors:

. . . will *recognize* the difference between big and little

. . . will *understand* the value of coins

. . . will *develop* an appreciation of Chaucer

. . . will *remain* on-task during reading group

Using such broad terms, any individual who works with a student is, at best, inferring the existence of some behavior (guessing). These four descriptions would require yet another step to increase their specificity: the use of an **operational definition.** Operationally defining a vague term provides examples of what action(s) would confirm that the student is exhibiting the target behavior. In other words, the observer would look to the operational definition to determine the observable action(s) the student would perform to indicate he or she were "understanding" or "on-task."

The need for added explanation through an operational definition is reduced when the behavior analyst uses more precise verbs *in the objective.* Increased precision also promotes more accurate recording of data. A precise behavioral description such as "will circle," "will state orally," or "will point to" is less likely to be interpreted differently. With such a description, individuals can agree on whether an action did or did not take place without calling for a conference or lengthy written explanation. Here are some examples of precise behavioral descriptions:

... will point to the little car

... will verbally count the equivalent in pennies

... will write a translation of the prologue to *Canterbury Tales*

... will face his book or the speaker

One guide for selecting appropriate verbs has been offered by Deno and Jenkins (1967). Their classification of verbs is based upon agreement of occurrence between independent classroom observers. They arrived at the three sets of verbs shown in Table 3–1, categorized as *directly observable action verbs, ambiguous action verbs,* and *not directly observable action verbs.*

In addition to these verbs, which are appropriate for more general areas of instruction, individual discipline specialists may derive observable verbs specific to their needs and activities, as in these examples:

Fine Arts

blend, brush, carve, drill, fold, hammer, heat, melt, mix, paste, flour, roll, sand, saw, shake, sketch, trace, trim varnish.

Drama

clasp, cross, direct, display, enter, exit, move, pantomime, pass, perform, sit, start, turn.

Laboratory Science

calibrate, connect, convert, dissect, feed, insert, lengthen, limit, manipulate, operate, plant, remove, (re)set, straighten, time, transfer, weigh.

Language

abbreviate, accent, alphabetize, capitalize, define, edit, hyphenate, indent, print, pronounce, punctuate, recite, sign, spell, state, summarize, syllabicate, translate.

Music

blow, bow, clap, compose, finger, harmonize, hum, mute, play, pluck, practice, sing, strum, tap, whistle.

Physical Education

catch, climb, float, grasp, hit, jump, kick, lift, pitch, pull, push, run, skip, somersault, swim, swing, throw.

TABLE 3–1
Observability
classification of
verbs

Action Verbs That Are Directly Observable		
to cover with a card	to draw	to place
to mark	to lever press	to cross out
to underline	to point to	to circle
to repeat orally	to walk	to say
to write	to count orally	to read orally
to shade	to put on	to name
to fill in	to number	to state
to remove	to label	to tell what

Ambiguous Action Verbs		
to identify in writing	to check	to construct
to match	to take away	to make
to arrange	to finish	to read
to play	to locate	to connect
to give	to reject	to select
to choose	to subtract	to change
to use	to divide	to perform
to total	to add	to order
to measure	to regroup	to supply
to demonstrate	to group	to multiply
to round off	to average	to complete
to inquire	to utilize	to summarize
to acknowledge	to find	to borrow
to see	to convert	to identify

Action Verbs That Are Not Directly Observable		
to distinguish	to be curious	to solve
to conclude	to apply	to deduce
to develop	to feel	to test
to concentrate	to determine	to perceive
to generate	to think	to create
to think critically	to discriminate	to learn
to recognize	to appreciate	to discover
to be aware	to become competent	to know
to infer	to wonder	to like
to realize fully	to analyze	to understand

In order to evaluate a description of a target behavior, Morris (1976) suggests using his IBSO (Is the Behavior Specific and Objective) Test questions:

1. Can you count the number of times that the behavior occurs in, for example, a 15-minute period, a one-hour period, or one day? Or, can you count the number of minutes that it takes for the child to perform the behavior? That is, can you tell someone that the behavior occurred "X" number of times or "X" number of minutes today? (Your answer should be yes.)

2. Will a stranger know exactly what to look for when you tell him/her the target behavior you are planning to modify? That is, can you actually see the child performing the behavior when it occurs? (Your answer should be yes.)

3. Can you break down the target behavior into smaller components, each of which is more specific and observable than the original target behavior? (Your answer should be no.) (p. 19)

Identify the Conditions of Intervention

Conditions are antecedent stimuli tied to the target behavior.

The statement of conditions, the third component of a behavioral objective, is a description of the antecedent (behavior-preceding) stimuli in the presence of which the target behavior is expected to occur. This part of the objective affirms that the learning experience can be consistently reproduced.

The teacher may set up an occasion for the appropriate response using any or all of several categories of antecedent stimuli:

1. Verbal requests or instructions
Sam, point to the little car.
Debbie, add these numbers.
Jody, go back to your desk.

2. Written instructions
Diagram these sentences.
Find the products.
Draw a line from each word to its definition.

3. Demonstration

The teacher must be sure that the verbal or visual cue planned does in fact provide an opportunity for the desired response in the student. That is, a teacher should deliver an unambiguous request or instruction to the student. The teacher who holds a flashcard with the word "get" and says, "Give me a sentence for get," is likely to hear, "I for get my milk money," or "I for get my homework."

Providing appropriate antecedent stimuli will be discussed in Chapter 9.

The materials described in the objective should ensure stimulus consistency for the learner and reduce the chance for inadvertent, subtle changes in the learning performance being requested. For example, presenting a red, a blue, and a green block and asking the student to "Point to red" is a less complex task than presenting a red car, a blue sock, and a green cup and making the same request. Giving the student a page with written instructions to fill in the blanks in sentences is less complex when a list of words is provided that includes the answers. Asking the student to write a story based on a stimulus picture is different from asking the student to write a story without visual stimulus.

The following are examples of condition statement formats:

Given an array of materials containing . . .

Given a textbook containing 25 division problems with single-digit divisors . . .

Given the manual sign for toilet . . .

Given the use of a thesarus and the written instruction . . .

Given a pullover sweater with the label cued red and the verbal cue . . .

Given a ditto sheet with 20 problems containing improper fractions having unlike denominators and the written instruction to "Find the products" . . .

Without the aid of . . .

Careful statement of the conditions under which the behavior is to be performed may prevent problems like the one encountered by Ms. Samuels in the vignette that follows.

Ms. Samuels Teaches Long Division

Ms. Samuels was once again in trouble with a regular classroom teacher. She and Mr. Watson, the sixth-grade math teacher, had agreed that she would work on long division with Harvey. Ms. Samuels had carefully checked to be sure that the method she taught Harvey to use in the resource room was the same as Mr. Watson's. She made dozens of practice worksheets and Harvey worked long division problems until he could do them in his sleep. Ms. Samuels was predictably horrified then, when Mr. Watson asked her if she ever planned to start working on division with Harvey. Investigation revealed that in the regular classroom, Harvey was expected to copy the problems from the math book to notebook paper and that he made so many copying errors that he seldom got the correct answer. The conditions under which the task was to be performed were thus significantly different.

As part of a plan of instruction for students with learning handicaps, teachers may need to include extra support in the form of supplementary cues, such as a model of a completed long division problem for the student to keep at his desk. It is important to include a description of such supplementary cues in the condition component of the behavioral objective so that no misunderstanding will occur. When the cue is no longer needed, the objective may be rewritten.

Identify Criteria for Acceptable Performance

Criterion statements set minimum performance standards.

In the criterion statement included in a behavioral objective, the teacher sets the standard for minimally acceptable performance. This statement indicates the level of performance the student will be able to achieve as a result of the intervention. The performance itself has been defined; the criterion sets the standard for evaluation. Throughout the intervention process, this criterion is used to measure the effectiveness of the intervention strategy selected to meet the behavioral objective.

The basic criterion statement for initial learning or acquisition indicates the *accuracy* of a response or the response's *frequency of occurrence*. Such statements are written in terms of the number of correct responses, the student's accuracy on trial presentations, the percentage of accurate responses, or some performance within an error of limitation. Here are some sample criterion statements:

. . . 17 out of 20 correct responses

... (label) all 10 objects correctly

... with 80% accuracy

... 20 problems must be answered correctly (100% accuracy)

... four out of five trials correct

... on five consecutive successful trials

... complete all steps in the toileting program independently

... (list) all four of the main characters within a book report of no less than 250 words

... with no more than five errors in spelling

See Chapter 4 for an extended discussion of duration and latency.

Two additional types of criteria may be included when time is a critical dimension of the behavior. *Duration* is a statement of the length of time the student performs the behavior. *Latency* is a statement of the length of time that elapses before the student begins performing the behavior.

Criterion statements addressing duration:

... for at least 20 minutes

... for no more than ½ hour

... with 2 weeks

Statements addressing latency:

... within 10 seconds after the flashcard is presented

... within 1 minute after a verbal request

In selecting criteria for acceptable performance, the teacher must be careful to set a reasonable level of performance. Selection should reflect the nature of the content, the abilities of the student, and the learning opportunity being provided.

Writing criterion statements in terms of percentage requires care. How many problems, for example, would a student have to work correctly to satisfy a 90% criterion on a 5-item math test?

Certain types of content require particular criterion levels. When a student is acquiring basic skills upon which other skills will be built, a criterion of 90% may not be high enough. For example, learning "almost all" of the multiplication facts may result in a student's going through life never knowing what 8×7 is. There are other skills as well that require 100% accuracy. Remembering to look both ways before crossing the street only 90% of the time may result in premature termination of the opportunity for future learning!

For certain students with physical handicaps, a disability may influence the force, direction, or duration of the criterion set by the teacher. For example, a student may not be able to hammer a nail all the way into a piece of wood; range-of-motion limitations may influence motor capability for reaching; hypotonic muscles (those with less-than-normal tone) may limit the duration of walking or sitting; or a spastic condition may limit the perfection of cursive handwriting.

In addition to considering the number or percentage correct and the accuracy of response, writers of behavioral objectives must also determine the number of times a student must meet a criterion to demonstrate mastery. For example, how often must Jane perform a behavior successfully on 8 out of 10 trials before the teacher will be convinced of mastery and allow her to move on to the next level of learning or to the next behavioral objective?

It may be inferred from an open-ended criterion statement that the first time a student reaches 85% accuracy, the skill will be considered "learned" or that from now until the end of the school year, the teacher will continually test and retest to substantiate the 85% accuracy. Either inference could be false. Therefore, a statement such as one of the following should be included in the behavioral objective to provide a point of closure and terminal review:

... 85% accuracy *for 4 consecutive sessions*

... 85% accuracy *for 3 out of 4 days*

... on 8 out of 10 trials *for 3 consecutive teaching sessions*

... will return within 10 minutes *on 3 consecutive trips to the bathroom*

FORMAT FOR A BEHAVIORAL OBJECTIVE

A management aid for the teacher in writing behavioral objectives is the adoption of a standard format. A consistent format helps the teacher include all the components necessary for communicating all intended information. No single format is necessarily superior to others; a teacher should simply find one that is compatible with his or her writing style. Here are two such formats:

Format 1

Conditions:	Given 20 flash cards with preprimer sight words and the instruction "Read these words,"
Student:	Sam
Behavior:	will read the words orally
Criterion:	within 2 seconds for each word with 90% accuracy on 3 consecutive trials.

Format 2

Student:	Marvin
Behavior:	will write in cursive handwriting 20 fourth-grade spelling words
Conditions:	from dictation by the resource teacher
Criterion:	with no more than 2 errors for 3 consecutive weeks.

The following behavioral objectives may be derived from the educational goals previously set for students Bill and Thom.

See pp. 71–72 for Bill and Thom's long-term goals.

Arithmetic

Goal: Bill will master basic computation facts at the first-grade level.

Objective: Given a worksheet of 20 single-digit addition problems in the form 6 + 2 and the written instruction "Find the sums," Bill will complete all problems with 90% accuracy for three consecutive math sessions.

Social Studies

Goal: Bill will demonstrate knowledge of the functions of the three branches of the federal government.

Objective: Given the assignment to read pages 23–26 in the text *Our American Heritage*, Bill will list the 10-step sequential process by which a bill becomes a law. This list will have no more than one error of sequence and one error of omission. This will be successfully accomplished on an in-class exercise and on the unit-end test.

Reading

Goal: Bill will be able to identify relevant parts of a story he has read.

Objective: Given the short story "The Necklace," Bill will write a minimum 200-word paper that (1) lists all the main characters and (2) lists the sequence of main actions, with no more than two errors.

Science

Goal: Bill will demonstrate knowledge of the structure of the solar system.

Objective: Given a map of the solar system, Bill will label each planet in its proper position from the sun with 100% accuracy on two consecutive sessions.

Language arts

Goal: Bill will increase the creative expression of his oral language.
Objective: Given an array of photos of people, objects, and locations, Bill will tell a 5-minute story to the class that makes use of a minimum of seven items, on 3 out of 5 days.

Physical education

Goal: Bill will increase his skills in team sports.
Objective: Given a basketball, Bill will throw the ball into the hoop from a distance of 10′, 8 out of 10 trials for four consecutive gym classes.

Recall from our earlier discussion that while Bill is a mildly handicapped student, Thom is much more severely handicapped. Here are some objectives and corresponding goals for Thom:

Cognitive

Goal:	Thom will be able to categorize objects according to their function.
Objective:	Given 12 Peabody cards (4 foods, 4 clothing, 4 grooming aids), a sample stimulus card of each category, and the verbal cue "Where does this one go?", Thom will place the cards on the appropriate category pile with 100% accuracy, for 17 out of 20 trials.

Language

Goal:	Thom will demonstrate increased receptive understanding of functional labels.
Objective:	Given an array of three objects found in his environment (cup, spoon, dish) and the verbal cue "Give me the . . .," Thom will hand the teacher the named object 9 out of 10 times for 4 consecutive sessions.

Motor

Goal:	Thom will develop gross motor capability of his upper extremities.
Objective:	Given a soft rubber ball suspended from the ceiling and the verbal cue "Hit the ball," Thom will hit the ball causing movement 10 out of 10 times for 5 consecutive days.

Social

Goal:	Thom will learn to participate appropriately in group activities.
Objective:	When sitting with the teacher and two other students during story time, Thom will make an appropriate motor or verbal response to each of the teacher's questions when called upon a minimum of three times in a 10-minute period for 5 consecutive days.

Self-help

Goal:	Thom will demonstrate the ability to dress himself independently.
Objective:	Given a pullover sweater with the back label color-cued red and the verbal cue "Put on your sweater," Thom will successfully complete all steps of the task without physical assistance 2 out of 3 trials for 4 consecutive days.

Maladaptive behavior

Goal:	Thom will decrease out-of-seat behavior.
Objective:	Given the period from 9:00 to 9:20 (functional academics), Thom will remain in his seat, unless given permission by the teacher to leave, for 5 consecutive days.

Professor Grundy's Class Writes Behavioral Objectives

It was the time of the semester for Professor Grundy's 8 o'clock class to learn about behavioral objectives. After presenting a carefully planned lecture (remarkably similar to the first part of this chapter), Grundy asked if there were any questions. Dawn Tompkins stopped filing her nails long enough to ask, with a deep sigh, "Yes, Professor, would you please tell me what a behavioral objective is, exactly?"

"I was under the impression, young lady, that I had done just that," replied Grundy. "Is anyone else confused?"

A chorus of muttering and rumbling ensued from which Grundy was able to extract clearly only two questions.

"Is this covered in the book?" and "Will it be on the test?"

After once more presenting a drastically abbreviated description of the components of a behavioral objective, Grundy announced that each member of the class was to write a behavioral objective for the curriculum area of science and present it to him for checking before leaving class. This announcement, followed by a chorus of groans and considerable paper shuffling, also brought forth a flurry of hands:

"You mean list the components?"

"No," said Grundy, "Write an objective."

"You mean define a behavioral objective?"

"No," said Grundy, "Write one."

"But you never *said* anything about writing them."

"What," Grundy retorted, "did you think was the purpose of the lecture?"

After everyone who lacked these tools had been provided with paper and pencil, silence descended upon the class. DeWayne was the first one finished and proudly presented his objective to the Professor:

> To understand the importance of the digestive system.

"Well, DeWayne," said the Professor, "That's a start, but do you not remember that a behavioral objective must talk about *behavior*? Remember the list of verbs I gave you . . . When DeWayne continued to look blank, Grundy rifled through his briefcase and found a copy. (See Table 3–1).

"Look here," the professor said, "Use one of these directly observable verbs."

DeWayne returned some time later with his rewritten objective:

> To label the parts of the digestive system.

"Good, DeWayne," sighed the Professor. "That's a behavior, all right. Now, do you recall the components of a behavioral objective?" Once again, DeWayne looked blank. Grundy carefully wrote:

Conditions   Student   Behavior   Criteria

across the top of the gasoline credit card receipt DeWayne had evidently found in his wallet. DeWayne returned to his desk.

An hour and a half later, as Grundy was regretting ever having made this assignment, DeWayne returned again:

Given an unlabeled diagram of the human digestive system, fourth-grade students will label the major parts of the digestive system (mouth, esophagus, stomach, small intestine, large intestine) with no errors.

Grundy read DeWayne's objective with interest, because his own digestive system was beginning to be the major focus of his attention. "Excellent, DeWayne," said the professor, "I suppose it's too late to get lunch in the cafeteria. Why didn't you do this in the first place?"

"Well, Professor," answered DeWayne, "I didn't really understand what you wanted. I'm still not sure I could do another one."

After getting some crackers from the vending machine, Grundy returned meditatively to his office. He found a piece of paper and began to write as he munched:

Given a worksheet listing appropriate verbs and the components of a behavioral objective, students enrolled in Education 411 will write five behavioral objectives including all components.

"Do you have any idea what his objective for this class is?"

After musing for a few minutes, he added:

. . . in less than half an hour.

"Perhaps," Grundy muttered to himself, "if I had been sure what I wanted and told the students at the front end, they would have had less trouble figuring it out."

EXPANDING THE SCOPE OF THE BASIC BEHAVIORAL OBJECTIVE

Earlier discussion in this chapter outlined the process of developing a standard behavioral objective for acquiring a new behavior. In this process, the teacher describes a behavior that had not previously been in the student's repertoire. Mastery of such a behavior is admirable, but the teacher has yet to provide the student with a functional response, that is, one that can be used outside the confines established in the objective as written. The student has demonstrated the behavior only under controlled conditions. Many teachers select new target behaviors as soon as this degree of mastery is demonstrated. Moving on to a new objective at this point is often inappropriate and may interfere with or halt learning at a critical stage. Before replacing one objective with the next one in the learning sequence, the teacher should *ensure the student's functional use of the behavior.* Two possible perspectives on expanded use are

1. Programming according to a hierarchy of response competence
2. Programming according to a hierarchy of levels of learning

Hierarchy of Response Competence

Go beyond acquisition: make behaviors functional.

A measure of response accuracy (8 out of 10 correct, for example) is only one dimension for evaluating performance. It represents the **acquisition** level of response competence. At this level, we merely verify the presence of the ability to do something the student wasn't previously able to do and the ability to do it with some degree of accuracy. Moving to measures of competence in performance beyond accuracy, beyond this acquisition level, requires alterations in the statements of criteria and conditions. Such alterations reflect a hierarchy of response competence. Once a child can perform the behavior, we are then concerned with **fluency**, or rate, of performance, as well as performance under conditions other than those imposed during the initial teaching process.

A response hierarchy should contain the following minimum levels:

\ Generalizaton /
\ Maintenance /
\ Fluency /   ↑   Increasing functional
\Acquisition/   use of a response

As an example of the use of this hierarchy, let us assume that John has reached the acquisition level on the following objective:

> Given one quarter, two dimes, two nickels, and one penny and the verbal cue "John, give me your bus fare," he will hand the teacher coins equalling 35¢, 8 out of 10 trials for three consecutive sessions;

and that Lauren has reached acquisition on this objective:

> Given a ditto of 20 division problems with two-digit dividends and single-digit divisors, Lauren will write the correct answer in the appropriate place on the radical with 90% accuracy for 4 consecutive days.

After John and Lauren have met these stated criteria for their performances, the teaching concern should turn to their fluency of performance, or the rate at which they perform the behavior. *Fluency* refers to the appropriateness of the rate at which the student is accurately performing this newly acquired response. In John's case, we know that he can select the appropriate coins to make 35¢, but this does him little good if when we take him to the bus, it takes him 5 minutes to do it. The bus driver cannot wait this long. In Lauren's case, we know she can now solve division problems, but it takes her so long that either we interrupt her when her reading group is scheduled or she misses part of her reading lesson so she can finish her problems.

In both instances, the students are demonstrating accurate performance at an inappropriate rate. Recognizing the necessity for an appropriate rate of performance, a teacher can indicate an acceptable fluency when the behavioral objective is written. This is accomplished by adding a time limit to the statement of criteria, as found in parentheses, in the following objectives:

> Given one quarter, two dimes, two nickels, and one penny and the verbal due "John, give me your bus fare," he will hand the teacher coins equalling 35¢ (within 90 sec.) 8 out of 10 trials for three consecutive sessions.

> Given a mimeographed sheet of 20 division problems with two-digit dividends and single-digit divisors, Lauren will write the correct answer in the appropriate place on the radical (within 20 minutes) with 90% accuracy for 4 consecutive days.

In cases of typical or high-functioning learners, rate is often included within the initial objective, thus combining acquisition and fluency within a single instructional procedure.

Programming for maintenance will be discussed in Chapter 10.

It is not necessary to adjust the original behavioral objective to include the level of competence labeled *maintenance*. **Maintenance** is the ability to perform a response over time without reteaching. Maintenance-level competence is confirmed through the use of postchecks or probes, during which the teacher rechecks the skill to be sure the student can still do it. Maintenance may be promoted through building in the opportunity for

Programming for overlearning.

Distributed practice is a more efficient way of learning for long-term maintenance.

overlearning trials and distributed practice (Biehler, 1974; Denny, 1966; Gruenenfelder & Borkowski, 1975). *Overlearning* refers to repeated practice after an objective has been initially accomplished. An optimum number of overlearning opportunities is approximately 50% of the number of trials required for acquisition of the behavior. If it takes John 10 teaching sessions to learn to tie his shoes, we should ideally provide 5 additional sessions for overlearning. *Distributed practice* is practice that is spread out over time (as opposed to *massed practice*, which is compressed in time). An example of massed practice familiar to college students is cramming for an exam. The material may be learned between 10 P.M. and 6 A.M. the night before the test, but most of it will be rapidly forgotten. If maintenance is desired, the preferable approach is studying for short periods every evening for several weeks before the exam, using distributed practice. Another means of providing for maintenance, alteration of schedules of reinforcement (Skinner, 1968), will be discussed in Chapter 6.

The level of competence labeled **generalization** is of prime importance in assuring a functional response. Generalization indicates that the student can perform the behavior under conditions different from those under which it was acquired. An acquired response may be generalized across four basic dimensions: across cues, across materials or formats, across trainers, and across environments (settings). Programming for generalization will be discussed at length in Chapter 9.

Hierarchy of Levels of Learning

It may seem that writing behavioral objectives inevitably focuses teacher attention on very concrete, simple forms of learning. Indeed, this has been one of the most frequent criticisms of a behavioral approach. However, it is not necessary to confine behavioral objectives to lower levels of learning. Bloom (1956) has proposed hierarchies of learning in cognitive, affective, and psychomotor areas. These hierarchies classify possible learning outcomes in terms of increasingly abstract levels. They are helpful in writing objectives in behavioral terms because they suggest observable, measurable behaviors that may occur as the result of both simple and complex learning.

The cognitive hierarchy, which will serve as our example, contains six levels of learning, as shown in the following diagram (Bloom, 1956).

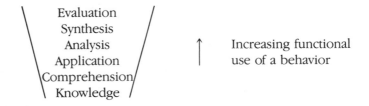

Many behavioral objectives are written in terms of the knowledge level of the hierarchy—we simply want students to demonstrate that they know or remember something we have taught them. Once the student has achieved mastery on this lowest of the six levels, the teacher can shift programming toward higher levels of learning by preparing subsequent objectives that alter the target behavior and criterion statements. As an aid in this process, Gronlund (1978) prepared a table that illustrates behavioral terms appropriate to describe target behaviors at each level of learning. (See Table 3–2).

Knowledge

Bloom defines learning at the *knowledge level* as the recall or recognition of information ranging from specific facts to complete theories (Bloom, 1956). These memory functions are the only behavior to be demonstrated at this basic level of cognitive learning. The following acquisition objectives are examples written for students at this level:

> Upon reading and completing the exercise in Chapter 2 of *Biology for Your Understanding*, Virginia will list the biological categories of the Linnaean system, in their order of evolutional complexity, without error during two class sessions and on a unit-end exam.

> Given the symbols for the arithmetic processes of addition, subtraction, multiplication, and division, Sandra will respond with 90% accuracy on a multiple-choice test of their labels and basic functions.

> Given the study guide to Shakespearian tragedies, Deborah will recite the names of all 10 plays with no more than one error of omission.

Comprehension

Once the student has reached the performance criterion at the knowledge level, the teacher moves to the *comprehension level,* the understanding of meaning. The student may demonstrate comprehension by paraphrasing and providing examples. Here are some sample objectives at this level:

> Given the Linnaean system of biological classification, Virginia will provide a written description of an organism within each category. The description will include at least one factor that distinguishes the category from others.

> Given a worksheet of 40 basic arithmetic examples requiring addition, subtraction, multiplication, and division, Sandra will complete the sheet with 90% accuracy.

> Given the metaphoric passage, "Oh that this too, too solid flesh would melt . . ." from *Hamlet,* Deborah will write an essay describing the literal intent of the passage. The essay will be a minimum of 300 words.

Application

Programming at Bloom's *application level* requires the student to use the method, concept, or theory in various concrete situations. Consider these objectives:

TABLE 3–2   Examples of general instructional objectives and behavioral terms for the cognitive domain of the taxonomy.

Illustrative general instructional objectives	Illustrative behavioral terms for stating specific learning outcomes
Knows common terms Knows specific facts Knows methods and procedures Knows basic concepts Knows principles	Defines, describes, identifies, labels, lists, matches, names, outlines, reproduces, selects, states
Understands facts and principles Interprets verbal material Interprets charts and graphs Translates verbal material to mathematical formulas Estimates future consequences implied in data Justifies methods and procedures	Converts, defends, distinguishes, estimates, explains, extends, generalizes, gives examples, infers, paraphrases, predicts, rewrites, summarizes
Applies concepts and principles to new situations Applies laws and theories to practical situations Solves mathematical problems Constructs charts and graphs Demonstrates correct usage of a method or procedure	Changes, computes, demonstrates, discovers, manipulates, modifies, operates, predicts, prepares, produces, relates, shows, solves, uses
Recognizes unstated assumptions Recognizes logical fallacies in reasoning Distinguishes between facts and inferences Evaluates the relevancy of data Analyzes the organizational structure of a work (art, music, writing)	Breaks down, diagrams, differentiates, discriminates, distinguishes, identifies, illustrates, infers, outlines, points out, relates, selects, separates, subdivides.
Writes a well organized theme Gives a well organized speech Writes a creative short story (or poem, or music) Proposes a plan for an experiment Integrates learning from different areas into a plan for solving a problem Formulates a new scheme for classifying objects (or events, or ideas	Categorizes, combines, compiles, composes, creates, devises, designs, explains, generates, modifies, organizes, plans, rearranges, reconstructs, relates, reorganizes, revises, rewrites, summarizes, tells, writes
Judges the logical consistency of written material Judges the adequacy with which conclusions are supported by data Judges the value of a work (art, music, writing) by use of internal criteria Judges the value of a work (art, music, writing) by use of external standards of excellence	Appraises, compares, concludes, contrasts, criticizes, describes, discriminates, explains, justifies, interprets, relates, summarizes, supports

Given the names of five organisms and the Linnaean system, Virginia will place each in its proper category and write a list of rationales for placement. Each rationale will contain a minimum of two reasons for placement.

Given a set of 10 paragraphs that present problems requiring an arithmetic computation for solution, Sandra will write the correct answer, showing all computations with 100% accuracy.

After reading *Hamlet*, Deborah will be able to explain the parallels between Hamlet's ethical dilemma and the problem of abortion and to cite an additional current parallel example of her own choosing.

Analysis

Analysis is the ability to break down material into its constituent parts in order to identify these parts, discuss their interrelationship, and understand their organization as a whole. The following objectives are analytically oriented:

Given a list of five organisms, Virginia will use appropriate references in the library to investigate and report to the class the role of the organisms in either the food chain or in the ecological stability of their habitat.

Given a written statement of the Associative Property, Sandra will be able to explain accurately to the class, using examples at the chalkboard, the property's relation to the basic additive and multiplicative functions.

After having read *Hamlet* and *Macbeth*, Deborah will guide the class in a discussion of the play's plot development. This discussion will be based upon a schematic representation of each scene that she will provide in written form.

Synthesis

At the cognitive level of *synthesis*, the student should demonstrate the ability to bring parts together resulting in a different, original, or creative whole:

Given a list of reference texts, Virginia will write a 1000-word summary paper that explains biological classifications within Darwin's theory of evolution. The paper will be evaluated on the basis of accuracy, completeness, organization, and clarity.

Given the numerical systems of base 10 and base 2, Sandra will orally demonstrate the use of the functions of addition, subtraction, multiplication, and division within each system.

Given the study of the Shakespearian tragedy *Macbeth*, Deborah will rewrite the end of the play in iambic pentameter, assuming that the murder of the king was unsuccessful.

Evaluation

The highest level of learning demonstrated within this hierarchy is *evaluation*. The student is asked to make a judgment of value:

Based upon the principles of mutual exclusion, Virginia will devise a taxonomy for the classification of means of transportation and provide a justification for the categories created and their constituent parts.

Given a set of unknown values and a given arithmetic computational function, Sandra will explain the probability of differing answers that may be correct.

Given plays by Shakespeare and Bacon, Deborah will state a preference for one, and justify her preference in a 500-word essay based upon some element(s) of style.

Learning Levels for the Limited Learner

Even limited learners can acquire higher-level cognitive skills.

In most instances of planning for expanded instructional intent, we tend to focus on a hierarchy of response competence for the more limited learner and a hierarchy of levels of learning for the typical or above-average learner. This dichotomy is not necessarily warranted simply by the level of the student's functioning. Consider the following examples of how we may write behavioral objectives for the limited learner in conjunction with levels of learning:

Knowledge	Given a common coin and the verbal cue, "What is the name of this," George will state the appropriate label on 18 out of 20 trials for 5 consecutive sessions.
Comprehension	Given a common coin and the verbal cue "What is this worth," George will count out the coins equivalent in pennies and state something to the effect that "A dime is worth 10 pennies" on 8 out of 10 trials for each coin.
Application	When presented with 10 pictures of food items, each with its cost written on it, George will count out coins equal to the amount written upon the verbal cue "Show me the amount" on 18 out of 20 trials.
Analysis	When presented with pictures of items each with its cost printed on it, a $1 bill, and a verbal cue such as "Can you buy a pencil and a newspaper?" George will respond correctly on 18 out of 20 trials.
Synthesis	Given a $1 bill and the instruction to buy various priced items, George will simulate the buying exchange and decide whether he was given correct change without error on 10 trials.
Evaluation	Given 35¢ and a 5-mile ride from the sheltered workshop to his home, George will use the 35¢ for the bus ride rather than a candy bar.

BEHAVIORAL OBJECTIVES AND THE IEP

The development of educational goals and behavioral objectives (short-term objectives) for students in need of special education services was included as one of the mandates of The Education for All Handicapped Chil-

*PL 94–142 requires an
IEP for every
handicapped student.*

dren Act of 1975 (P.L. 94–142). Among the results of this legislation have been the formalization of the planning aspects inherent in the writing of goals and objectives and the provision for active parental participation in the educational planning process for their child. This planning process ultimately results in the development of an Individualized Education Program (IEP).

The rules and regulations of P.L. 94–142 require five elements to appear as part of the IEP:

The individualized education program for each child must include:

1. A statement of the child's present levels of educational performance;

2. A statement of annual goals, including short term instructional objectives;

3. A statement of the specific special education and related services to be provided to the child, and the extent to which the child will be able to participate in regular education programs;

4. The projected dates for initiation of services and the anticipated duration of the services; and

5. Appropriate objective criteria and evaluation procedures and schedules for determining, on at least an annual basis, whether the short term instructional objectives are being achieved. (121a 346, 1977)

This excerpt shows the parallel in procedural format between the development of behavioral objectives and the content of the IEP adopted by the legislative authors. Both processes include the accumulation of data to determine the student's current levels of functioning, the development of behavioral objectives (short-term objectives) for attaining the goals, and the review of objective mastery.

On the sample form (see Figure 3–1), the components that state the target behavior and the conditions are placed in the column labeled *Short-term instructional objectives*. The component that states the criterion for mastery is placed in the correspondingly labeled column. Because each IEP is developed for an individual student, the student's identity need not be restated within each objective on the form.

The behavioral objective on the following page has been transferred onto the sample form:

Student: _____　　Person to provide service to: _____

Short-term instructional objectives	Person responsible	Criteria for mastery	Date reviewed	Mastery Yes No
1. Complete single-digit addition problems in the form 6 + 2 given written instructions to "find the sum."		90% accuracy for 3 consecutive math sessions.		

FIGURE 3–1　Sample IEP

Given a worksheet of 20 single-digit addition problems in the form 6 + 2 and the written instruction, "Find the sums," Bill will complete all problems with 90% accuracy for 3 consecutive math sessions.

In order to manage an IEP and monitor its constituent objectives, teachers should observe the following recommendations:

1. Short-term objectives should be directly related to goal statements (long-term objectives) in a sequential manner.

2. In the case of mildly handicapped students, the short-term objective should deal directly with the reason for their referral for special education services.

3. For moderately and severely handicapped students, two or three short-term objectives per learning domain should be placed on the IEP. These objectives should be included following the consideration of prerequisite skill requirements across the various curricular areas.

4. New short-term objectives should not be added until maintenance has been achieved on current objectives and generalization instruction has begun. Once the planned generalizations have been reached, the objective may be removed from the IEP.

5. The management of the IEP should be a continuing process. Teachers and administrators should not overlook the regulations stating that a review should be conducted "on at least an annual basis," not "only on an annual basis." The annual review should include the full committee.

When Bill's math skills are at grade level, he is no longer handicapped.

(a) The objectives of mildly handicapped students should be reviewed as soon as achievement has been verified to assess whether or not the original need for special education services still exists.

(b) Reasonable dates of review should be set for objectives of moderately and severely handicapped students. As objectives are met, the teacher should add new short-term objectives and notify the committee chairperson and parent(s) in writing, with full data justification made at the annual review. Such a procedure will foster the student's progress, while prohibiting stagnation of instruction until the full committee can be gathered.

Thom's objective should be reviewed frequently so that he will make the maximum possible progress.

6. Review dates should be set bearing in mind the need for instruction at the higher levels of learning in order to promote full functional use of the skill.

7. If personnel other than the special education teacher (such as the physical therapist, speech pathologist, or regular class teacher) take part in the instruction of a particular objective, each individual's name and signature should appear on the IEP along with that of the teacher.

SUMMARY

In this chapter, we described the process of writing behavioral objectives and the relationship between such objectives and the IEP required for handicapped students. This process is an integral part of any program for behavior change, whether the program is directed toward academic or social behavior. A program for changing behavior is unlikely to be successful unless we are sure what constitutes success. Behavioral objectives facilitate communication, so that everyone knows the goal of instruction. They also provide for evaluation, so that everyone knows whether the goal has been reached.

REFERENCES

BAYLEY, N. 1969. *Bayley scales of infant development.* New York: The Psychological Corp.

BENDERESKY, M. (Ed.). 1978. *Uniform performance assessment system.* Seattle: University of Washington.

BIEHLER, R.F. 1974. *Psychology applied to teaching.* Boston: Houghton Mifflin.

BLOOM, B.S. (Ed.). 1956. *Taxonomy of educational objectives handbook I: Cognitive domain.* New York: David McKay Company, Inc.

BRIGANCE, A. 1977. *Inventory of basic skills.* Newton, Mass.: Curriculum Assoc.

DENNY, M.R. 1966. A theoretical analysis and its application to training the mentally retarded. In N.R. Ellis (Ed.), *International review of research in mental retardation* (vol. 2). New York: Academic Press.

DENO, S., & JENKINS, J. 1967. *Evaluating preplanning curriculum objectives.* Philadelphia: Research for Better Schools.

DUBOSE, R., & LANGLEY, M. 1977. *Developmental activities screening inventory.* New York: Teaching Resources Corp.

DUNN, L.M., & MARKWARDT, F.C. 1970. *Peabody individual achievement test.* Circle Pines, Minn.: American Guidance Service.

GAGNE, R. 1970. *The conditions of learning.* New York: Holt, Rinehart & Winston.

GRONLUND, N. 1978. *Stating objectives for classroom instruction.* New York: Macmillan.

GRUENENFELDER, T., & BORKOWSKI, J. 1975. Transfer of cumulative rehearsal strategies in children's short-term memory. *Child Development, 46,* 1019–1024.

MAGER, R. 1962. *Preparing instructional objectives.* Palo Alto, Calif.: Fearon Publishers.

MERCER, J. 1979. *System of multicultural pluralistic assessment.* New York: The Psychological Corp.

MORRIS, R. 1976. *Behavior modification with children.* Cambridge, Mass.: Winthrop Publications.

NIHIRA, K., FOSTER, R., SHELLHAAS, M., & LELAND, H. 1975. *Adaptive behavior scales,* 1975 rev. ed. Washington, D.C.: American Association on Mental Deficiency.

QUAY, H.C., & PETERSON, D.R. 1975. *Manual for the behavior problem checklist.* Champaign: Children's Research Center, University of Illinois.

SANTA CRUZ COUNTY. 1973. *Behavioral characteristics progression.* The Office of the Santa Cruz County Superintendent of Schools. Palo Alto, Calif.: Vort Corp.

SKINNER, B.F. 1968. *The technology of teaching.* New York: Appleton-Century-Crofts.

TURNBULL, A.P., STRICKLAND, B.B., & BRANTLEY, J.C. 1978. *Developing and implementing individualized education programs.* Columbus, Ohio: Charles E. Merrill.

WECHSLER, D. 1974. *The revised Wechsler intelligence scale for children (WISC-R).* New York: The Psychological Corp.

4 Collecting and Graphing Data

Teachers react strongly to suggestions that they collect data in their classrooms:

- "I don't have time to write down everything anyone does."

- "I just don't think I can manage shuffling all those sheets of paper, handling stopwatches, wrist counters, giving proper cues. When am I supposed to concentrate on teaching?"

- "This data collection adds an extra hour a day, at least, in summarizing the data, putting it on graphs, and so on. Where's that time supposed to come from?"

- "How utterly ridiculous."

Most teachers regard the kind of data-collection procedures that we shall discuss in this chapter with the same enthusiasm they reserve for statistics. In some cases, their comments are thoroughly justified. Some of the systems we are going to discuss are not practical for everyday classroom use. Classroom teachers may never use some of the more complex systems. However, understanding how these systems work helps in understanding the research published in this area. That teachers read research and attempt to apply it in their classrooms is a fond belief we are unwilling to abandon. This chapter does describe, then, the most common systems and shows how many of them can be adapted for classroom use. Because these systems require recording and graphing data, the chapter focuses on both of these tasks.

A RATIONALE

Even after accepting the feasibility of data collection in the classroom, many teachers see very little value in it. Beyond recording grades on tests, most teachers have traditionally kept very few records of their student's academic and social behaviors. There are, nevertheless, two excellent reasons for collecting classroom data. The first is that precise observation and measurement of behavior may enable teachers to determine the best way to change it. The second is that observation and measurement enable very accurate determination of a particular intervention's effect.

By writing behavioral objectives, teachers communicate their intent to change particular behaviors. They also state the criteria they will use to judge whether change procedures have been successful. In many classroom situations, the intervention's effect upon the students' original level of performance would be evaluated by administering a pretest and posttest. However, the precision desired within a behavioral approach to instruction and in program evaluation demands additional data.

Behavioral evaluation requires observation of students' current functioning and ongoing progress.

Behavioral evaluation has two requirements. First, a detailed observation of the student's current functioning is needed. This observation should reflect the conditions and description of the behavior stated in the objective. For example, a behavioral objective stating that students should solve 25 long division problems in 30 minutes requires that the teacher determine how many long division problems the students can already solve in 30 minutes. Second, evaluation of an instructional program must facilitate ongoing monitoring of the teaching/learning process as well as provide a system for terminal evaluation. Thus, evaluation must be continuous so that programs can be adjusted as instruction progresses. As the students in our example receive instruction in long division, for instance, the teacher could record daily how many problems they could solve in 30 minutes, thus providing continuous evaluation. The monitoring process can provide guidelines for continuing or altering the application of instructional techniques and help to avoid false assumptions about student progress. Such false assumptions are unfortunately very common, as illustrated by the following vignette.

Ms. Waller Goes Electronic

Ms. Waller was ecstatic. After months of complaining that she had no materials to use to teach reading to her "Tortoise" group, she had received a computerized teaching machine. The salesman demonstrated the machine proudly and pointed out the features that justified the hundreds of dollars invested.

"All you have to do," he assured her, "is hook the little, er, students up to these here headphones, drop in a casette, and turn this baby on. Everything else is taken care of . . . you don't do a thing."

Ms. Waller briskly administered the pretest included in the materials, scheduled each Tortoise for 15 minutes a day on the machine, and assumed that her worries were over.

At the end of the school year, Ms. Waller administered the posttest. Imagine her distress when, although several members of the group had made remarkable progress, some students had made none at all.

"I don't understand," she wailed. "The machine was supposed to do everything. How was I supposed to know it wasn't working?"

"Perhaps," suggested her principal kindly, as he wished her success in her career as an encyclopedia salesperson, "you should have checked before now."

CHOOSING A SYSTEM

The first step in the evaluative process of ongoing measurement of behavior is selection of a system of data collection. The characteristics of the system selected must be appropriate to the behavior being observed and to the kind of behavior change desired.

Behavior may be measured and changed on a number of dimensions (White & Haring, 1980):

Dimensions for observation of behavior.

1. *Rate.* When frequency data are expressed in a ratio with time, we have a measure of rate.
John made 7 trips per hour to the bathroom.
Harold did 11 math problems per minute.
Marvin had 8 tantrums per week.

2. *Duration.* When we describe how long a behavior lasts, we are measuring duration.
John stayed in the bathroom for 3 hours.
Harold worked on his math for 20 minutes.
Marvin's tantrum lasted for 1 hour and 40 minutes.

3. *Latency.* A measure of how long it takes before a student starts performing a behavior is a measure of latency.
After I told John to come out of the bathroom, it took him 5 minutes to appear at the door.
After the teacher said, "Get to work," Harold stared into space for 5 minutes before he started his math.
It took 20 minutes for Marvin to stop having a tantrum after I put him in the time-out room.

4. *Topography.* Topography refers to the "shape" of the behavior—what it looks like.
Harold writes all the 4's backwards on his math paper.
Marvin screams, kicks his heels on the floor, and pulls his hair during a tantrum.

5. *Force.* When we are concerned with how strongly (loudly, for example) a behavior is performed, we measure force.
Harold writes so heavily that he makes holes in his math paper.

Marvin screams so loudly that the teacher three doors down the hall can hear him.

6. *Locus.* A concern with where a behavior occurs is described as locus.
I refused to let John go to the bathroom, so he wet his pants.
Harold wrote the answers to his math problems in the wrong spaces.
Marvin had a tantrum in the lunchroom.

The decision to use a particular system of data collection will be made partly on the basis of the dimension of behavior that is of concern, as well as on the basis of convenience. Systems for collecting data can be classified into three general categories. First is the recording and analysis of written reports that ideally include a full record of the sample behavior and all behaviors emitted during the observation period. Second is the recording of a sample of the behavior as it occurs. Third is the observation of tangible products resulting from a behavior. These systems may be categorized as shown in the accompanying table.

Analyzing written records	Anecdotal reports
Observing tangible products	Permanent product recording
Observing a sample of behavior	Event recording
	Interval recording
	Time sampling
	Duration recording
	Latency recording

ANECDOTAL REPORTS

Anecdotal reports are written to provide as complete a description as possible of a student's behavior in a particular setting or during an instructional period. Anecdotal reports do not identify a predefined or operationalized target behavior. Rather, this approach anticipates that, as an end product of writing and analyzing anecdotal data, a specific behavior may be identified as the source of a general disturbance or lack of learning by the student, therefore requiring some type of modification. Anecdotal records are useful primarily for analysis not for evaluation.

An anecdotal system of data collection is frequently used by teachers, parents, and therapists to describe some general disturbance that is taking place or a lack of academic progress for a reason that cannot be determined. For example, it might be reported that "Shelia constantly disrupts the class and does not complete her own work," or that "During therapy sessions, I cannot seem to get Shelia under control in order to do the needed speech remediation," or that "Shelia keeps getting up from the dinner table and throwing her food."

Reports such as these are common and should prompt the applied behavior analyst to ask the series of questions offered in Chapter 3 on page

68. Should the specific behavior continue to elude identification, even after those questions have been asked, the analyst will have to further isolate and identify a target behavior that may be the source of the complaint in the natural setting of the behaviors—such as at the dinner table, in the schoolyard during recess, in the classroom during reading period—and attempt to write down everything that occurs.

This system of data collection produces a written description of nearly everything that occurred in a specific time period or setting. Therefore, this procedure does not yield isolated markings on a data sheet (such as + and −), but rather a report or narrative, written in everyday language, describing individuals and interactions. Wright (1960) has provided some suggested guidelines for the writing of anecdotal reports:

Guide to writing anecdotal records.

1. Before beginning to record anecdotal data, write down the setting as you initially see it, the individuals in the setting and their relationships, and the activity occurring as you are about to begin recording (e.g., lunch, free play).

2. Include in your description everything the target student(s) says and does and to whom or to what.

3. Include in your description everything said and done to the target student(s) and by whom.

4. As you write, clearly differentiate fact (what is actually occurring) from your impressions or interpretations of cause or reaction.

5. Provide some temporal indications so as to be able to judge duration of particular responses or interactions.

Structuring an Anecdotal Record

After observations have been made, an anecdotal report must be analyzed to determine the behavior(s), if any, that should be the subject of a behavior-change program. The observations in this initial anecdotal format are difficult to separate into individual behaviors and relationships, so it is helpful to present the anecdotal data in a more schematic manner for review. Bijou, Peterson, and Ault (1968) employed a system for sequence analysis in which they redrafted an anecdotal report into a form that reflects a behavioral view of environmental interactions. By this system, the contents of the report are arranged into columns divided to indicate antecedent stimuli, specific responses, and consequating stimuli. This table format clearly represents the temporal relationship among individual behaviors, the antecedents that stimulate them, and the consequences that maintain them.

The anecdotal report in Figure 4–1 was prepared by the staff of the Judevine Center in St. Louis, Missouri. It was taken in the home and records a period of interaction between a young boy named Chris and his mother.

FIGURE 4–1
Excerpt from an
anecdotal report
taken at the
Judevine Center

. . . Chris is walking around the kitchen. Mother is fixing something for dinner, working at the sink. Chris opens a cabinet and throws the pans out onto the floor. Mother turns the fire down on the stove and picks up the pans. She tells Chris to stop. Chris picks up a lid and begins to spin it on the floor. Mother restacks the pans and closes the cabinet door. She goes back to the sink. Chris throws the pan lid across the room. It hits the table and the salt shaker spills. Mother stops her cooking and cleans up the spilt salt. Chris picks up the pan lid and spins it again. Mother opens the refrigerator. Chris winces and pulls on the refrigerator door. She holds it open with her hip and gets something out. Chris screams. She moves and the refrigerator slams shut. Mother bends down and talks to Chris. He pushes her away, gets up and walks out of the room. Mother follows him into the living room. He is throwing some clothes on the floor which were stacked on the ironing board. She picks up the clothing and scolds Chris. He watches quietly. The clothes are folded again. Mother takes Chris over to a small desk. He sits and watches her. She puts some building toys on the top of the desk and guides his hands to play with them. Now he is stacking the toys by himself. Mother picks up the clothing and leaves the room. Chris gets up and follows her to the linen closet. He grabs her hand and pulls it back toward the living room. She gets the clothes in one stack on the shelf in the linen closet and follows Chris to a bookcase in the hall. She picks out a picture book and asks him if he wants to hear a story. Chris begins to scream and throws all of the books onto the floor. She picks up the books and Chris crawls into the living room. She goes into the kitchen. Chris screams. She runs into the living room. Chris has pulled the ironing board down onto himself and is lying on the floor. . . .

Using the approach suggested by Bijou et al., the beginning of this report could be transposed into columns as begun in Figure 4–2. The antecedents, behaviors, and consequences are numbered to indicate the time sequence. Note that transposing the report makes it apparent that, in several instances, consequences of a given response can become the antecedents for a succeeding response.

When the content of an anecdotal report has been arranged in a format that clearly presents the sequence of and the relationships among behavioral events, the source of the problem behavior may be determined. The following questions help in analysis:

Questions for analyzing anecdotal information.

1. What are the behaviors that can be described as *inappropriate*? The behavior analyst should be able to justify labeling the behaviors *inappropriate* given the setting and the activity taking place.

2. Is this behavior occurring frequently, or has a unique occurrence been identified?

3. Can reinforcement or punishment of the behavior be identified? The reinforcement may be delivered by the teacher, parent, another child, or by some naturally occurring environmental consequence.

Review the sequential analysis of the anecdotal record for Chris. Can you answer any of these questions?

4. Is there a pattern to these consequences?

5. Can antecedents to the behavior(s) be identified?

6. Is there a pattern that can be identified for certain events or stimuli (antecedents) that consistently precede the behavior's occurrence?

FIGURE 4–2
Structuring of an
anecdotal record

Time	Antecedent	Response	Consequence
5:15 P.M.	1. Mother working at sink.		
		2. Chris throws pans onto floor.	
			3. M. picks up pans, tells C. to stop.
		4. C. spins lid.	
			5. M. restacks pans, closes cabinet.
	6. M. returns to sink.		
		7. C. throws lid across room, spills salt shaker.	
			8. M. stops cooking, cleans salt.
		9. C. spins pan lid.	
	10. M. opens refrigerator.		
		11. C. pulls refg. door.	
			12. M. holds door.
		13. C. screams.	
			14. M. slams door, talks to Chris.

7. Are there recurrent chains of certain antecedents, behaviors, and consequences?

8. Given the identified inappropriate behavior(s) of the student and the patterns of antecedents and consequences, what behavior really needs to be modified, and who is engaging in the behavior (for example, the referred student, or the teacher, or the parent)?

The use of anecdotal records is seldom practical for regular classroom teachers. However, special education teachers who work with students only part of the day in a special classroom or resource room may be called upon to observe referred students in the regular classroom. For such observation, skill in recording and analyzing anecdotal data is extremely valuable. Anecdotal records can enable these teachers to determine accurately what factors in the classroom are occasioning or maintaining appropriate and inappropriate behaviors.

PERMANENT PRODUCT RECORDING

Teachers have been using permanent product recording since the first time a teacher walked into a classroom. A teacher uses **permanent product recording** to grade a spelling test, check arithmetic problems, verify the creation of a chemical emulsion, or count the number of pegs a student has placed in a container. *Permanent products* are tangible items or environmental effects that result from a behavior. Permanent products are outcomes of behavior; thus, this method is sometimes called *outcome recording*. This type of recording is an *ex post facto* method of data collection, meaning it takes place after the behavior has occurred.

In order to collect permanent product data, the teacher reviews the statement of the behavior as written in the behavioral objective and determines what constitutes an acceptable outcome of the behavior. For example, if the behavior is building a tower of blocks, the objective will state whether the student is required to place one block on top of another or whether the blocks should be arranged in a certain sequence of colors. If the behavior is academic, conditions will also be specified. In each case, the teacher reviews the operational definition of the behavior. After evaluating the products of the required behavior, the teacher simply notes how many of the products were produced and how many were acceptable according to the definition.

Because the concrete results of a behavior are being evaluated and recorded, the teacher does not have to observe the student directly engaged in the behavior. Nor does the teacher have to sit and watch while the student takes the spelling test, writes a composition, or reads a language sample into a tape recorder. Convenience is the explanation for the frequent use of permanent product recording in the classroom—it causes minimal interference with a classroom schedule.

Permanent product recording is the easiest to use, but do all behaviors have a permanent product?

The versatility of permanent product recording makes it useful in a variety of instructional procedures and settings. In the home, permanent product recording has been used to record data on bedwetting (Hansen, 1979) —some permanent products are less appealing than others—and room cleaning (Wood & Flynn, 1978). In sheltered workshops and vocational settings, this method has been used to record increases in work production (Martin, Pallotta-Cornick, Johnstone, & Goyos, 1980) and the acquisition of janitorial skills (Cuvo, Leaf, & Borakove, 1978). In educational settings, it has been employed to record the expanded complexity of sentence writing (Heward & Eachus, 1979), to check spelling accuracy (Neef, Iwata, & Page, 1980), and to note development of fine motor control in cursive writing (Trap, Milner-Davis, Joseph, & Cooper, 1978). More generally, permanent product recording has been used in such diverse projects as controlling blood pressure (Cincirpini, Epstein, & Martin, 1979), assessing litter-control techniques (Bacon-Prue, Blount, Pickering, & Drabman, 1980), and increasing student self-recording skills (Hundert & Bucher, 1978).

Applications of permanent product recording.

The main advantage of permanent product recording is the durability of the sample of behavior obtained. The permanent product is not apt to disappear before its occurrence can be recorded. In light of this, the teacher may keep an accurate file of the actual products of certain target behaviors (such as test papers) or a report of the products for further review or verification at a later time.

With the increased availability of audiovisual equipment in schools and institutions, permanent product recording may include the use of audio-tape and videotape systems. With such devices, the teacher can take samples of specific transitory behaviors that would not ordinarily produce permanent products, such as a child's gait or a student's social interactions in play group. Taping makes a permanent recording of a behavior sample that might otherwise have been transitory.

What permanent products or outcomes might be observed for each of the behavioral dimensions discussed in the section on choosing a system of data collection?

Behavioral dimensions for which permanent product recording may be used.

Rate	The number of written products of any academic behavior per unit of time
Duration or latency	These behavioral dimensions unfortunately do not lend themselves to permanent product recording unless recording equipment is available
Topography	The correct formation of letters or numerals
	Following a pattern in such activities as pegboard designs, bead-stringing, or block-building
Force	Too light, too heavy, or uneven pressure when writing or typing
	Holes kicked in a classroom wall by a student having a tantrum
	Correct assembly of parts in a sheltered workshop

This list of examples is by no means exhaustive. Because permanent product recording is relatively simple and convenient, teachers can be imaginative in defining behaviors in terms of their outcomes. We have known teachers who operationally defined

Test anxiety as the number of visible erasures on a test paper

Sloppiness as the number of pieces of scrap paper on the floor within 2 feet of a student's desk

Hyperactivity as the number of table tennis balls still balanced in the pencil tray of a student's desk (the classroom was carpeted, which cut down on the noise).

The following vignette examines one use of permanent product recording.

Mr. Martin Observes Room Cleaning

Mr. Martin, while majoring in special education, was a night-shift aide at a residential institution for severely emotionally disturbed students. One of his duties was to see that each bedroom (shared by two residents) was cleaned before bedtime. He had decided that he would establish some system for reinforcing room-cleaning but was uncertain about what he should measure. When he tried measuring and reinforcing the time students spent cleaning up their rooms, he found that, while there was a great deal of scurrying around, the rooms were still very messy. Because the major problem appeared to be clothes, toys, and trash scattered on the floors, beds, and other furniture, he decided to use the number of such objects as his measure. Each evening before lights-out, he stood at the door of each bedroom with a clipboard containing a sheet of paper with each pair of residents' names and a space for each day of the week. He rapidly counted the number of separate objects scattered in inappropriate places and entered the total in the space on his data sheet.

OBSERVATIONAL RECORDING SYSTEMS

Whereas the permanent product method of data collection records the outcome of a behavior, **observational recording systems** are used to record behavior samples as the behavior is actually occurring. A data collector may choose from a pool of basic observational recording systems. Teachers who are interested in recording the number of times a behavior is occurring may select **event recording**. Those who want to find the proportion of a specified time period the behavior occurs may select **interval recording** or **time sampling. Duration recording** allows the teacher to determine the length of time the student spends performing some behavior. **Latency recording** measures the length of time it takes a student to start doing something. An illustration of the relationship between observational recording procedures and the components of a behavioral, stimulus-response sequence may be noted in Figure 4–3.

FIGURE 4–3
Observational data collection systems as related to the basic behavioral paradigm

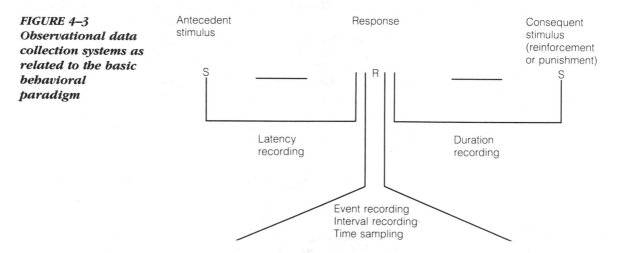

Event Recording

Event recording is a frequently used observational recording procedure because it most directly reflects a behavior's frequency. In event recording, a notation is made every time the student engages in the target behavior. The tally data are an exact record of how often the behavior occurred. A count of the target behavior is made within a specified sample-observation period—for example, during the reading period, from 10:00–10:30 A.M., or during lunch in the cafeteria. Recording how often the behavior occurs in this time period documents its **frequency** or rate of occurrence.

Event recording provides an exact count of how many times a behavior occurs.

Event recording is usually the method of choice when the objective is to increase or decrease the number of times a student engages in a certain behavior. Event recording may be used to record an increase in an appropriate social behavior (raising hands prior to talking in the classroom) or an increase in an academic response (number of correctly identified sight vocabulary words) or a decrease in an inappropriate behavior (such as hitting other children, cursing, or spitting).

Event recording can be used only for discrete behaviors.

Because the teacher attempts to record exactly the number of times the behavior occurs, event recording must be used with behaviors that are discrete. **Discrete behaviors** have an obvious beginning and an obvious end. With such a behavior, the teacher can clearly judge when one occurrence ends and the next begins, thus enabling an accurate frequency count. Event recording has been used in counting and recording a range of discrete behaviors, such as the number of manual signs a student uses (Carr, Binkoff, Kologinsky, & Eddy, 1978; Duker & Morsink, 1984), the numbers repeated in digit span (Farb & Throne, 1978), the number of math problems completed (Stevenson & Fantuzzo, 1984), the use of adjectives in verbal sentence production (Brody, Lahey, & Combs, 1978), the frequency of responses during small group conversations (Gajar, Schloss, Schloss, & Thompson, 1984), the use of yes or no responses (Neef, Walters, & Egel, 1984), letter recognition and labeling (Kincaid & Weisberg, 1978), echolalic responses (Schreibman & Carr, 1978), picture naming (Olenick & Pear, 1980), steps in a tooth-brushing procedure (Fisher, 1979), occurrences of aggressive behavior such as scratching and hitting (Carr, Newsom, & Binkoff, 1980), and nondelusional responses by schizophrenic children (Schraa, Lautmann, Luzi, & Screven, 1978).

Certain behaviors, however, may not be adequately measured using event recording. This data collection procedure is not appropriate in the following instances:

1. *Behavior occurring at such a high frequency that the number recorded may not reflect an accurate count.* Certain behaviors, such as the number of steps taken while running, some stereotypic behaviors (such as hand-flapping or rocking by severely handicapped students), and eye blinking may occur at such high frequencies that it becomes impossible to count them accurately (especially if the data collector blinks).

2. *Cases in which one behavior or response can occur for extended time periods.* Examples of such behaviors might be thumb-sucking or attention to task. If out-of-seat behavior were being recorded, for instance, a record showing that the student was out of her seat only one time during a morning would give an inaccurate indication of what the student was

FIGURE 4–4
Basic data recording sheet for event recording

Event Recording Data Sheet

Student: __PATRICIA__
Observer: __MRS. COHEN__

Behavior: __INAPPROPRIATE TALK-OUTS (NO HAND RAISED)__

	Time Start Stop	Notations of occurrences	Total occurrences
5/1/80	10:00 10:20	LHT LHTII	12
5/2/80	10:00 10:20	LHT IIII	9

FIGURE 4–5
Adapted data collection sheet for event recording

Student: __JEREMY__
Observer: __MS. GARWOOD__

Behavior: __ERRORS IN ORAL READING__

	CAROL	5/1/80	EXERC. #2	READER PG. 7.
SUBSTITUTIONS				
MISPRONOUNCIATIONS				
INSERTIONS				
REPETITIONS				

actually doing if this one instance of the behavior lasted from roll call to lunch time.

An advantage of event recording, in addition to accuracy, is the relative ease of the data collection itself. The teacher does not need to interrupt

FIGURE 4–6
Adapted data
collected sheet for
event recording

Event Recording Data Sheet

Student: _GAIL_
Observer: _MRS. ELLIOT_

Behavior: _VERBALLY STATING SIGHT WORDS_

	1	2	3	5/1	5/2	CAROL	ROBERT
OUT							
IN							
GIRL							
BOY							
EXIT							

the lesson in order to take data. The teacher may simply make a notation on an index card or paper on a clipboard, make slash marks on a piece of tape around a wrist (requiring stoicism upon removal), or transfer paper clips from one pocket to another. Following the lesson, this information may be tallied and transferred to a data sheet similar to the one presented in Figure 4–4.

Event recording is also easily used for many academic behaviors. Figure 4–5 is a data sheet used for recording the errors made during oral reading exercise. The teacher simply places a mark in the appropriate row as a particular type of error is made. The top of the columns may record the day of week, child who's reading, dates, page number from the reader where the mistake was made, and so on.

On Figure 4–6's data sheet, the teacher is able to record the correct oral naming of sight words (listed in the left column) during or following instruction. A simple method for recording mistakes is to mark directly on the back of the flashcard being used. Marks can later be tallied on a summary data sheet.

For the more mechanically inclined, counting devices are commercially available. While these make data collection easier and more accurate, they also entail some expense and may break. An inexpensive counter sold for tallying purchases in a grocery store may be useful, but it is noisy and possibly distracting. Stitch counters designed to fit on the end of a knitting needle come in sizes large enough to fit on a pen. Other commercial devices include

Wrist golf-counters or Wrist Tally Board
Behavior Research Company
P.O. Box 3351
Kansas City, KS 66103

Hand Tally Digital Counters (counter for single behavior)
Cambosco Scientific Co., Inc.
342 Western Ave.
Boston, MA 01235

Hand Tally Digital Counters (counter for multiple behaviors)
Lafayette Radio & Electronics
111 Jerico Turnpike
Syosset, Long Island, NY 11791

The wrist tally board is an arrangement of beads on a leather strap that functions as an abacus. It is a rather attractive device somewhat reminiscent of jewelry made at summer camp.

Recording of Controlled Presentations

One variation on the event recording technique is the use of **controlled presentations** (Ayllon & Milan, 1979). In this method, the teacher structures or controls the number of opportunities the student will have to perform the behavior. Most often this method consists of presenting a predetermined number of opportunities or trials in each instructional session. A trial may be viewed as a discrete occurrence, because it has an identifiable beginning and ending. A trial is defined by its three behavioral components: an antecedent stimulus, a response, and a consequating stimulus (S-R-S). The delivery of the antecedent stimulus (usually a verbal cue) marks the beginning of the trial, and the delivery of the consequating stimulus (reinforcement or punishment) signifies the termination of the trial. For example, in a given session the teacher may decide that a student will be given ten opportunities, or trials, to respond by pointing to a specified object upon request. Each trial is then recorded as correct or incorrect. Controlled presentation allows the teacher to monitor progress simply by looking at the number of correct responses for each session.

Figures 4–7, 4–8, and 4–9 present three variations of data sheets that have been used in the collection of trial or controlled presentation data. The data sheet in Figure 4–7 (Saunders & Koplik, 1975) is arranged to allow for up to 20 trials per instructional session. The teacher records dichotomous data (correct or incorrect) using the following simple procedure:

Following each trial

1. Circle the trial number that corresponds to a correct response.

2. Slash (/) through the trial number that corresponds to an incorrect response.

Following each session

1. Total the number of correct trials (those circled).

2. Place a square around the corresponding number in the session column that corresponds to the number of correct trials.

3. In order to graph directly on the data sheet connect the squared numbers across the sessions to yield a learning curve.

Figure 4–8 (Bellamy, Horner, & Inman, 1979) is arranged for recording dichotomous data on the instruction of complex tasks. One column is provided for listing antecedents to be provided (S^D) and another for a description of the subbehaviors or steps being requested (Response). Each trial consists of an opportunity for the student to perform the entire chain of steps (1–13) that make up the complex behavior. For each trial, the teacher may record the accuracy of the student's performance of each step within the chain, using the same circle and slash procedure. As with the previously described data sheet, this format also allows for graphing directly on the raw data sheet. (*Raw data* is the original record taken by the observer—it has not yet been converted or processed.) The number corresponding to the number of correctly performed steps is indicated by a square, and the squares are connected across trials.

Figure 4–9 (Alberto & Schofield, 1979) is systematized to permit recording of the type of interactions between teacher and student. The data sheet depicts five sessions, each containing ten trials. For each trial, the teacher can indicate the type of assistance that was necessary to enable the student to perform the response. Following each trial, the trial number is marked on the row corresponding to the type of assistance provided (e.g., the use of a gesture or a prompt). This data sheet may also serve as a self-graphing sheet if the marked trial numbers are connected within and across sessions, thus yielding a curve that displays the increasing independence of the student's performance of the response.

Classroom teachers can improve instruction by using controlled presentations. For example, a teacher might want to be sure to ask each member of a seminar group five questions during a discussion on early Cold War events and the Berlin Wall. A very simple data sheet with the names of the students and a space to mark whether answers were correct or incorrect would provide valuable information for analysis and evaluation.

Event recording (including controlled presentation) lends itself to the observation of the rate or frequency of behavior, for example:

Behavioral dimensions appropriate to event recording.

Number of times Mel talks out in an hour

Number of times Charlie hits another student in a 20-minute recess

Number of questions Melissa answers correctly during a 15-minute world geography review

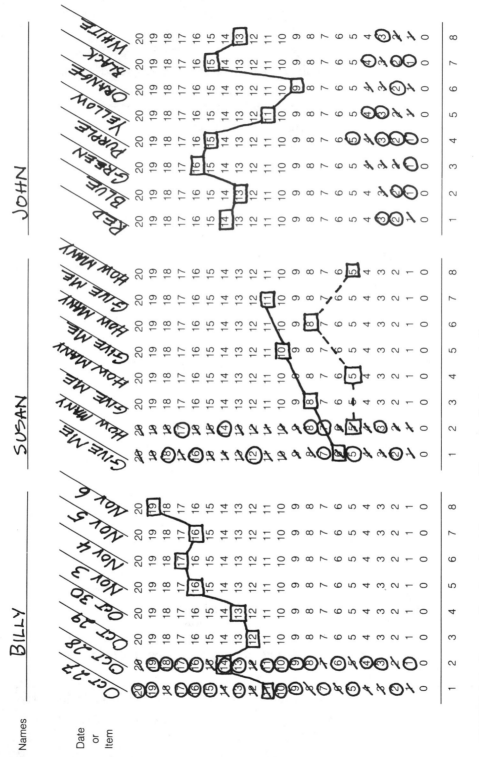

FIGURE 4-7 *Data collection sheet for use with controlled presentations*

Note. From "A multi-purpose data sheet for recording and graphing in the classroom," by R. Saunders and K. Koplik. *AAESPH Review*, 1975. *1*. 1. Copyright 1975 by The Association for the Severely Handicapped. Reprinted with permission.

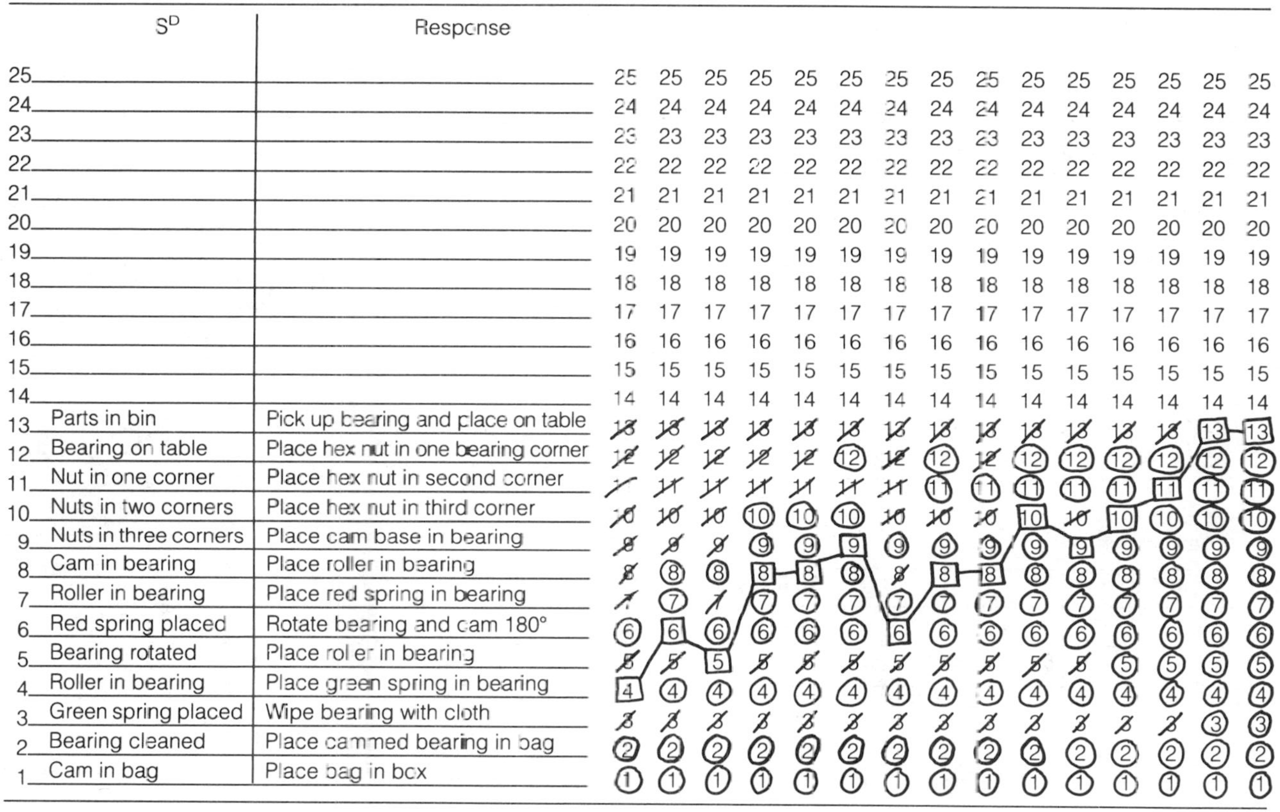

FIGURE 4–8 is a data collection sheet with a grid for recording trial-by-trial responses.

S^D	Response
13. Parts in bin	Pick up bearing and place on table
12. Bearing on table	Place hex nut in one bearing corner
11. Nut in one corner	Place hex nut in second corner
10. Nuts in two corners	Place hex nut in third corner
9. Nuts in three corners	Place cam base in bearing
8. Cam in bearing	Place roller in bearing
7. Roller in bearing	Place red spring in bearing
6. Red spring placed	Rotate bearing and cam 180°
5. Bearing rotated	Place roller in bearing
4. Roller in bearing	Place green spring in bearing
3. Green spring placed	Wipe bearing with cloth
2. Bearing cleaned	Place cammed bearing in bag
1. Cam in bag	Place bag in box

FIGURE 4–8 Data collection sheet for use with controlled presentations

Note. From *Vocational habilitation of severely retarded adults. A direct service technology*, by G. Bellamy, R. Horner & D. Inman (Baltimore University Park Press, 1979). Copyright© 1979 University Park Press, Baltimore. Reprinted by permission.

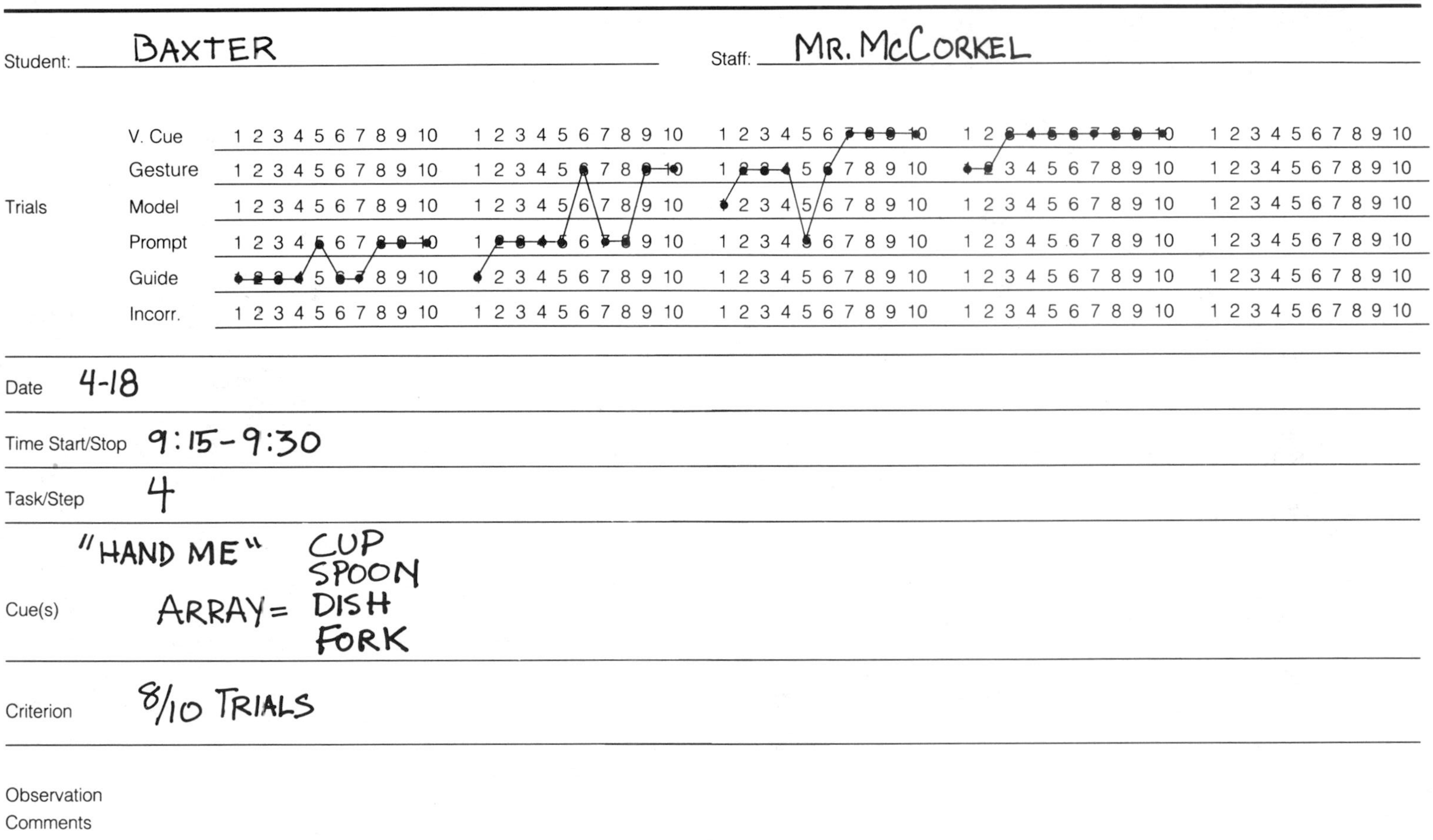

Student: __BAXTER__   Staff: __MR. McCORKEL__

	Trials			
V. Cue	1 2 3 4 5 6 7 8 9 10	1 2 3 4 5 6 7 8 9 10	1 2 3 4 5 6 7 8 9 10	1 2 3 4 5 6 7 8 9 10
Gesture	1 2 3 4 5 6 7 8 9 10	1 2 3 4 5 6 7 8 9 10	1 2 3 4 5 6 7 8 9 10	1 2 3 4 5 6 7 8 9 10
Model	1 2 3 4 5 6 7 8 9 10	1 2 3 4 5 6 7 8 9 10	1 2 3 4 5 6 7 8 9 10	1 2 3 4 5 6 7 8 9 10
Prompt	1 2 3 4 5 6 7 8 9 10	1 2 3 4 5 6 7 8 9 10	1 2 3 4 5 6 7 8 9 10	1 2 3 4 5 6 7 8 9 10
Guide	1 2 3 4 5 6 7 8 9 10	1 2 3 4 5 6 7 8 9 10	1 2 3 4 5 6 7 8 9 10	1 2 3 4 5 6 7 8 9 10
Incorr.	1 2 3 4 5 6 7 8 9 10	1 2 3 4 5 6 7 8 9 10	1 2 3 4 5 6 7 8 9 10	1 2 3 4 5 6 7 8 9 10

Date __4-18__

Time Start/Stop __9:15-9:30__

Task/Step __4__

Cue(s) __"HAND ME"   CUP   SPOON   DISH   FORK   ARRAY=__

Criterion __8/10 TRIALS__

Observation
Comments

FIGURE 4-9 *Data collection sheet for use with controlled presentations (Alberto & Schofield, 1979)*

Note. From "An instructional interaction pattern for the severely handicapped" by P.A. Alberto and P. Schofield, *Teaching Exceptional Children*, 1979, 12, 16-19. Copyright 1979 by the Council for Exceptional Children. Reprinted with permission.

Number of times Sam answers questions in a whisper

Number of times Mary throws trash on the floor

Number of stairs Eliot climbs putting only one foot on each step

---

Ms. Bragman Counts Tattling

The students in Ms. Bragman's third-grade class seemed to be spending most of their time telling her what other students were doing wrong. Ms. Bragman was concerned about this for two reasons: the students were not working efficiently and they were driving her up the wall. When she asked her friend Ms. Barbe for advice, Ms. Barbe suggested that the first thing to do was to find out how often the students were tattling.

"Otherwise," she said, "you wouldn't know for sure how well what you decide to do is working."

Ms. Bragman decided to count an instance of tattling each time a student mentioned another student's name to her and described any number of inappropriate behaviors. Thus, "Johnny's not doing his work and he's *bothering* me," was counted as one instance, but "Harold and Pete are talking," was counted as two instances. She then went back to Ms. Barbe.

"How can I write down something every time they do it?" she asked. "I move around my room all the time and I don't want to carry paper and a pencil."

Ms. Barbe laughed. "Believe it or not," she said, "what I use is macrame yarn. I fasten some of it around my waist with the ends hanging down and tie a knot every time I observe the behavior. It's easy and accurate and have you seen my new belt?"

INTERVAL RECORDING AND TIME SAMPLING

The data collection system of **interval recording** and **time sampling** are means of recording an estimate of the actual number of times behavior occurs. Instead of counting each occurrence of the behavior, the teacher counts the number of intervals of time within a sample observation period in which the behavior has occurred. With these methods, it is therefore possible to record continuous behaviors, as well as behaviors of a high frequency that may be incompatible with event recording.

The observer counts intervals, not discrete behaviors.

In terms of making the closest representation of the actual occurrence of the behavior, event recording is the most accurate, interval recording is the next, and time sampling the least exact (Repp, Roberts, Slack, Repp, & Berkler, 1976). Each system, however, has its own advantages and disadvantages.

Interval Recording

In interval recording, the teacher defines a specific time period in which the behavior will be observed. This observation period is then divided into equal intervals. Typically, the intervals are no longer than 30 seconds in length (Cooper, 1981). In order to record these data, the teacher draws a series of boxes representing the intervals of time. Within each box or in-

FIGURE 4–10
*Interval recording
for one minute*

	10″	20″	30″	40″	50″	60″
1 min	−	+	+	−	−	−

terval the teacher simply notes whether the behavior occurred (+) or did not occur (−) *at any time during the interval*. Therefore, each interval has only one notation. The example of a 1-minute observation period shown in Figure 4–10 has been divided into 10-second intervals. The sample target behavior occurred during two of the intervals in this observation period, namely the second and third 10-second intervals.

Because of the manner in which these interval data are recorded, only limited conclusions can be drawn from the record of the behavior's occurrence. Regardless of whether the behavior occurred once or five times during the interval, it is recorded by a single notation. Therefore, the actual number of occurrences is not included in the record. If, in the preceding example, cursing was being recorded, all the teacher could say was that, during two of the intervals, the student cursed. There were at least two instances of this behavior, but there may have been more. Even if the student cursed 11 times during the second interval, only one notation would have been made. Recording occurrences of discrete behaviors, such as cursing or hitting, is known as *partial-interval* recording (the behavior does not consume the entire interval).

Interval recording does not provide an exact count of behaviors, but is especially appropriate for continuous behaviors.

Behaviors like walking around the room or being off-task may begin in one interval and continue unbroken into the next interval. Such timing would appear as two instances, because it would be recorded in two intervals in this instance, but the same duration of behavior would appear as only one instance if it fell within a single interval. Recording ongoing behaviors that may continue for several intervals is known as *whole-interval* recording (the behavior consumes the entire interval).

An additional problem encountered with interval data collection is created by the shortness of each interval in which the notation is to take place. It is very difficult to teach and collect interval data simultaneously. The teacher must keep an eye on a student or students, observe a stopwatch or second hand on a watch, and note the occurrence or nonoccurrence of the target behavior all within matters of seconds; therefore, a third-party observer is sometimes required. It is fairly easy to teach even elementary school students to collect interval data. Most of them enjoy it thoroughly.

Even with an observer, however, the necessity of looking down at the data sheet to make a recording might cause an observer to miss an occurrence of the behavior, therefore resulting in inaccurate data. A cassette tape with timed beeps for the chosen intervals at least eliminates the timing difficulty. Another way of simplifying the task is to allow for recording inter-

FIGURE 4–11
Interval recording
with 5-second
scoring interludes

10″	5″	10″	5″	10″	5″	10″	5″	10″	5″	10″
Observe	Score	Observe	Score	Observe	Score	Observe	Score	Observe	Score	Observe

ludes as part of the data collection schedule. Using this system, a specific time is allotted both for observation and scoring, as seen in the sample row from a data-collection sheet shown in Figure 4–11.

Figures 4–12 and 4–13 are examples of interval recording sheets for a

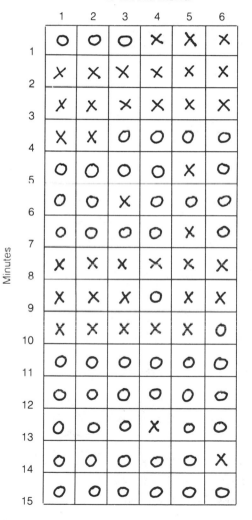

Student ___MALCOLM___
Date ___2/2/82___
Observer ___MR. RILEY___
Time Start ___9:15___
Time End ___9:30___
Behavior ___OFF-TASK___

Note occurrence within
the 10-second interval.

X = occurrence
O = non occurrence

FIGURE 4–12   Interval recording of off-task behavior

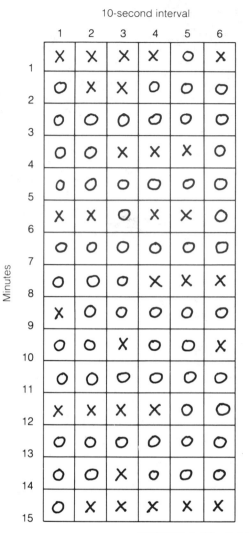

10-second interval

Student __LEROY__
Date __9/19/82__
Observer __MS HEILBRUNER__
Time Start __11:00__
Time End __11:15__
Behavior __OFF-TASK__

Note occurrence within
the 10-second interval.

X = occurrence
O = non occurrence

FIGURE 4–13 *Interval recording of off-task behaviors*

15-minute period that is divided into 10-second intervals. Looking at the notations of occurrence and nonoccurrence, the data collector can infer certain information:

1. Approximate number of occurrences of the behavior

2. Approximate duration of the behavior within the observation period

3. Distribution of the behavior across the observation period

Assuming that off-task behavior during a written arithmetic assignment is recorded in both examples, in Figure 4–12 the behavior appears to have occurred in 38 of the 90 intervals. When looking at the successive intervals in which the behavior occurred, it can be seen that the off-task behavior occurred for long durations (3 minutes each) but it appears to have been confined primarily to two periods. When reviewing such data, teachers should analyze the situation for some indication of what seem to be the immediate precipitating factors. In this example, off-task behavior may have been due to the two sets of written instructions on the worksheet, which prompted the student to ask a neighbor what to do.

In Figure 4–13, the off-task behavior is distributed throughout the time period. The teacher might find that the off-task behavior coincided with times when another student asked a question. Therefore, this student's problem may be distractibility rather than inability or unwillingness to read instructions.

Time Sampling

Time sampling allows for only one observation per interval.

In order to use time sampling, the data collector selects a time period in which to observe the behavior and divides this period into equal intervals. This process is similar to that employed with interval recording; however, the intervals in time sampling are often minutes rather than seconds. Such a format allows for observing the behavior over longer periods of time (Ayllon & Michael, 1959). In order to record these data, the observer draws a series of boxes representing the intervals. Within each box (interval), the observer simply notes whether the behavior was occurring (+) or not (−) when the student was observed *at the end of the interval*. Each interval therefore has only one notation. Note that the time sampling procedure differs from interval recording in that the student is observed only at the end of the interval rather than throughout it.

The 30-minute observation period in Figure 4–14 has been divided into 5-minute intervals. The target behavior occurred at the end of four of the intervals in this observation period: the first, second, fourth, and fifth intervals. A fairly simple way of recording time sampling data is to set a kitchen timer to ring at the end of the interval and to observe the behavior when the timer rings. To prevent students' figuring out the schedule and performing (or not performing) some behavior only at the end of the interval, the intervals may be made of random length. For example, a 5-minute time sampling recording system might have intervals of 2, 6, 5, 4, and 3 minutes. The average interval duration would be 5 minutes, but the students would never know when they were about to be observed. Common sense indicates the need to hide the face of the timer.

Because of the method of recording time sampling data, only limited conclusions can be drawn about the behavior recorded. As with interval

FIGURE 4–14
Time sampling at 5-minute intervals

	5′	10′	15′	20′	25′	30′
30 min.	+	+	−	+	+	−

An observer may miss a lot of behavior when using time sampling.

The longer the time-sampling interval, the less accurate the data will be.

Behavioral dimensions appropriate to interval and time-sampling recording.

recording, the behavior may have occurred more than once within the 5-minute observation interval. A particularly serious drawback with time sampling occurs when a single instance of a behavior occurs just before or just after the observer looks up to record the occurrence, resulting in a record of nonoccurrence.

When time sampling intervals are divided into segments by minutes as opposed to seconds, the procedure allows for longer periods between observations. It is therefore more convenient for simultaneous teaching and data collection. Indeed, the interval may be set at 15, 30, 60 minutes or further apart, allowing for observation throughout an entire day or class period. As the interval gets longer, however, the similarity between the data recorded and the actual occurrence of the behavior probably *decreases*. Time sampling, therefore, is suitable primarily for recording behaviors that are frequent or of long duration: for example, attention to task, out-of-seat behavior, or thumbsucking.

Data collected using interval recording or time sampling can be used to measure behavior along the dimension of frequency but *cannot be converted to rate*. You cannot say that a certain behavior occurred at the rate of two per minute when what you have recorded is that the behavior occurred during two 10-second intervals in a 60-second period. Interval and time sampling data are most often expressed in terms of the percentage of intervals during which the behavior occurred. The procedure for converting raw data into percentages will be discussed later in his chapter.

Measurement of duration can be approximated using interval recording, but this procedure does not lend itself to measures of latency. Force, locus, and topography may be measured, as with event recording, if they are included in the operational definition.

Ms. Simmons Observes Pencil Tapping

Ms. Simmons is an elementary school resource teacher for students with learning disabilities. One of her students, Arnold, tapped his pencil on the desk as he worked. He completed a surprising number of academic tasks but tapped whenever he was not actually writing. Ms. Simmons' concern was that this behavior would make it difficult for Arnold to be accepted in a regular classroom. She had tried counting pencil taps but found that Arnold tapped so rapidly that the pencil was a blur.

With another student, Shane, she was working on paying attention and concentrating. She decided that recording interval data would be an excellent task for Shane. She carefully defined the behavior, provided Shane with a recording sheet on a clipboard, gave him a stopwatch, and told him to mark a + in the box if Arnold tapped during a 10-second interval and a − if he didn't. She left Shane to observe and went to teach a small group. Very soon, she heard the sound of the clipboard hitting the floor as well as an expression of annoyance totally unacceptable in the classroom.

"I'm sorry, Ms. Simmons," said Shane, "but how the (expletive deleted) am I 'posed to watch the kid, the watch, and the sheet at the same time?"

Realizing that she had asked too much of a student who had difficulty concentrating, Ms. Simmons revised her procedure. She took home her cassette recorder and prepared to make a tape that would "bleep" every 10 seconds. She set up the recorder next to her new electronically controlled microwave oven and touched the "clear" button every 10 seconds as she watched the stopwatch. She had to start over only once when her teenage son, who had learned to expect dinner after a small number of "bleeps" (see Chapter 8 for a discussion of stimulus control), came in to ask when and what he was going to be fed.

The next day Ms. Simmons provided Shane with earphones and a 20-minute tape divided into 10-second intervals. Shane was able to record pencil-tapping this way very efficiently, even though he couldn't hear it.

After several days of observation and analysis of data, it occurred to Ms. Simmons that she might have used time sampling to observe Arnold's pencil tapping, because it occurred so frequently.

Variations on Collection Sheet

These data recording procedures are very flexible. They can be used to record data on several students simultaneously.

The basic data collection sheet format for interval recording and time sampling is an extremely flexible tool that can easily be adapted to meet a variety of instructional situations that arise in the classroom.

Cases of Multiple Students

In many instances, a teacher encounters several students in the same classroom who exhibit the same behavior within the same observation period. In such cases, data must be collected for each student. The basic data collection can be adapted to meet this need by adding additional row(s) of intervals for each student, as shown in Figure 4–15.

Another way of adapting the system for use with a group is a round-robin format (Cooper, 1981) of interval recording or time sampling. With this format, the behavior of a single group member per interval is observed and recorded. For example, when instructing a group of students in a DISTAR language lesson, the teacher might choose to monitor each student's attending behavior. The language period would then be divided into

FIGURE 4–15
Data sheet for multiple students

	10″	20″	30″	40″	50″	60″
Sara						
Max						
Sara						
Max						
Sara						
Max						

Event Recording Data Sheet

FIGURE 4–16
Round-robin format
of interval recording

	1st 15-sec. interval	2nd 15-sec. interval	3rd 15-sec. interval	4th 15-sec. interval
	Kate	Michael	Harry	Jody
1				
2				
3				
4				

equal 15-second intervals to accommodate the group of four students, with the name of each group member assigned to an interval. As shown in Figure 4–16, Kate was observed for occurrence or nonoccurrence of attending during the first 15-second interval of each minute; Michael's attending was observed and recorded during the second; Harry's, during the third; and Jody's, during the fourth.

A method of data collection similar to time sampling and useful for recording group behavior is PLACHECK—The Planned Activity Check (Doke & Risley, 1971). Doke and Risley (1972) used the PLACHECK procedure to measure and record participation of children in a day-care center in required-versus-optional activities. At 3-minute intervals, observers recorded the number of children participating in an appropriate manner in the alternative activities.

The PLACHECK procedure is similar to time sampling except that several students are observed at the end of the interval. The number of students engaged in activity can be compared to the number of students present.

A classroom teacher might use a PLACHECK procedure to record how many of her students were actively engaged, for example, in an art activity. She could set the timer, in a manner similar to that used in time sampling, and then count the number of participants each time the timer rang.

Cases of Multiple Behaviors

They can be used to record data on several behaviors simultaneously.

In some instances, a teacher may be focusing on a behavior that has a number of possible responses. The crucial task in this case is recording, with some detail, the nature or topography of the behavior as it occurs. Such a refinement of the recording system is possible through the use of a coding procedure (Bijou, Peterson, & Ault, 1968).

A data collection sheet for coding has a legend at the top indicating the behaviors being observed and their codes. The example in Figure 4–17 shows a few rows of data from an observation period in which occur-

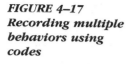

FIGURE 4–17
Recording multiple behaviors using codes

Interval Recording

Student _____ Time Start _____ H = Hitting
Date _____ Time End _____ T = Talking
Observer _____ P = Pinching

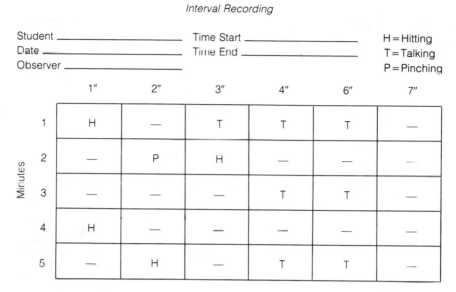

Minutes	1"	2"	3"	4"	6"	7"
1	H	—	T	T	T	—
2	—	P	H	—	—	—
3	—	—	—	T	T	—
4	H	—	—	—	—	—
5	—	H	—	T	T	—

rences of "disturbing one's neighbor" were recorded. The legend includes hitting (H), talking (T), and pinching (P).

When the behaviors to be coded occur during the observation period (interval recording) or at the end of the interval (time sampling), the code for the behavior is recorded for that interval. Coding enables the observer to record a category of behavior or several different behaviors.

An alternative system for coding more than one behavior or for coding a behavior whose topography ("shape") varies is the use of a tracking format (Bijou, Peterson, & Ault, 1968). In this format, each interval is divided into the number of rows corresponding to the possible behaviors or operationally defined responses that are to be recorded. Each row is reserved for the notation of that behavior. Figure 4–18 shows an example of data collected on stereotypic behavior where the exact responses to be rec-

H = head rolling
F = hand flapping
C = finger contortions
V = vocalizations

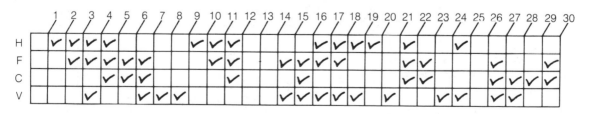

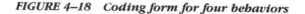

FIGURE 4–18   Coding form for four behaviors

FIGURE 4–19
Coding form with additional rows for nontarget behavior

	10"	10"	10"	10"	10"	10"	10"	10"	10"	10"
On-task										
Verbal off-task										
Motor off-task										
Passive off-task										

orded were head-rolling, hand-flapping, finger contortions, and high-pitched vocalizations, behaviors sometimes observed in people with autism, severe or profound mental retardation, or other severe handicapping conditions.

The format in Figure 4–18 targets specific responses for observation; however, it is also possible to track occurrences of more general behaviors. In Figure 4–19's data sheet (Cooper, 1981), on-task behavior was tracked. The teacher also wanted to know, however, the general nature of any off-task behavior. Therefore, in addition to noting the occurrence or nonoccurrence of on-task behavior, she provided track rows to indicate the general nature of any off-task behavior that occurred. The observer would simply put a check mark in the appropriate box(es) to indicate which behavior(s) occurred in which intervals.

A third type of coding, represented in Figure 4–20 (Alberto, Sharpton, & Goldstein, 1979), uses letter codes for specific responses within each interval. With this prepared format the observer simply places a slash through the appropriate letter(s) indicating the behavior that occurred. During the observation period, the student's social interactions were monitored for

FIGURE 4–20
Format for coding specific behaviors within intervals

	10"		20"		30"		40"		50"		60"	
1 min.	WG	AI	WG	AI	WG	AI	WG	AI	WG	AI	WG	AI
	A		A		A		A		A		A	
	CG	NI	CG	NI	CG	NI	CG	NI	CG	NI	CG	NI
2 min.	WG	AI	WG	AI	WG	AI	WG	AI	WG	AI	WG	AI
	A		A		A		A		A		A	
	CG	NI	CG	NI	CG	NI	CG	NI	CG	NI	CG	NI
3 min.	WG	AI	WG	AI	WG	AI	WG	AI	WG	AI	WG	AI
	A		A		A		A		A		A	
	CG	NI	CG	NI	CG	NI	CG	NI	CG	NI	CG	NI

within-group interaction (WG), cross-group interaction (CG), adult interaction (AI), no interaction (NI), or aggressive behavior (A).

DURATION AND LATENCY RECORDING

Duration and latency recording emphasize measures of time rather than instances of behavior.

Event recording, interval recording, and time sampling techniques of collection focus primarily on exact or approximate counts of the occurrence of a behavior. **Duration** and **latency recording** differ from these systems in that the focus is on a temporal rather than a numerical dimension of the behavior.

Duration Recording

Duration recording is used when the primary concern is the length of time a student engages in a particular behavior. For example, if a teacher wants to know about a student's out-of-seat behavior, either event recording or duration recording might be appropriate. Event recording would provide information about the number of times a student left her seat.

See pages 104–108.

However, if the teacher's concern is *how long* she stays out of her seat, the most appropriate data collection method would be duration recording. In this example, event recording would mask the temporal nature of the target behavior. While event data might indicate, for example, that the number of times the student got out of her seat had decreased substantially, it would not reveal that the length of time spent out-of-seat might actually have increased.

Duration recording, like event recording, is suitable for discrete behaviors—those that have an easily identifiable beginning and end. It is important to define clearly the onset of the behavior and its completion. The observer may time the duration of the behavior using the second hand of a watch or wall clock, but a stopwatch makes the process much simpler.

You could measure average duration of tantrums, of time spent on-task, of recreational reading.

There are two basic ways to collect duration data: by recording average duration or total duration. The *average duration* approach is used when the student performs the target behavior routinely or with some regularity. In a given day, the teacher measures the length of time consumed in each occurrence (its duration) and then finds the average duration for that day. If the behavior occurs at regular but widely spaced intervals (for example, only once per day or once per class period), the data may be averaged for the week. One behavior that can be measured by the duration data is time spent in the bathroom. Perhaps his teacher feels that each time John goes to the bathroom he stays for an unreasonable length of time. In order to gather data on this behavior, she decides to measure the amount of time he takes for each trip. On Monday, John went to the bathroom three times. The first trip took him 7 minutes, the second 11 minutes, and the third 9 minutes. If she continued to collect data in this manner during the rest of the week, the teacher would be able to calculate John's average duration of bathroom use for the week.

Total duration could be used to record time spent talking, reading, or playing with toys.

Total duration recording measures how long a student engages in a behavior within a limited time period. This activity may or may not be continuous. As an example, the target behavior "appropriate play" might be observed over a 15-minute period. The observer would record the number of minutes within this period the student was engaging in appropriate play. The child might, for example, have been playing appropriately from 10:00–10:04 (4 minutes), from 10:07–10:08 (1 minute), and from 10:10–10:15 (5 minutes). Although such a behavior record is clearly noncontinuous, these notations would yield a total duration of 10 minutes of appropriate play during the 15-minute observation period.

Latency Recording

Refer to Figure 4–3.

Latency recording measures how long a student takes to begin performing a behavior once its performance has been requested. This procedure measures the length of time between the presentation of an antecedent stimulus and the initiation of the behavior. For example, if a teacher says, "Michael, sit down" (antecedent stimulus) and Michael does, but so slowly that 5 minutes elapses before he is seated, the teacher would be concerned with the latency of the student's response.

Latency Recording Data Sheet		

Student ___EDITH___   Observer ___MR. HALL___
Behavior ___TIME ELAPSED BEFORE TAKING SEAT___

Operationalization of behavior initiation _____
_____
_____

Date	Time		Latency
	Delivery of S^D	Response initiation	

Duration Recording Data Sheet		

Student ___SAM___   Observer ___Ms. JAMES___
Behavior ___TIME SPENT IN BATHROOM___
_____

Behavior initiation _____
Behavior termination _____

Date	Time		Duration
	Response initiation	Response termination	

FIGURE 4–21   Basic formats for latency and duration data collection sheets

FIGURE 4–22
Self-graphing
duration data
collection sheet

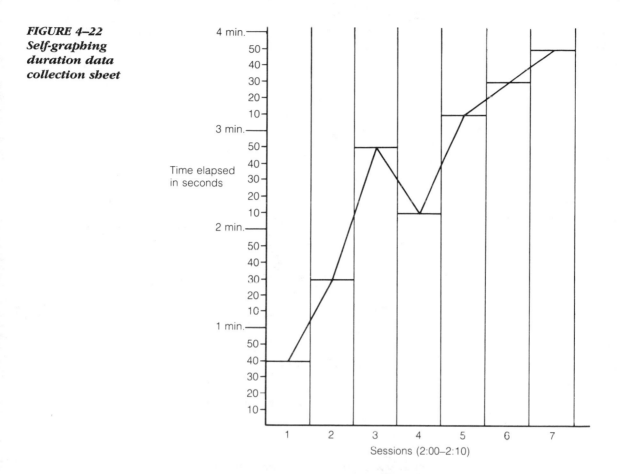

As seen in Figure 4–21, a basic duration, or latency, data collection sheet should provide space for noting when the stimulus is presented and when the response begins. The data sheet in Figure 4–22 can be made to be self-graphing. After timing each occurrence of the behavior, place a data point at the appropriate intersection of seconds and session and then simply connecting these data points.

Duration and latency recording are closely matched to the behavioral dimensions of duration and latency. However, consideration of topography, locus, and force may also apply here. For example, a teacher might want to measure

How long Calvin can maintain perfectly a position in gymnastics

How long Roberta talks to each of a number of other students

How long after being given a nonverbal signal to lower her voice Ellen actually does so

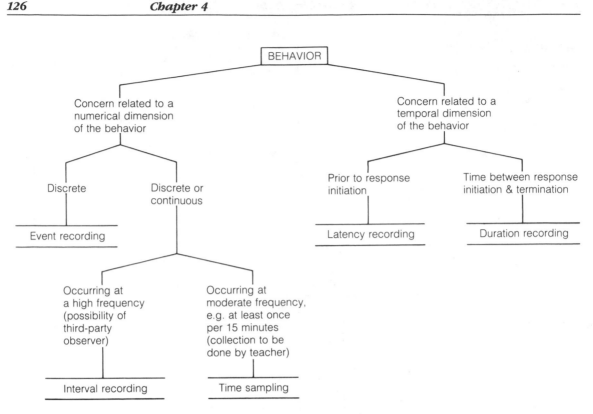

FIGURE 4–23 *Selecting observational recording procedures*

The five specific observational systems available to the data collector are *event recording, interval recording, time sampling, duration recording,* and *latency recording.* Figure 4–23 summarizes the decision-making process involved in selecting the system appropriate for a particular target behavior. This decision process is based on a series of questions to be answered by the data collector:

Questions for the data-collector. How to choose a recording system.

1. Is the target behavior numerical or temporal in nature?

2. If it is numerical

    a. Is the behavior discrete or continuous?

    b. Is the behavior expected to occur at a high, moderate, or low frequency?

    c. Will I be able to collect data during intervention/instruction, or will I need a third party to collect the data so as not to interrupt instruction?

3. If it is temporal, do I want to measure the time before initiation of the response or time elapsed during performance of the response?

RELIABILITY

When data collection depends upon human beings (teachers are included within this category), there is always the possibility of error. Even in the case of permanent product data, which are easiest to record, mistakes may happen. Teachers occasionally count math problems as incorrect even when they are correctly done or overlook a misspelled word in a paragraph. However, because there is something tangible, the teacher can easily go back to check the accuracy or **reliability** of her observations of the behavior. However, in using an observational recording system the teacher does not have this advantage. The behavior occurs and then disappears, so she cannot go back and check her own accuracy. In order to be sure that the data are correct, or reliable, it is wise to periodically have a second observer record the same behavior of the same student(s) at the same time. When this is done, the two observations can be compared and a coefficient or percent of *interobserver reliability* may be computed.

Computing reliability for permanent product data.

To check event recording, the teacher and a second observer, a paraprofessional or another student, simultaneously watch the student and record each instance of the target behavior. After the observation period, the teacher calculates the coefficient of agreement or reliability by dividing the smaller number of recorded instances by the larger number of recorded instances. For example, if the teacher observed 20 instances of talking-out in a 40-minute session, and the second observer recorded only 19, the calculation would be $19/20 = .95$. Therefore the coefficient of interobserver agreement would be .95. Coefficients of agreement are often reported as percent of agreement. The percent of agreement is calculated by multiplying the coefficient by 100. So in this example the percent of agreement is $.95 \times 100 = 95\%$.

Computing reliability for event recording.

However, this method of calculating reliability for research purposes lacks a certain amount of precision and has, therefore, been referred to as a gross method of calculation. "The problem is that this method does not permit the researcher to state that both observers saw the same thing or that the events they agreed on were all the same events" (Tawney & Gast, 1984, p. 138). In other words, there is no absolute certainty that the 19 occurrences noted by the paraprofessional were the same ones noted by the teacher.

Computing reliability for duration and latency recording.

The reliability of duration and latency data is determined by a procedure similar to that of event recording, except that the longer time is divided into the shorter, as in

$$\frac{\text{Shorter number of minutes}}{\text{Longer number of minutes}} \times 100 = \text{percent of agreement.}$$

When using interval recording or time sampling the basic formula for calculating reliability is

$$\frac{\text{Agreements}}{\text{Agreements} + \text{Disagreements}} \times 100 = \% \text{ of agreement.}$$

If the data below represent 10 intervals during which the teacher and the paraprofessional were recording whether or not Lauren was talking to her neighbor, we see that their data agree in 7 intervals (i.e., intervals 1, 2, 3, 4, 6, 7, and 8); their data are not in agreement in 3 intervals (i.e., 5, 9, and 10). Therefore, using the basic formula the calculation for reliability would be

$$\frac{7}{7 + 3} \times 100 = 70\%$$

	1	2	3	4	5	6	7	8	9	10
Teacher	×	×	—	—	×	×	—	—	×	—
Paraprofessional	×	×	—	—	—	×	—	—	—	×

Under certain research circumstances an additional, more rigorous determination of reliability should be considered. This should be a calculation of *occurrence reliability* or of *nonoccurrence reliability*. When the target behavior is recorded to have occurred in *less* than 75% of the intervals, occurrence reliability should be calculated. When the target behaviorl is recorded to have occurred in *more* than 75% of the intervals, nonoccurrence agreement should be computed (Tawney & Gast, 1984). These coefficients are determined with the same basic formula (agreements/ agreements + disagreements × 100) except that only those intervals in which the behavior occurred (or did not occur) are used in the computation.

In general, applied behavior analysts aim for a reliability coefficient of around .90. Anything less than .80 is a signal that something is seriously wrong. The problem is most often in the definition of the behavior; sloppy definitions result in low reliability, because the observers have not been told with sufficient exactitude what they are observing.

Kazdin (1977) has suggested four additional sources of bias that can affect interobserver agreement. First, just knowing another person is present collecting reliability data can influence the accuracy of the primary observer. Such knowledge has influenced reliability data by as much as 20–25%. It is suggested that reliability checks should be unobtrusive or covert or that the second observer collect data on several students including the target student. These suggestions may not be practical in the classroom. However, limiting communication between the teacher and the second observer during the observation period can reduce their influence on one another's observations.

The second concern is referred to as observer drift. This is the tendency of observers to change over time the stringency with which they apply the operational definitions of behavior. They may begin to record as "instances" behaviors that do not exactly conform to the operational definition. It is recommended that the observers periodically review definitions together and practice during the course of the program.

A third influence on the reliability of data concerns the complexity of the observational coding system. The more complex the system, the more in jeopardy the reliability. Complexity refers to either the number of different types of a response category being recorded (e.g., types of disruptive behavior), the number of different students being observed, or the number of different behaviors being scored on a given occasion. In a classroom, the teacher might mitigate the effects of complexity by limiting the number of behaviors or students to observe at any given time. The fourth bias is that of expectancies. Observers who look for behavior change are likely to find it. The reverse is also true; a teacher who has decided that nothing can be done with a student is not likely to find behavior change in her data.

Kazdin (1977) provides suggestions for limiting complexity bias in research studies.

The procedures just described are adequate for determining reliability for most teachers. More stringent standards are sometimes applied in research studies. The student who is interested in learning more about inter-observer reliability should consult the article by Hawkins and Dotson (1975) listed in the references and the series of invited articles on this topic in Volume 10 (1977) of the *Journal of Applied Behavior Analysis*.

FUNDAMENTALS OF GRAPHING DATA

In order to be useful, data obtained through observational techniques must be presented in an easily readable form. Tallies or coded entries on data collection forms (raw data) are difficult if not impossible to interpret for information about what is happening to the target behavior. The most common method of data presentation used by the applied behavior analysts is graphing. The translation of raw data into a graph is not difficult. It may involve two important steps: conversion of data and plotting the graph.

Conversion of Data

Before data can be graphed, it must be converted into a form that will allow for consistent graphing. The data may be reported as a number correct, as a percentage, or as a rate. The type of measurement conversion used depends upon the type of collection system that was employed and the kind of data obtained.

Converting permanent product data.

Permanent product data are reported as a number of items or a percentage of items resulting from the behavior. For example, a teacher might record the number of math problems completed, the percentage of correctly spelled words, the number of beads strung, or the number of dirty clothes placed in the hamper. If the number of opportunities for responding remains constant, as in spelling tests that always have 20 items or in a

FIGURE 4–24
Choosing
measurement
conversion for
permanent product
data

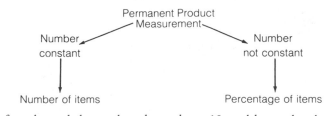

series of math worksheets that always have 10 problems, the data may be graphed simply as the number of items. If, on the other hand, the number of opportunities varies—different numbers of test items or math problems —the analyst will need to calculate percentages. (See Figure 4–24.)

Calculating percentages of correct responses requires only dividing the number of correct responses by the total number of responses and multiplying the result by 100:

$$\frac{\text{Number of correct responses}}{\text{Total number of responses}} \times 100 = \text{Percentage of correct responses.}$$

Percentages do not take time (speed) into account as a factor in describing the behavior.

Calculation of Rate

A conversion to rate data is required when the behavior analyst is concerned with the occurrence of the target behavior *within a certain time period* as well as with the number and accuracy of responses. Rate data reflect accuracy and speed, or fluency of performance, and thereby allow judgments about the development of proficiency. It is important to note that if the time allowed for the response(s) is the same across all sessions, a simple reporting of frequency is all that is necessary. Such is the case where each day the student has 20 minutes to complete a set of math problems. However, in instances where the time allocated for responding, or that required by the student to complete a task, varies from session to session, a rate calculation is needed for comparison.

Calculating rate.   A *rate of correct responding* is computed by dividing the number of correct responses by the time taken for responding:

$$\text{Correct rate} = \frac{\text{Number correct}}{\text{Time}}.$$

For example, if on Monday Kevin completed 15 problems correctly in 30 minutes, his rate of problems correct would be .5 per minute. If, on Tuesday, he completed 20 correct problems in 45 minutes, his rate per minute would be .45. If Kevin's teacher had merely recorded that Kevin completed 15 problems on Monday and 20 problems on Tuesday, the teacher might think that Kevin's math was improving. In reality, though the number of math problems increased, the rate decreased, and Kevin did not do as well on Tuesday as on Monday.

Monday:
$$\frac{15 \text{ correct problems}}{30 \text{ minutes}} = .5 \text{ correct problems per minute.}$$

Tuesday:
$$\frac{20 \text{ correct problems}}{45 \text{ minutes}} = .45 \text{ correct problems per minute.}$$

Calculating and plotting a rate of error may be done by dividing the number of errors by the time:

$$\text{Error rate} = \frac{\text{Number of errors}}{\text{Time}}.$$

These calculations will provide the teacher with the numbers of correct and incorrect behaviors per minute (or second, or hour).

Session 1:
$$\frac{12 \text{ spelling errors}}{20 \text{ minutes}} = .60 \text{ errors per minute.}$$

Session 2:
$$\frac{10 \text{ spelling errors}}{30 \text{ minutes}} = .33 \text{ errors per minute.}$$

Converting other kinds of data

Event recording data may be presented in terms of the number of occurrences if the opportunities for responding are consistent each time data are collected. If opportunities for responding are not consistent—as might happen if we were counting the number of times a student followed an instruction from the teacher—the data may be converted to a percentage. Controlled presentation would eliminate this necessity. For example, the teacher might decide to issue exactly ten instructions in a 1-hour period. If the fluency of responding is an issue, conversion may also be made to rate.

Interval or time sampling data are reported as the number or percent of intervals in which the behavior occurred. They are most often presented as percentages.

Duration data are reported as the number of minutes or seconds during which the behavior occurred within a specified period of time.

Latency data are reported as the number of minutes that lapsed prior to the initiation of the behavior following a request. A summary of data conversion procedures appears in Table 4–1.

Basic Components of a Graph

The basic model for graphing is a simple *line graph*, drawn to include two *axes*. The horizontal axis is the abscissa or *x*-axis. The vertical axis is the ordinate or *y*-axis. Each data point is placed at the intersection of the ses-

TABLE 4–1   *Summary of data conversion procedures*

Type of recording	Data conversion	
Permanent product recording Event recording	Report number of occurrences . . .	if both time and opportunities to respond are constant.
	Report percentage . . .	if time is constant (or not of concern) and opportunities vary.
	Report rate . . .	if both time (which is of concern) and opportunities vary. *OR* if time varies and opportunities are constant.
Interval recording Time sampling	Report number . . .	if constant.
	Report percentage of intervals . . .	during or at the end of which behavior occurred.
Duration	Report number of seconds/ minutes/hours . . .	for which the behavior occurred.
Latency	Report number of seconds/ minutes/hours . . .	between antecedent stimulus and onset of behavior.

sion during which the behavior was recorded and the quantity or level of the behavior performance. (See Figure 4–25).

The time dimension is indicated along the abcissa. Each unit along this axis often represents an observation period during which data were collected. The observation period might be units of days or sessions (of 15 minutes, from 10:00–10:30), or sets of trials (one session equaled 20 trials), or opportunities for response. The first principle of graphing, then, is that *behavior* goes on the ordinate, *time* on the abscissa.

FIGURE 4–25
Basic line graph

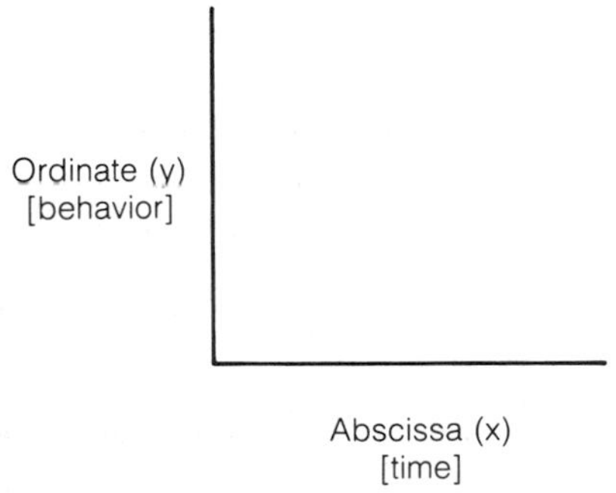

Ordinate (y)
[behavior]

Abscissa (x)
[time]

FIGURE 4–26
Sample graph
labeling

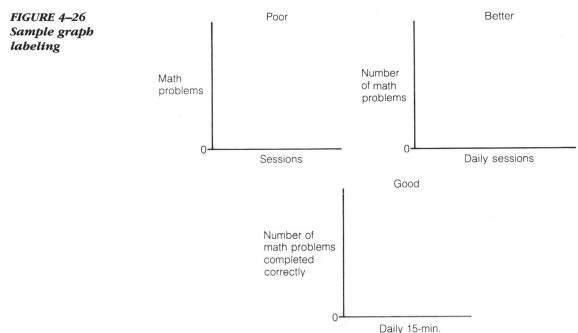

After labeling, the lines on the graph are then numbered in equal increments to show the passage of time and the number, percentage, or rate of
the behavior. Notice in Figure 4–27 that the intersection of the axes is the
zero point—the numbers start on the first line of the graph.

A more specific description of the time units and behavior is included
in the titles written on each axis. It is a challenge to include as much information as possible without crowding the graph. For example, see the labeling on the graphs in Figure 4–26.

Each data point, converted from the record sheets, is next plotted in the
form of a dot at the appropriate intersection of the two points on the axes.
If Harold completed five problems the first day, eight the second, and four
the third, the graph would look like the one in Figure 4–28. To complete
the graph, the dots are then connected with a solid line, as in Figure 4–29.
This is the basic form for presenting applied behavior analysis data.

Alternative Graphing Formats

Several alternative methods are used for displaying data graphically: cumulative graphs, ratio graphs, and bar graphs.

*A cumulative graph
shows the total number
of responses for all
sessions including the
one being plotted.*

Cumulative graph.   On a line graph, data points are plotted at the appropriate intersections without regard to the performance level of the previous session. For a **cumulative graph**, on the other hand, the number of
occurrences observed in a given session are graphed *after being added* to

FIGURE 4–27
Sample graphs with
units numbered

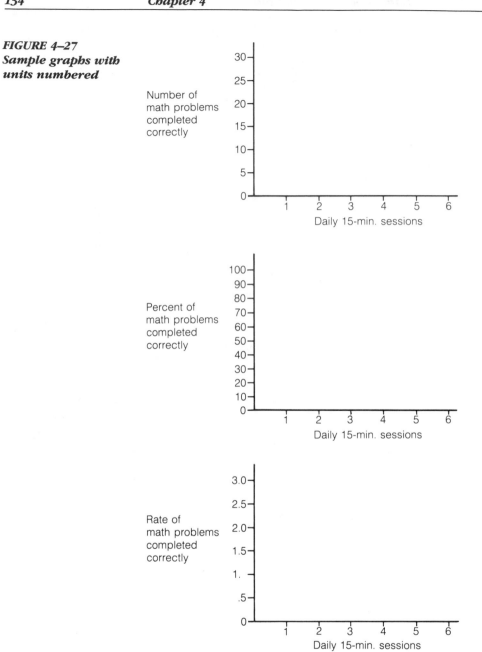

the number of occurrences of the previous session. This approach provides a count of the total number of responses. Such graphs must always
demonstrate an upward curve provided that any behavior at all is being
recorded (Ferster, Culbertson, & Boren, 1975). A cumulative graph presents an additive view of a behavior across the intervention sessions.

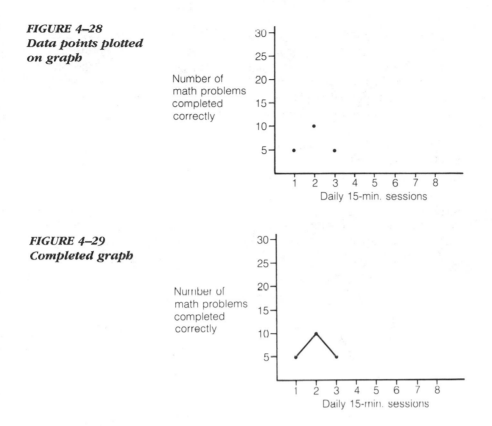

FIGURE 4–28
Data points plotted on graph

FIGURE 4–29
Completed graph

This form of data illustration provides a continuous line with a slope that indicates the rate of responding. A steep slope indicates rapid responding, a gradual slope indicates slow responding, and a plateau or straight line indicates no responding. (See Figure 4–30).

A teacher might want to use a cumulative graph to trace a student's progress toward performing a predetermined number of behaviors. For example, in a prevocational class, the teacher might decide that students

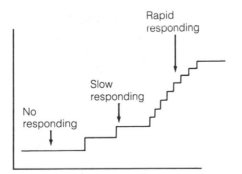

FIGURE 4–30
Slope and rate of responding

should assemble the parts of a simple appliance 50 times in order to demonstrate mastery. A cumulative graph would show clearly when this objective had been accomplished. The slope of the graph would also give the teacher informal data about how rapidly students were moving toward mastery. The hypothetical graphs in Figure 4–31 show the same raw data plotted on a line graph (a) and a cumulative graph (b).

Ratio Graphs.   A form of graphing particularly suited to charting rate data has been described by White and Haring (1980). Their **ratio graphs** are designed to show proportional change. The procedure for plotting data points for this kind of graph is the same as that for line graphs; only the form of the chart or graph itself is different. All data for ratio graphing are converted to rate per minute, and the chart can be used to plot behaviors that occur at the rate of 1000 per minute down to behaviors that occur at

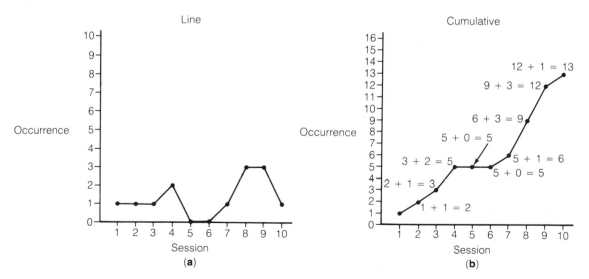

Raw Data

Session	Occurrences
1	1
2	1
3	1
4	2
5	0
6	0
7	1
8	3
9	3
10	1

FIGURE 4–31 Comparing line and cumulative graphing

the rate of .000695 per minute (once in 24 hours). Such graphs are a major tool in an outgrowth of applied behavior analysis known as *precision teaching*. Precision teaching involves recording students' responses during a specified time period and plotting these data in the form of rate of responding.

Teachers may initially be put off by the appearance of a ratio or logarithmic graph, but with a little practice, these tools are easy to use and interpret. Ratio graphs provide the basis for making decisions about teaching based on students' rate of performance. Figure 4–32 shows an example of a ratio or logarithmic chart.

Bar Graph. A **bar graph**, or histogram, is yet another means of displaying data. As its name implies, a bar graph uses vertical bars rather than horizontal lines to indicate levels of performance.

A bar graph may be the preferred means of displaying data in situations where clear interpretation of the pattern of behavior plotted on a line graph is difficult. Such confusion may result when several lines are plotted

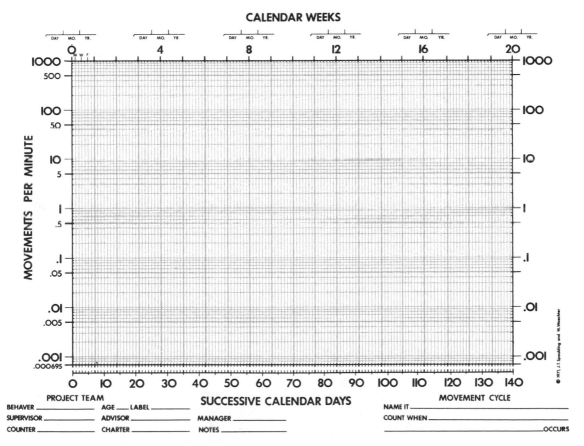

FIGURE 4–32   Chart for recording ratio data

FIGURE 4–33
Comparing line and
bar graphing

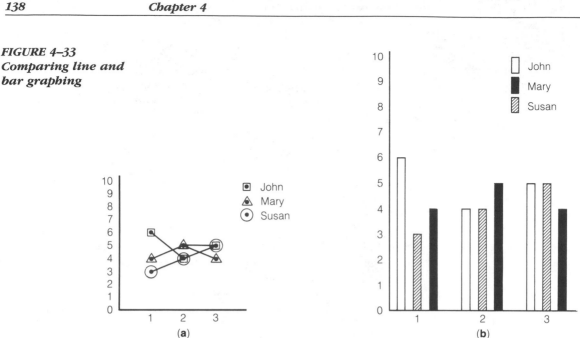

(a)   (b)

on a single line graph, as when a teacher chooses to include data from several students or from multiple behaviors. In such cases, plotted lines may overlap or appear extremely close together because data points fall at the same intersections. Figure 4–33 offers an example of the same data plotted on a line graph and on a bar graph. The bar graph is plainly much clearer. A classroom teacher might use a bar graph to display daily the number of correct responses from each member of a small group.

A Day in the Life of Professor Grundy

The alarm clock beside Professor Grundy's bed rang loudly. He groped for the off button and silenced the noise. 6:30 A.M. He put the pillow over his head and contemplated his day. He remained prone for several minutes, then sat up, turned on the bedside lamp and swung his feet to the floor. On the alarm clock was taped a data record sheet on which he recorded the latency of his getting up behavior.

"Ten minutes," he groaned. "Yesterday, I made it in five, but that was before behavioral objectives. Oh Lord, today I have to do observational procedures with the graduate class." Professor Grundy felt himself pulled again toward the pillow but forced himself into the bathroom instead, where he shed his pajamas and stepped on the scale. He noted the (very) permanent product thus measured on a graph handily placed on the bathroom wall.

"Oh well," sighed Professor Grundy, "I have big bones."

He then proceeded to the kitchen where his wife, Minerva, was preparing breakfast. He watched his stopwatch carefully.

"I compliment you, my dear," he exclaimed, when she served him. "Your facial twitch is much better. It occurred in only 35% of the 10-second intervals during the 15 minutes it took you to cook today. That's down 12% from yesterday."

"Which data sheet is it, Oliver?"

Professor Grundy proceeded to the university. During the day, he picked up the phone in his office at 15-minute intervals, recording a time sample of his garrulous colleague's talking on the party line they shared.

"Just as I thought," he gloated. "That psychoanalytic fruitcake is on the phone 65% of the time. I knew he was hogging the line."

At 4:30, Professor Grundy went to teach his graduate class. All the students were employed as teachers during the day, and it was a challenge merely to keep them awake. He presented a crisp, factual description of behavioral observation procedures (remarkably similar to the one presented in this chapter) and then assigned an observation project.

One of the students still awake snorted, "How utterly ridiculous . . .," apparently waking up the rest of the class, all of whom chimed in: "Yes, I have to teach, I don't have time . . .," "You must be kidding . . ."

Professor Grundy had activated his stopwatch when the first comment was made. He waited calmly until the uproar ceased, then made a notation in his grade book.

"Eleven minutes this semester," he told the class. "Last semester the duration of complaints was only eight minutes."

After repeating the instructions for the assignments, Professor Grundy returned home where he recorded and graphed

The latency of Minerva's getting dinner on the table after he announced that he was hungry

The rate per hour of television commercials

The fact that once again Minerva went to bed early with a blinding headache, brought on, she announced, by the sight of graph paper all over the house

SUMMARY

This chapter described procedures for collecting data that will form the basis for decisions about intervention strategies. We discussed a number of types of data collection: permanent product, observational techniques, and anecdotal records. One section examined basic procedures for determining interobserver reliability. The chapter concluded with information on graphing data, including a description of line graphs, cumulative graphs, bar graphs, and ratio graphs. You'll find tools presented in this chapter very useful in working with data.

REFERENCES

ALBERTO, P.A., & SCHOFIELD, P. 1979. An instructional interaction pattern for the severely handicapped. *Teaching Exceptional Children, 12*, 16–19.

ALBERTO, P.A., SHARPTON, W., & GOLDSTEIN, D. 1979. Project Bridge: Integration of severely retarded students on regular education campuses. Atlanta: Georgia State University.

AYLLON, T.A., & MICHAEL, J. 1959. The psychiatric nurse as a behavior engineer. *Journal of the Experimental Analysis of Behavior, 2*, 323–334.

AYLLON, T., & MILAN, M. 1979. *Correctional rehabilitation and management: A psychological approach*. New York: Wiley.

BACON-PRUE, A., BLOUNT, R., PICKERING, D., & DRABMAN, R. 1980. An evaluation of three litter control procedures—Trash receptacles, paid workers, and the marked item technique. *Journal of Applied Behavior Analysis, 13*, 165–170.

BELLAMY, G., HORNER, R., & INMAN, D. 1979. *Vocational habilitation of severely retarded adults: A direct service technology*. Baltimore: University Park Press.

BIJOU, S.W., PETERSON, R.F., & AULT, M.H. 1968. A method to integrate descriptive and experimental field studies at the level of data and empirical concepts. *Journal of Applied Behavior Analysis, 1*, 175–191.

BRODY, G.H., LAHEY, B.B., & COMBS, M.L. 1978. Effects of intermittent modelling on observational learning. *Journal of Applied Behavior Analysis, 11*, 87–90.

CARR, E.G., BINKOFF, J.A., KOLOGINSKY, E., & EDDY, M. 1978. Acquisition of sign language by autistic children. I: Expressive labelling. *Journal of Applied Behavior Analysis, 11*, 489–501.

CARR, E.G., NEWSOM, C.D., & BINKOFF, J.A. 1980. Escape as a factor in the aggressive behavior of two retarded children. *Journal of Applied Behavior Analysis, 13*, 101–117.

CINCIRIPINI, P.M., EPSTEIN, L.H., & MARTIN, J.E. 1979. The effects of feedback on blood pressure discrimination. *Journal of Applied Behavior Analysis, 12*, 345–353.

COOPER, J. 1981. *Measurement and analysis of behavioral techniques*. Columbus, Ohio: Charles E. Merrill.

CUVO, A.J., LEAF, R.B., & BORAKOVE, L.S. 1978. Teaching janitorial skills to the mentally retarded: Acquisition, generalization, and maintenance. *Journal of Applied Behavior Analysis, 11*, 345–355.

DOKE, L., & RISLEY, T.R. 1971. The PLACHECK evaluation of group care. Unpublished paper presented at Annual Meeting of the Kansas Psychological Association, Overland Park, Kan., April.

DOKE, L.A., & RISLEY, T.R. 1972. The organization of day-care environments: Required vs. optional activities. *Journal of Applied Behavior Analysis, 5*, 405–420.

DUKER, P., & MORSINK, H. 1984. Acquisition and cross-setting generalization of manual signs with severely retarded individuals *Journal of Applied Behavior Analysis, 17,* 93–103.

FARB, J., & THRONE, J.M. 1978. Improving the generalized mnemonic performance of a Down's syndrome child. *Journal of Applied Behavior Analysis, 11*, 413–419.

FERSTER, C.B., CULBERTSON, S., & BOREN, M.C.P. 1975. *Behavior principles* (2nd ed.). Englewood Cliffs, N.J.: Prentice-Hall, Inc.

FISHER, E.B., Jr. 1979. Overjustification effects in token economies. *Journal of Applied Behavior Analysis, 12*, 407–415.

GAJAR, A., SCHLOSS, P., & THOMPSON, C. 1984. Effects of feedback and self-monitoring on head trauma youths' conversational skills. *Journal of Applied Behavior Analysis, 17*, 353–358.

HALL, C.S. 1954. *A primer of Freudian psychology*. Cleveland: World Publishing.

HALL, R. 1971. *Behavior modification: The measurement of behavior*. Lawrence Kan.: H & H Enterprises.

HANSEN, G.D. 1979. Enuresis control through fading, escape, and avoidance training. *Journal of Applied Behavior Analysis, 12*, 303–307.

HAWKINS, R.P., & DOTSON, V.S. 1975. Reliability scores that delude: An Alice in Wonderland trip through the misleading characteristics of inter-observer agreement scores in interval recording. In E. Ramp & G. Semp (Eds.), *Behavior analysis: Areas of research and application* (pp. 359–376). Englewood Cliffs, N.J.: Prentice-Hall.

HERSEN, M., & BARLOW, D.H. 1976. *Single-case experimental designs: Strategies for studying behavior change*. New York: Pergamon.

HEWARD, W.L., & EACHUS, H.T. 1979. Acquisition of adjectives and adverbs in sentences written by hearing impaired and aphasic children. *Journal of Applied Behavior Analysis, 12*, 391–400.

HUNDERT, J., & BUCHER, B. 1978. Pupils' self-scored arithmetic performance: A practical procedure for maintaining accuracy. *Journal of Applied Behavior Analysis, 11*, 304.

KAZDIN, A. 1977. Artifact, bias, and complexity of assessment: The ABCs of reliability. *Journal of Applied Behavior Analysis, 10,* 141–150.

KINCAID, M.S., & WEISBERG, P. 1978. Alphabet letters as tokens: Training preschool children in letter recognition and labelling during a token exchange period. *Journal of Applied Behavior Analysis, 11*, 199.

MARTIN, G., PALLOTTA-CORNICK, A., JOHNSTONE, G., & GOYOS, A.C. 1980. A supervisory strategy to improve work performance for lower functioning retarded clients in a sheltered workshop. *Journal of Applied Behavior Analysis, 13*, 183–190.

NEEF, N.A., IWATA, B.A., & PAGE, T.J. 1980. The effects of interspersal training versus high density reinforcement on spelling acquisition and retention. *Journal of Applied Behavior Analysis, 13*, 153–158.

NEEF, N., WALTERS, J., & EGEL, A. 1984. Establishing generative yes/no responses in developmentally disabled children. *Journal of Applied Behavior Analysis, 17*, 453–460.

OLENICK, D.L., & PEAR, J.J. 1980. Differential reinforcement of correct responses to probes and prompts in picture-name training with severely retarded children. *Journal of Applied Behavior Analysis, 13*, 77–89.

REPP, A.C., ROBERTS, D.M., SLACK, D.J., REPP, C.F., & BERKLER, M.S. 1976. A comparison of frequency, interval, and time-sampling methods of data collection. *Journal of Applied Behavior Analysis, 9*, 501–508.

SANFORD, F.L., & FAWCETT, S.B. 1980. Consequence analysis: Its effects on verbal statements about an environmental project. *Journal of Applied Behavior Analysis, 13*, 57–64.

SAUNDERS, R., & KOPLIK, K. 1975. A multi-purpose data sheet for recording and graphing in the classroom. *AAESPH Review, 1*, 1.

SCHRAA, J.C., LAUTMANN, L., LUZI, M.K., & SCREVEN, C.G. 1978. Establishment of nondelusional responses in a socially withdrawn chronic schizophrenic. *Journal of Applied Behavior Analysis, 11*, 433–434.

SCHREIBMAN, L., & CARR, E.G. 1978. Elimination of echolalic responding to questions through the training of a generalized verbal response. *Journal of Applied Behavior Analysis, 11*, 453–463.

SIDMAN, M. 1960. *Tactics of scientific research*. New York: Basic Books.

STEVENSON, H., & FANTUZZO, J. 1984. Application of the "generalization map" to a self-control intervention with school-age children. *Journal of Applied Behavior Analysis, 17*, 203–212.

TAWNEY, J., & GAST, D. 1984. *Single subject research in special education*. Columbus, Ohio: Charles E. Merrill.

TRAP, J.J., MILNER-DAVIS, P., JOSEPH, S., & COOPER, J.O. 1978. The effects of feedback and consequences on transitional cursive letter formation. *Journal of Applied Behavior Analysis, 11*, 381–393.

WHITE, O.R., & HARING, N.G. 1980. *Exceptional teaching* (2nd ed.). Columbus, Ohio: Charles E. Merrill.

WOOD, R., & FLYNN, J.M. 1978. A self-evaluation token system *versus* an external evaluation token system alone in a residential setting with predelinquent youth. *Journal of Applied Behavior Analysis, 11*, 503–512.

WRIGHT, H. 1960. Observational study. In P. H. Mussen (Ed.), *Handbook of research methods in child development*. New York: Wiley.

Single-subject Designs

Did you know that . . .

- There's no such thing as an average student?
- What you think is working may not be?
- Classroom teachers can conduct research?
- Classroom teachers must conduct research if their instruction is to be effective?

When teachers collect data on student progress, they can make systematic observations of student behavior change. Data collection and observation certainly make possible statements about the direction and magnitude of behavioral changes. They do not, however, provide sufficient information to indicate a cause-and-effect relationship between an intervention and the behavior in question.

In order to make assumptions about cause and effect, data collection and recording must be carried out under more rigorous conditions. There are certain patterns, or *designs,* for collecting data that enable professionals to determine the relationship between an intervention and a behavior. Only when such a relationship can be established may teachers make confident statements about the effect of an intervention on behavior.

In this chapter we will describe a number of designs that enable teachers and researchers to determine the relationship between intervention procedures and behaviors. We include research designs for two reasons: (1) to enable teachers to design classroom intervention procedures that can be systematically evaluated; and (2) to familiarize teachers with experimental designs used in applied behavior analysis. Actual research applications, taken from professional journals, accompany the text describing each research design. Each design discussed in the chapter is also applied to a classroom problem to stress the utility of applied behavior analysis designs in the classroom. After reading this chapter, teachers should be able both to conduct simple research and to read and understand experimental research reported in professional journals.

VARIABLES

Before discussing specific designs, we shall describe some terms basic to an understanding of experimental investigation. The term **variable** may be used to refer to any of a number of factors involved in research. These may include attributes unique to the individual involved in the study or conditions associated with the environment in which the study is carried out. Experimental design uses two types of variables: dependent and independent. The term *dependent variable* refers to the behavior being measured, while *independent variable* refers to the treatment or procedure under experimenter control. In the following sentences from a research synopsis, the independent variable is italicized and the dependent variable appears in parentheses.

> *Verbal praise* is contingently presented when (oral questions are answered correctly).
> *Time-out* is contingently presented when (a temper tantrum occurs).

If a design is able to demonstrate that manipulation of the independent variable results in changes in the dependent variable, a tentative cause-and-effect relationship may be assumed. This link is termed a **functional relationship.** There is a functional relationship, in other words, if the behavior analyst can demonstrate that the dependent variable systematically changes as a result of changes in the independent variable. The changes in the dependent variable "depend" on changes in the independent variable.

BASIC CATEGORIES OF DESIGNS

Group designs vs. single-subject designs.

Two major categories of designs are **group designs** and **single-subject,** or single-case, **designs.** As indicated by the names, group designs focus on data related to a number of individuals, whereas single-subject designs focus on a single individual in a research sample. Group designs are used to compare the performance of groups of people (or subjects) on some dependent variable. The comparison is most often made using an average of each group's performance. Most applied behavior analysts prefer to record specific information about an individual rather than information about the average performance of a group. The use of averages may obscure important information as illustrated in the following anecdote.

Ms. Witherspoon Orders Reading Books

Ms. Witherspoon, a third-grade teacher, was urged by her principal to order new reading books at the beginning of the school year. Being unfamiliar with her class, Ms. Witherspoon decided to use a reading test to determine which books to order. She administered the test and decided to average the scores in order to determine the most appropriate reader. She came up with an average of exactly third grade, first month, and ordered 30 readers on that level.

When the books arrived, she found that the reader was much too hard for some of her students and much too easy for others. Using an average score had concealed the fact that, although the class average was third grade, some students were reading at first-grade level and others at sixth.

Single-subject Designs

Single-subject designs compare the effects of different conditions on the same individual.

In order to avoid the problems associated with the use of averages, most applied behavior analysis researchers prefer to use single-subject designs. These designs monitor the performance of individual subjects during manipulation of the independent variable(s). Certain techniques, which will be described later in the chapter, are used to ensure that changes in the dependent variable result from experimental manipulations and not from chance or coincidental factors.

Single-subject designs require repeated measures of the dependent variable. The performance of the individual whose behavior is being monitored (the subject) is recorded weekly, daily, or even more frequently over an extended period of time. That subject's performance can then be compared under different *experimental conditions,* or manipulations of the independent variable. Each subject is compared only to himself or herself, an approach that contrasts with group-design research, in which groups of subjects are compared to each other. Certain components are common to all single-subject designs. These include a measure of baseline performance and at least one measure of performance under an intervention or treatment condition.

Baseline Measures

The first phase of single-subject design involves the collection and recording of **baseline data.** Baseline data are measures of the level of behavior (the dependent variable) as it occurs naturally. Kazdin (1982) states that baseline data serve two functions. First, baseline data serve a *descriptive function.* These data describe the existing level of student performance. When data points are graphed, they provide a picture of the student's behavior—his current ability to solve multiplication problems or his current rate of talk-outs. This objective record can assist the teacher in verifying the existence, and extent, of the behavior deficit (lack of ability to do multiplication) or behavior excess (talking-out) that she has decided to alter.

Second, baseline data serve a *predictive function.* "The baseline data serve as the basis for predicting the level of performance for the immediate future if the intervention is not provided" (Kazdin, 1982, p. 105). In order to evaluate the success of an intervention (the independent variable) the teacher must know what student performance was like before the intervention. Baseline data serve a purpose similar to that of a pretest. "The predication is achieved by projecting or extrapolating into the future a continuation of baseline performance" (Kazdin, 1982, p. 105). It is against this projection that the effect of an intervention is judged.

The baseline phase continues for several days before beginning the intervention phase(s). In most instances, at least five baseline data points are collected and plotted. However, the extent of baseline data collection is affected by certain characteristics of these data points.

Baselines should be stable.

Stability.   Because baseline data are to be used to judge the effectiveness of the teacher's intervention, it is important that the baseline be *stable,* providing a representative sample of the natural occurrence of the behavior. The stability of a baseline is assessed by two characteristics: variability of the data points and trends in the data points.

Variability of data refers to fluctuations in the student's performance. "As a general rule, the greater the variability in the data, the more difficult it is to draw conclusions about the effects of the intervention" (Kazdin, 1982, p. 109) and to make projections about future performance. When baselines are unstable, the first thing to examine is the definition of the target behavior. A lack of stability in the baseline may suggest that the operational definition of the target behavior is not sufficiently descriptive to allow for accurate and consistent recording or that the data collector is not being consistent in the procedure used for data collection.

See Chapter 3 for suggestions on writing operational definitions.

In laboratory settings, other sources of variability can very often be identified and controlled. In classrooms, attempts to control variability are desirable if the sources of variability can be identified—for example, if fluctuations are caused by inconsistent delivery of medication. In cases of temporary fluctuations caused by such unusual events as a fight or a problem at home, the teacher may just wait for the fluctuation to pass. However, in classrooms, unlike laboratories, "variability is an unavoidable fact of life," and in such settings there are seldom "the facilities or time that would be required to eliminate variability" (Sidman, 1960, p. 193).

Where variables can be rigorously controlled, a research-oriented criterion for the existence of variability would be data points within a 5% range of variability (Sidman, 1960). A therapeutic criterion of 20% has been suggested (Repp, 1983). However, in classrooms where pure research concerns might be less important than rapid modification of the behavior, we suggest a more lenient parameter of 50% variability. If variability exceeds 50%, statistical techniques for performance comparisons must be used (Barlow & Hersen, 1984). A baseline is often considered stable if no data point of the baseline varies more than 50% from the mean, or average, of the baseline. Figure 5–1 illustrates a procedure for computing the stability of a baseline based upon this criterion.

FIGURE 5–1
Computing baseline
stability

Session	Data points
1	14
2	10
3	20
4	16
5	12

Baseline mean (arithmetic average) = 14.4 = 14
50% of mean = 7
Acceptable range of data points = 7 – 21 (14 ± 7)
This baseline is stable, because no data point varies more than 50% from the mean.

A *trend in the data* refers to an indication of a distinctive direction in the performance of the behavior. A trend is defined as three successive data points in the same direction (Barlow & Hersen, 1984). A baseline may show no trend, an increasing trend, or a decreasing trend. Figures 5–2 and 5–3 illustrate two types of trends—increasing and decreasing. Figure 5–4 does not depict a trend, because only two points show increases in the behavior.

An *ascending baseline* denotes an increasing trend. Teachers should initiate intervention on an ascending baseline only if the objective is to decrease the behavior. Because the behavior is already increasing, the effects of an intervention designed to increase behavior will be obscured by the baseline trend.

FIGURE 5–2
Increasing trend
(ascending baseline)

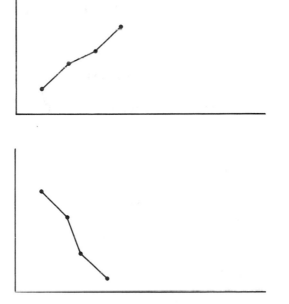

FIGURE 5–3
Decreasing trend
(descending baseline)

FIGURE 5–4 Data that do not reflect a trend

Take baseline trends into account before intervention.

A *descending baseline* includes at least three data points that show a distinctive decreasing direction or trend in the behavior. Teachers may initiate intervention on a descending baseline only if the objective is to increase the behavior.

Intervention Measures

The second component of any single-subject design is a series of repeated measures of the subject's performance under a treatment or intervention condition. The independent variable (treatment or intervention) is introduced and its effects on the dependent variable (the subject's performance) are monitored. Trends in treatment data indicate the effectiveness of the treatment and provide the teacher or researcher with guidance in determining the need for changes in intervention procedures.

Experimental Control

Experimental control refers to the researcher's efforts to ensure that changes in the dependent variable are in fact related to manipulations of the independent variable—that a functional relationship exists. The researcher wants to eliminate to the greatest extent possible the chance that other, confounding variables are responsible for changes in the behavior. *Confounding variables* are those environmental events or conditions that are not controlled by the researcher but may affect behavior. For example, if a teacher institutes a behavioral system for reducing disruptive behavior in a class after the three most disruptive students have moved away, she really cannot be sure that the new system is responsible for lower levels of disruption. Removal of the three students is a confounding variable.

Professor Grundy encounters a confounding variable when he again visits Ms. Harper later in this chapter.

The designs discussed in this chapter provide varying degrees of experimental control. Some, called here *teaching designs,* do not permit confident assumption of a functional relationship. They may, however, provide sufficient indication of behavior change for everyday classroom use, particularly if the teacher remains alert to the possibility of confounding variables. Other designs, called *research designs,* provide for much tighter experimental control and allow the teacher or researcher to presume a functional relationship. Research designs may be used in classrooms when a teacher is particularly concerned about possible confounding variables and wants to be sure that intervention has the desired effect on behavior. The teacher who is interested in publishing or otherwise sharing with other professionals the results of an intervention would also use a research design if at all possible. In the following sections on specific research designs, both a research and a classroom application are described for each design whenever possible.

AB DESIGN

The AB design is a teaching design.

The AB design is the most basic of all single-subject designs. Each of the more sophisticated designs is actually an expansion on this simple one. The designation *AB* refers to the two phases of the design: the A, or baseline, phase and the B, or intervention, phase. During the A phase, baseline data are collected and recorded. Once a stable baseline has been established, the intervention is introduced, and the B phase begins. In this phase, intervention data are collected and recorded. Trends in the intervention data provide information about the effectiveness of an intervention. Using this information, the teacher can make decisions about continuing, changing, or discarding the intervention.

Implementation

Table 5–1 shows data collected using an AB design. The teacher in this instance was concerned about the few correct answers a student gave to questions about a reading assignment. For 5 days, she collected baseline data. She then made 2 minutes of free time contingent on each correct answer and continued to record the number of correct responses. As can be seen in Table 5–1, the number clearly increased during the intervention phase. The teacher could make a tentative assumption here that her intervention has been effective.

TABLE 5–1
Sample data from an AB design (to evaluate correct responses on reading assignments)

Baseline Data	
Day	Number of correct responses
Monday	2
Tuesday	1
Wednesday	0
Thursday	2
Friday	1

Intervention Data	
Day	Number of correct responses
Monday	5
Tuesday	6
Wednesday	4
Thursday	8
Friday	6

Graphic Display   Data collected using an AB design are graphed in two phases: A, or base-line, and B, or intervention. The two phases are then separated by a broken line on the graph and data points between phases are not connected. The graph (Figure 5–5) shows an even clearer indication of the effectiveness of an intervention than the data (Table 5–1).

Design Application   The basic AB design is not often found in the research literature because it is the weakest of the single-subject designs. The following example demonstrates the use of an AB design in a classroom setting.

Jack Learns to Do His French Homework

Mr. Vogl had difficulty working with Jack, a student in the fourth-period French class. Jack was inattentive when homework from the previous evening was reviewed. Closer investigation revealed that Jack ignored the review sessions because he was not doing the assignments. To increase the amount of homework completed, Mr. Vogl decided to use positive reinforcement. To evaluate the effectiveness of the intervention, he selected the AB design using the number of homework questions completed correctly as the dependent variable.

Over a baseline period of 5 days, Jack answered 0 out of 10 (0/10) questions correctly each day. Since Jack frequently asked if he could listen to tapes in the French lab, Mr. Vogl decided to allow Jack to listen to tapes for 2 minutes for each correct answer given during the review. Data collected during the intervention phase indicated an increase in the number of questions Jack answered correctly. Data analysis suggested that the intervention technique was an effective one.

Advantages and Disadvantages   The primary advantage of an AB design is its simplicity. It provides the teacher with a quick, uncomplicated means of comparing students' behavior before and after implementation of some intervention or instructional procedure, making instruction more systematic.

The disadvantage of the AB design is that it cannot be used to make a confident assumption of a functional relationship. While the design does

FIGURE 5–5
Graph of AB design data from Table 5–1

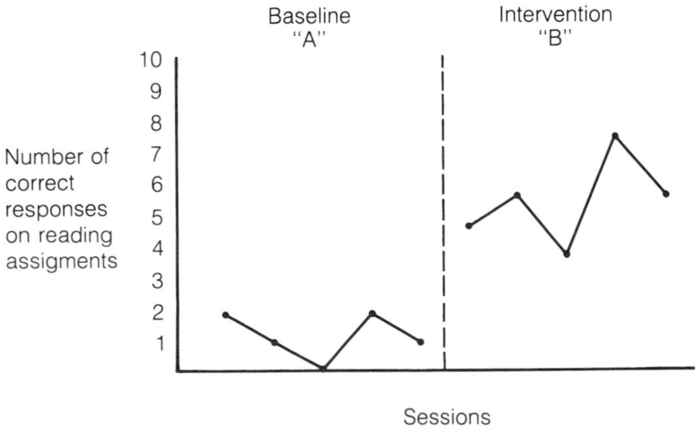

Many teachers use AB designs to evaluate their students' progress.

provide an indication of effectiveness, it does not control for confounding variables, as illustrated by the following example.

Ms. Harper Conducts Research

As part of her student teaching assignment, Ms. Harper was required to carry out a simple research project using an AB design. She had decided to use Ralph's staying in his seat as her dependent variable. (Remember Ralph from Chapter 1?) Ms. Harper collected baseline data for several days and determined that Ralph stayed in his seat for periods varying from 20 to 25 minutes during the 1-hour reading class. She prepared to intervene, choosing as her independent variable points exchangeable for various activities that Ralph enjoyed. When Professor Grundy made a visit soon after intervention began, Ms. Harper met him at the door in a state of high excitement.

"It's working, Professor!" gloated Ms. Harper. "Look at my graph! Ralph was absent the first two days of this week, but since he's been back and I've been giving him points, he's been in his seat 100% every day. Do you think I'll get an *A* on my project?"

Professor Grundy inspected Ms. Harper's graph and agreed that her procedure appeared to be effective. He then sat down in the rear of the classroom to observe. After a few minutes, during which Ralph indeed stayed in his seat, Professor Grundy attracted Ms. Harper's attention and called her to the back of the room.

"Ms. Harper," he asked gently, "did it not occur to you that the hip-to-ankle cast on Ralph's leg might have some effect on the amount of time he spends in his seat?"

"It's working, professor. He's staying in his seat!"

REVERSAL DESIGN

The **reversal design,** another type of single-subject design, is used to analyze the effectiveness of a single independent variable. Commonly referred to as the *ABAB design,* this design involves the sequential application and withdrawal of an intervention to test the intervention's effects upon a behavior. By comparing baseline conditions before and after application of the intervention strategy, the researcher can determine whether a functional relation between the dependent and independent variables exists.

Implementation

In a reversal design, the first step is to collect baseline data on the target behavior (dependent variable). After a stable baseline is achieved, the intervention is begun and appropriate data taken. Intervention continues until (1) the criterion for the target behavior is attained (for example, 95% accuracy) or (2) a definite trend in the desired direction of behavior change is noted in the data.

At either of these two points, the intervention is withdrawn, and baseline conditions are reinstituted in order to evaluate the effect of the change in conditions on the dependent variable. The teacher or researcher knows that the tested procedure is effective if the rate or magnitude of the target behavior is drastically altered under renewed baseline conditions.

ABAB is a research design. A functional relationship can be demonstrated.

Data collected using a reversal design should be examined for a functional relationship between the dependent and independent variables. The data in Figure 5–6 demonstrate a functional relationship between the dependent and independent variables, said to exist if the second set of baseline data returns to a level close to that in the original A phase or if a trend is evident in the second A phase in the opposite direction of the first B phase. Figure 5–7 graphs data that do not demonstrate the existence of a functional relationship.

It must be noted that a second B phase (intervention phase) is also desirable for verification of a functional relationship. Cooper (1981, p. 117) states that researchers need three pieces of evidence before they can say

FIGURE 5–6
Reversal design graph demonstrating a functional relationship between variables

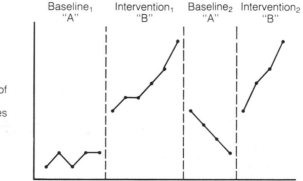

FIGURE 5-7
Reversal design that fails to demonstrate a functional relationship between variables

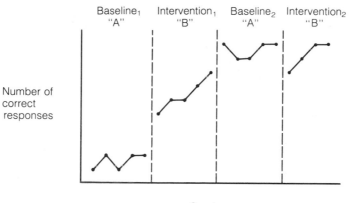

It would be unethical to withdraw, for purposes of research, an intervention that stopped head banging in a severely handicapped student.

that a functional relationship is demonstrated: (a) prediction: the instructional statement that a particular independent variable will alter the dependent variable, e.g., the contingent use of tokens to increase the number of math problems Michael completes; (b) verification of prediction: the increase (or decrease) in the dependent behavior during the first intervention phase, and the approximate return to baseline levels of performance in the second A phase; and (c) replication of effect: the reintroduction of the independent variable during a second B phase resulting again in the same desired change in behavior.

The reversal design is a research design that allows the teacher to assume a functional relationship between independent and dependent variables. It is unlikely that coincidental confounding variables would exist simultaneously with repeated application and withdrawal of the independent variable. The reversal design, however, is not always the most appropriate choice. The reversal design should not be used in these cases:

1. When the target behavior is of a dangerous nature, such as aggressive behavior directed toward other students (hitting) or self-injurious behavior. Because the reversal design calls for a second baseline condition to be implemented after a change in the target behavior rate, ethical considerations would prohibit withdrawing a successful intervention technique.

2. When the target behavior is not reversible. Many academic behaviors are not reversible, because the behavior change is associated with a learning process. Under such conditions, a return to baseline performance is not feasible: information cannot be "unlearned."

Graphic Display

The reversal design calls for four distinct phases of data collection. Figure 5-8 illustrates the basic reversal design. (Note that ABAB is derived from the labeling of each baseline period as an A phase and each intervention period as a B phase.)

FIGURE 5–8
Format of reversal
design

"A"	"B"	"A"	"B"
Baseline₁	Intervention₁	Baseline₂	Intervention₂

Design Variations   Variations of the reversal design can be found in current research. The first such variation does not involve a change in the structure of the design, but simply shortens the length of the initial baseline (A) period. This format of the design is appropriate when a lengthy baseline period is unethical or

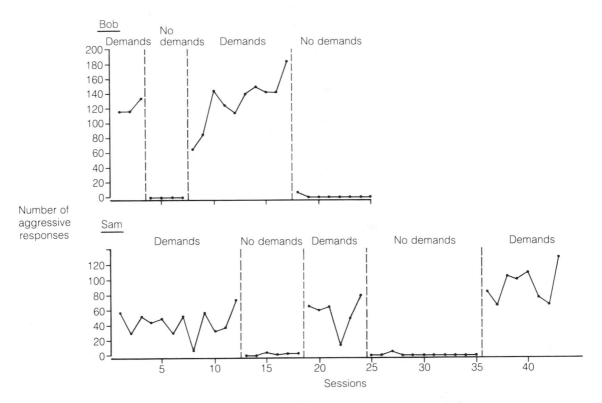

FIGURE 5–9   Research application of the reversal design

Note. From "Escape as a factor in the aggressive behavior of two retarded children" by E. Carr, C.D. Newsom, & J.A. Binkoff. *Journal of Applied Behavior Analysis.* 1980. *13.* 101–117. Copyright 1980 by Society for the Experimental Analysis of Behavior, Inc. Reprinted by permission.

not called for, as in the case of a student who is not capable of performing the target behavior to any degree.

A second variation of the reversal design omits the initial period entirely. This BAB variation is considered if the target behavior is obviously not in the student's repertoire. When this design is used, a functional relation between the dependent and independent variables can be demonstrated only in the second intervention (B) phase.

Research Application

There are many examples of the ABAB design applied to research. An ABAB design was employed by Carr, Newsom, and Binkoff (1980) as a means of determining the variables controlling the severely aggressive behavior of two retarded children. The experiment examined the effects of a simple "demand" situation: demands were alternately presented and withheld. Under the two demand conditions, the children's teacher selected problem situations as the dependent variable.

Figure 5–9 displays in graph form the results of this experiment. In Bob's case, the verbal cue "sit down" was chosen to require in-seat behavior. The demand situation for Sam involved practicing buttoning skills on an appropriate board with verbal cues provided at 10-second intervals. No demand situations imitated the nonstructured portions of a typical day. Note the clear relationship between the use of a verbal cue and the compliance of each child. More frequent demands resulted in more aggressive behavior.

Teaching Application

The following vignette illustrates the use of an ABAB design in the classroom.

Jill Learns Not to Suck Her Thumb

Ms. Kimball, a kindergarten teacher with 27 pupils, recently designed an effective intervention program to reduce her student Jill's thumb-sucking. She decided that a reversal design would allow her to determine the existence of a functional relation between the change in behavior and the selected intervention procedure. Ms. Kimball chose a time-sampling observation procedure. She looked at Jill every 10 minutes and marked a + on the data sheet if Jill was sucking her thumb and a − if she wasn't.

During the baseline condition (the first A phase), Ms. Kimball noted that Jill had her thumb in her mouth an average of 8 out of the 12 observations during a 2-hour period. Ms. Kimball decided to make a chart for Jill and to put a smelly sticker on the chart whenever Jill was not sucking her thumb at the end of an interval. After the intervention was applied (the first B phase), thumb-sucking occurred at the end of only 3 intervals on the average. To determine if a causal relation existed between the intervention and the behavior, Ms. Kimball returned to baseline (A) conditions. Jill no longer got smelly stickers, and the target behavior immediately returned to its previous level. Reinitiating the B condition immediately brought the behavior back to the lower level. Ms. Kimball felt confident that the intervention had changed Jill's behavior.

Advantages and a Disadvantage

ABAB designs allow for precise analysis, but require than an effective intervention be withdrawn.

As the preceding applications show, the reversal design offers the advantages of simplicity and experimental control. It provides for precise analysis of the effects of a single independent variable on a single dependent variable.

The primary disadvantage of this design is the necessity for withdrawing an effective intervention in order to determine whether a functional relationship exists. Even if the target behavior is neither dangerous nor irreversible, it often seems foolish to teachers to stop doing something that is apparently working.

CHANGING CRITERION DESIGN

The **changing criterion design** is used to evaluate a gradual and systematic increase or decrease in the performance level of the student by changing the criterion for reinforcement in a stepwise fashion (Hartmann & Hall, 1976). This design includes two major phases, much like the AB design. However, in the changing criterion design, the intervention phase is divided into subphases. Each subphase requires a closer approximation of the terminal behavior than the previous one. The subject's performance thus moves gradually from baseline levels to the terminal objective.

This design is particularly appropriate when the terminal goal of the behavior-change program requires a considerable length of time to reach. The changing criterion design is well suited for measuring the effectiveness of a shaping procedure (see Chapter 8). This design is also useful when the teacher wants to accelerate or decelerate behaviors that are measured in terms of frequency, duration, latency, or force.

Implementation

The first step in implementing the changing criterion design is to gather baseline data in the same manner used in other single-subject designs. After a stable baseline has been established, the teacher must determine the level of performance that will be required when the intervention phase is initiated. Two target levels of performance must be selected: an overall level of performance for the behavior-change program and an interim level of performance. The overall or terminal level of performance becomes the goal for the first subphase of intervention. The choice of the first interim level of performance may be determined using several techniques:

1. The interim criterion for performance can be set at and then increased by an amount equal to the mean of the stable portion of the baseline data. This technique is appropriate when the goal of the behavior-change program is to increase a level of performance and when the student's present level is quite low. For example, if a teacher wanted to increase the number of questions a student answered and the student's mean baseline level of correct responses was 2, that teacher might set two correct answers as a first interim criterion. Each subsequent subphase would then require two additional correct answers.

2. Interim levels of performance can be determined by taking 50% of the computed mean of baseline data (or the stable portion of the baseline data). For example, suppose that a student's number of problems correct was noted for 1 week and that the following data were recorded:

Monday	9
Tuesday	10
Wednesday	7
Thursday	6
Friday	8

The mean level of performance of the target behavior is 8. The first goal for the teacher would be to increase the number of correct math problems by 4 (50% of baseline mean) and continue to raise the criterion by 4 for each successive interim phase. Given the data above, the appropriate criterion for the first interim phase would be 12 (baseline mean + 50%, or 8 + 4).

3. Interim criteria can be based on selecting the highest (or lowest, depending on the terminal objective) level of baseline performance. This is probably most appropriate for use with social behavior, e.g., out-of-seat, positive peer interactions, rather than for an academic behavior. The assumption is that if the student was able to perform at that high (or low) level once, the behavior can be strengthened (or weakened) and maintained at that level.

4. Interim criteria can be based upon a professional estimate of the student's ability. This procedure is particularly appropriate when the student's present level of performance is 0.

Regardless of the technique a teacher uses to establish the initial goal of a behavior-change program, the data collected should be used to evaluate whether the selected goal is appropriate for the particular student.

The next step in implementing the changing criterion design is to begin the intervention phases. In each phase, if the student performs at least at the level of the interim criterion, the teacher provides reinforcement. It is important during the initial phase of the intervention for the teacher to analyze the appropriateness of the selected interim level of performance. If the student does not meet criterion after a reasonable number of trials, the teacher should consider decreasing the interim level of performance required for reinforcement. Conversely, the teacher should consider adjusting the interim level of performance required for reinforcement if the goal was reached too easily.

The teacher in our example would require 16, then 20, then 24 correct problems for reinforcement.

After the student has reached the established level of performance in two consecutive sessions, or in two out of three consecutive sessions of a phase, the level of performance required for reinforcement should be adjusted in the direction of the desired level of performance for the overall behavior-change program. Each successive interim level of performance should be determined using the same mathematical difference established at the first interim level of performance. That is, the behavior-change program should reflect a uniform step-by-step increase or decrease in criterion level. This process is continued until

1. The behavior is increased to a 100% or decreased to a 0% level of performance; or
2. The final goal established by the teacher in the behavioral objective is attained.

Changing criterion designs enable teachers and researchers to establish functional relationships.

A functional relationship between the dependent and independent variables can be demonstrated in the changing criterion design if the student's level of performance consistently changes when the criterion for reinforcement is adjusted. Generally, a student must meet the established criteria in a least three consecutive phases before the assumption of a functional relationship is valid.

Graphic Display

The basic display format for the changing criterion design is similar to the one used for the AB design. A baseline phase is followed by the intervention phase, with a dashed vertical line separating the two conditions. As seen in Figure 5–10, the data of the intervention phase are identified according to the level of performance selected for reinforcement. The procedure for graphing the data calls for connecting data points within each incremental value. Data points collected in different interim phases or subphases are never connected. The magnitude of student behavior necessary for consequation (delivery of reinforcement) should be clearly identified at each level of the intervention phase as seen in Figure 5–10.

Research Application

The changing criterion design was utilized by Hall and Fox (1977) to increase the number of math problems correctly solved by a child with a behavior disorder. Under baseline conditions, the student demonstrated a mean level of performance of one math problem.

The first interim level of performance was established at the next whole number greater than the mean baseline performance (2). If the student met this level of performance, he was allowed to play basketball. If the student failed to reach criterion he had to stay in the math session until the problems were solved correctly. As shown in Figure 5–11, this process was continued until 10 math problems were solved correctly.

Certain procedural elements may be included within the changing criterion design to increase its research credibility. In Figure 5–11, three sessions at each criterion were generally included; however, this number of sessions was altered in some cases. Varying the increase (or decrease) in performance required in subphases provides a more credible demonstration of experimental control than requiring the same performance level

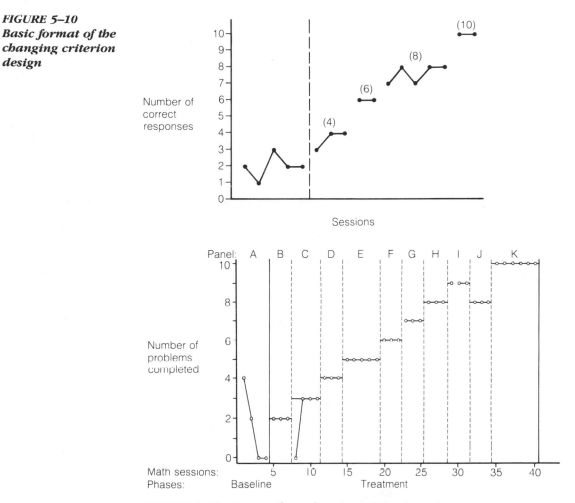

FIGURE 5–10
Basic format of the
changing criterion
design

FIGURE 5–11 Research application of the changing
criterion design

Note. From "Changing criterion designs: An applied behavior analysis procedure"
by R.V. Hall & R.G. Fox, in B.C. Etzel, J.M. LeBlanc, & D.M. Baer (Eds.), *New
developments in behavioral research: Theory, method and application.* Copyright
1977 by Lawrence Erlbaum Associates, Inc., Publishers. Reprinted with permission
of the authors and the publishers.

change across all subphases (Cooper, 1981). Finally, note that in subphase J
a change in the criterion for reinforcement was made in the direction opposite to the terminal objective. Returning the student's performance level
to a previously mastered criterion demonstrates a reversal effect similar to
that of a return to baseline conditions in an ABAB design.

**Teaching
Application**

Here's a teaching application of a changing criterion design to compare
with the research application.

Gay Learns to Sort by Color

Gay was a student in Mr. Carroll's intermediate class for the moderately retarded. Mr. Carroll was trying to teach Gay to sort objects rapidly by color. Gay
could perform the task, but she did it too slowly. Mr. Carroll decided to use a
changing criterion design to evaluate the effectiveness of a positive reinforcement procedure. He established that Gay's average baseline rate of sorting
was 4 objects per minute. He set 6 per minute as the first interim criterion and
30 per minute as the terminal goal. Gay earned a poker chip exchangeable for
a minute's free time when she met criterion. When Gay met criterion on two
consecutive trials or opportunities, Mr. Carroll raised the criterion required for
reinforcement by 2. He continued to do this until Gay sorted 30 objects per
minute in order to earn her poker chip. Mr. Carroll concluded that there was a
functional relationship between the dependent and independent variables, because Gay's behavior changed quickly each time the criterion was changed
but did not change until then.

**Advantage and
Disadvantage**

*Changing criterion
designs are unsuitable
with some behavior-
change programs.*

One advantage of the changing criterion design is that it can establish a
functional relationship while always changing the behavior in a positive direction. There is no need to withdraw a successful intervention. However,
using the changing criterion design necessitates very gradual behavior
change. It may, therefore, be inappropriate for behaviors that require or
lend themselves to rapid modification.

MULTIPLE BASELINE DESIGN

As indicated by its name, the **multiple baseline design** permits simultaneous analysis of more than one dependent variable. A teacher may exper-
imentally test the effects of intervention (the independent variable) on

1. Two or more behaviors associated with one student in a single setting:
for example, John's out-of-seat and speaking-out behaviors in social studies
class (*multiple baseline across behaviors*).

2. Two or more students exhibiting the same behavior in a single setting:
for example, the spelling accuracy of both Sara and Janet in English class
(*multiple baseline across individuals*).

3. Two or more settings in which one student is exhibiting the same
behavior: for example, Kurt's cursing during recess and also in the school
cafeteria (*multiple baseline across situations*).

*When to use a multiple
baseline design.*

The multiple baseline is the design of choice when the teacher is interested in applying an intervention procedure to more than one individual,
setting, or behavior. The multiple baseline design does not include a reversal phase; therefore, it may be used when the reversal design is not appropriate: when the target behavior includes aggressive actions or when
academic learning is involved.

*"Baseline,
directionality,
stability, control—is
this teaching or
tennis?"*

Implementation

A teacher using the multiple baseline design simultaneously takes data on each dependent variable. The teacher collects data under baseline conditions for each student, on each behavior, or in each setting. In setting up the data collection system, the teacher should select a scale for the ordinate that is appropriate for each of the variables involved in the program. In order to make data analysis possible, the same scale of measurement (for example, minutes involved in appropriate play, number of math problems completed correctly, percent of on-task behavior) should be used for each dependent variable.

After a stable baseline has been achieved on the first variable, intervention with that variable can be initiated. During the treatment period, baseline data collection continues for the remaining variables. Intervention on the second variable should begin when the first variable has reached the criterion established in the behavioral objective or when the data for the first variable show a trend in the desired direction as indicated by three consecutive data points. The intervention condition should be continued for the first variable and baseline data should still be collected for any additional variables. This sequence is to be continued until the intervention technique has been applied to all of the variables identified for the behavior-change program.

The data collected within a multiple baseline design should be examined for a functional relation between the independent variable and each of the dependent variables. This relationship is assumed if each dependent variable in succession shows a change when, and only when, the inde-

FIGURE 5–12
Data from a multiple baseline design that reflect functional relation

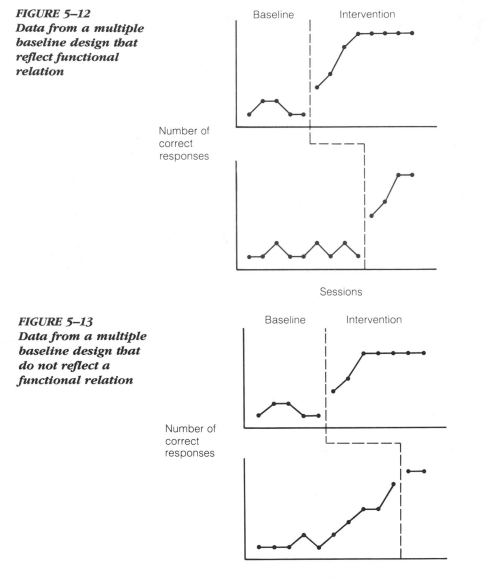

Baseline   Intervention

Number of correct responses

Sessions

FIGURE 5–13
Data from a multiple baseline design that do not reflect a functional relation

Baseline   Intervention

Number of correct responses

Sessions

pendent variable is introduced. Adjacent graphs should be examined to be sure that each successive intervention has an independent treatment effect on the appropriate dependent variable. Only the first independent variable should be affected by the first intervention. A change in the second and succeeding dependent variables should be seen only when the intervention is applied to them as well. Figure 5–12 shows an example of a functional relation, whereas Figure 5–13 does not. In Figure 5–13, the second dependent variable begins an upward trend when the intervention is intro-

FIGURE 5–14
*Basic format of the
multiple baseline
design*

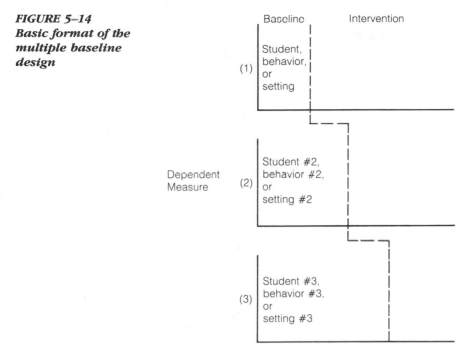

duced for the first variable, showing that the relationships between variables are not discrete, or independent.

Graphic Display

When using the multiple baseline design, the teacher should plot the data
collected using a separate axis for each of the dependent variables to
which intervention was applied (individuals, behaviors, or situations). Figure 5–14 shows a composite graph of a multiple baseline design.

Design Variation

*A problem with the
multiple baseline design
—and a suggested
solution.*

One problem associated with the multiple baseline design is the necessity
for collecting baseline data over extended periods of time, particularly
baseline data for the second and subsequent dependent variables. In cases
where the student cannot perform the behavior at all or where access to
additional settings is not available, collecting daily baseline data may take
more time than is actually warranted. The *multiple-probe technique* has
been suggested as a reasonable solution to this situation (Horner & Baer,
1978; Murphy & Bryan, 1980). Using this variation of the multiple baseline
design, data are not continuously collected on the behaviors (or students,
or settings) on which intervention is not being conducted. Rather a probe
trial (a trial under baseline conditions) or a probe session (more than one
trial) is conducted *intermittently* on these subsequent behaviors to verify
that the student can perform the behavior (following intervention), or that
he still cannot perform the behavior, or to record any changes in his abil-

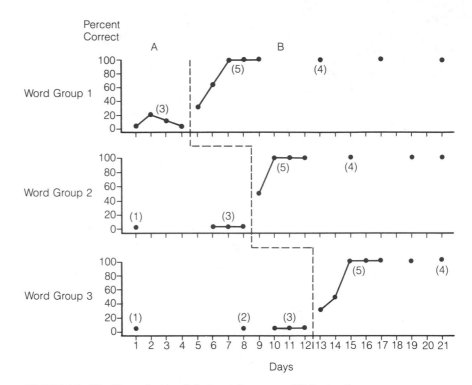

FIGURE 5–15   Hypothetical data using a multiple probe design across spelling word groups: (1) initial probe, (2) probe after criterion, (3) consecutive probe (baseline) sessions before intervention, (4) review probes, (5) continuous training data.

Note. From *Measuring behavior* 2nd ed. by J.O. Cooper, 144–145. Columbus, Ohio: Charles E. Merrill, 1981. Reprinted with permission.

ity before intervention. While intervening on behavior 1 (or with student 1 or in setting 1), the teacher intermittently probes behaviors 2 and 3. When behavior 1 reaches criterion, one or more probe sessions are conducted on all three behaviors. Then intervention is begun on behavior 2, while postcheck probes are conducted on behavior 1, and probes continue on behavior 3. When behavior 2 reaches criterion, one or more probe sessions are conducted on all three behaviors. Then intervention is begun on behavior 3 while postcheck probes are conducted on behaviors 1 and 2. Figure 5–15 illustrates this multiple-probe design using hypothetical data.

Research Applications

Across Behaviors

Nelson and Cone (1979) used a multiple baseline design across behaviors in a research study. These authors identified four groups or classes of behaviors—personal hygiene, personal management, ward work, and social skills—for training with 16 chronic psychiatric adult patients in a large institutional setting.

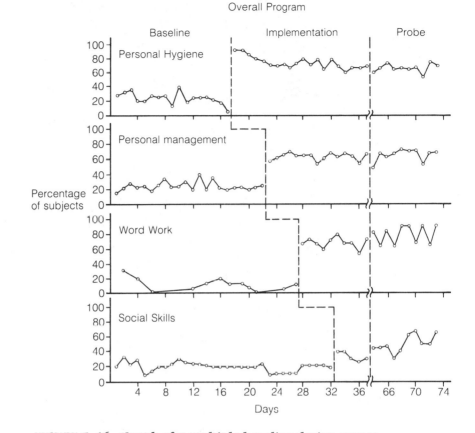

FIGURE 5–16   Graph of a multiple baseline design across behaviors

Note. From "Multiple-baseline analysis of a token economy for psychiatric inpatients" by G.L. Nelson & J.D. Cone, *Journal of Applied Behavior Analysis,* 1979, *12,* 255–271. Copyright 1979 by Society for the Experimental Analysis of Behavior, Inc. Reprinted by permission.

Figure 5–16 demonstrates the effect of the intervention on mean performance of all students in each of the four classes of behavior. The intervention used in this study consisted of introducing a token economy reinforcement program (see Chapter 6) for each class of behavior in a cumulative, sequential format. Results indicated abrupt and substantial increases in each target behavior when the independent variable was introduced.

Across Individuals

Stokes, Fowler, and Baer (1978) used a multiple baseline design across individuals to demonstrate the effects of children's cueing on delivery of social praise and teacher attention. Cues and praise were scored using a 10-second interval-recording procedure. Cues were defined as verbal com-

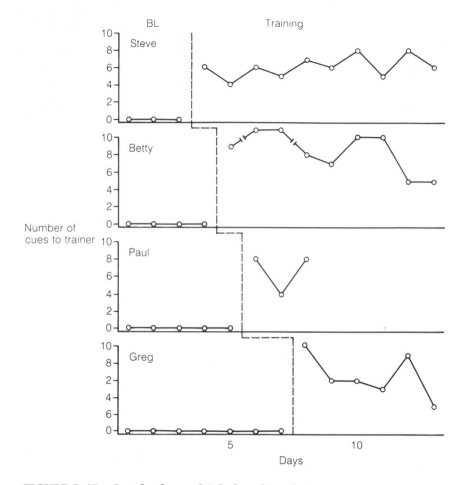

FIGURE 5–17   *Graph of a multiple baseline design across individuals*

Note. From "Training preschool children to recruit natural communities of reinforcement" by T.F. Stokes, S.A. Fowler, & D.M. Baer, *Journal of Applied Behavior Analysis, 1978, 11*, 285–303. Copyright 1978 by Society for the Experimental Analysis of Behavior, Inc. Reprinted by permission.

ments by the student that invited favorable comments or evaluations of the child's work or behavior by the teacher. Intervention involved the training of a number of specific cues (for example, "Have I worked well?" "How is this?") that were followed by feedback and praise. Results of the experiment showed an increase in students' cueing skill from baselines of 0 to a posttraining mean of 6.1. Figure 5–17 clearly illustrates the effectiveness of using the multiple baseline across individuals and the functional relationship between independent and dependent variables.

Across Situations

Goldstein, Minkin, Minkin, and Baer (1978) used a multiple baseline design across the dimension of setting to analyze the effects of free advertising on the number of "found" advertisements in three local newspapers. The rationale behind this experiment focused on the disproportionate number of "lost" advertisements in most newspapers, probably due to people's unwillingness to pay a fee to report found items. Intervention involved eliminating the fee for placing advertisements related to items found, allowing advertisements to be placed by telephone rather than requiring appearance at the newspaper business office, and listing on a daily basis this new "Free Ad" policy in each of the three newspapers in the study. The data in Figure 5–18 demonstrate that the intervention was successful in increasing the number of "found" ads in each of the three newspapers.

Teaching Application

The following vignette illustrates a classroom application of a multiple baseline design.

The Students Learn to Come to Class on Time

Ms. Raphael was a junior–high-school English teacher. The students in all three of her morning classes consistently came late. She began to record baseline data on all three classes. She recorded the number of students in their seats when the bell rang. She found that an average of five students in the first class, four in the second, and seven in the third class were in their seats. Ms. Raphael then began recording an extra-credit point in her grade book for each student in the first class who was in his or her seat when the bell rang. Within a week, 25 students were on time and in their seats. The baseline data for the other classes showed no change during his first intervention. When she began giving extra-credit points in the second class, the number of students on time increased immediately and dramatically. After a week, she applied the intervention in the third class with similar results. Ms. Raphael had accomplished two things: she had succeeded in getting her classes to arrive on time and she had established a functional relationship between her intervention (the independent variable) and her students' behavior (the dependent variable).

Advantages and Disadvantages

The multiple baseline design can establish a functional relationship without withdrawing the intervention, as is necessary in a reversal design, and without gradual alteration, as is required in a changing criterion design. These advantages make it a particularly useful design for classroom use. The multiple baseline design does, however, have some limitations. This type of design requires that the researcher apply the intervention to several students, behaviors, or settings. This may not always be practical.

Furthermore, the multiple baseline design is inappropriate in two specific situations:

1. When the target behavior calls for immediate action. The multiple baseline design calls for a considerable delay in delivery of the

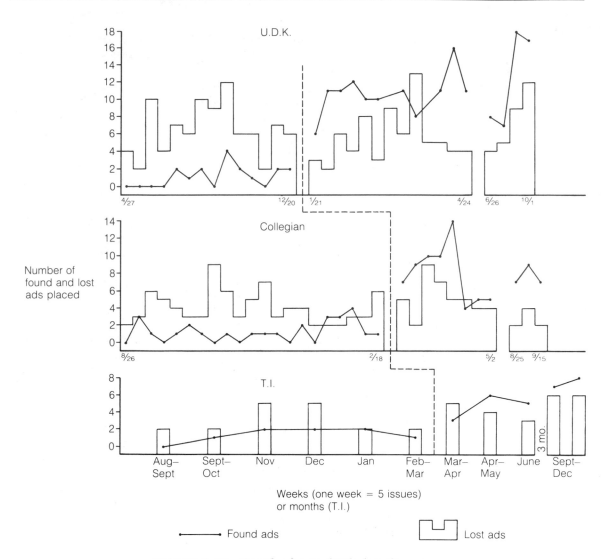

FIGURE 5–18   *Graph of a multiple baseline across situations*

Note. From "Finders, keepers?: An analysis and validation of a free-found-ad policy" by R.S. Goldstein, B.L. Minkin, N. Minkin, & D.M. Baer, *Journal of Applied Behavior Analysis,* 1978, *11,* 465–473. Copyright 1978 by Society for the Experimental Analysis of Behavior, Inc. Reprinted by permission.

intervention procedure for the second and subsequent dependent variables.

2. When the behaviors selected for intervention are not independent. In such a case, intervention with one behavior will bring about a change in the related behavior; therefore, the teacher will be unable to evaluate clearly the effects of the procedure. For example, if two behaviors targeted for a student are cursing and fighting, the teacher might find that after the

student's cursing decreases, there are fewer fights. In this case, the two behaviors are clearly not independent.

ALTERNATING TREATMENTS DESIGN

In contrast to the multiple baseline design that uses a single independent variable and multiple dependent variables, the **alternating treatments design** (Barlow & Hayes, 1979) allows comparison of the effectiveness of more than one treatment or intervention strategy on a single dependent variable. A number of different terms have been used for describe this design: **multiple schedule design** (Hersen & Barlow, 1976), **alternating conditions design** (Ulman & Sulzer-Azaroff, 1975), and **multi-element baseline design** (Sidman, 1960). This design can be used in two ways:

1. To compare the effectiveness of two or more procedures following a baseline condition.

2. To compare the effectiveness of two or more procedures (such as use of flashcards or participation in an instructional game) with a control condition (or no new treatment).

Implementation

Because the alternating treatments design may be used in two ways, the following sections describe implementation for each alternative.

Alternating Treatments Design with Baseline Condition

The first step in implementing data collection using the alternating treatments design with a baseline condition is identification of the dependent variable. A baseline is taken exactly as with other designs; then two or more interventions (independent variables) are identified. At this point, a schedule must be established for delivery of each intervention procedure. The student should be exposed to each treatment an equal number of times. The order for presentation of the independent variables (treatments) may be in a random order, e.g., ABBCAC, or may be rotated in block form. A block consists of one presentation of each treatment.

Consider that three treatments, A, B, and C, are being compared. The following order of presentation is possible: ABC, BCA, CAB, ACB, BAC, CBA. Each order of presentation should be employed at least once if data are collected for a sufficient length of time.

Alternating Treatments Design with Control Condition

Control means that no new treatment is implemented.

The alternating treatments design can also be used to compare two or more treatments with a no new treatment, or *control,* condition. *Control condition* is synonymous with *baseline condition.* The data on the baseline condition may be collected at the beginning of the study, as is done using other designs. However, when using an alternating treatments design the researcher often includes the control condition in the scheduling of inter-

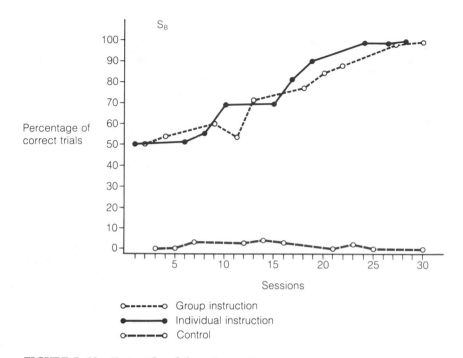

FIGURE 5–19   Example of the alternating treatment design with a control condition

Note. From "A comparison of individual and group instruction across response tasks" by P. Alberto, N. Jacobs, A. Sizemore, & D. Doran, *The Journal of the Association for the Severely Handicapped,* 1980. 5 285–293. Copyright 1980 by The Association for the Severely Handicapped. Reprinted by permission.

vention procedures. The baseline, in other words, is collected *throughout the study* rather than only before intervention begins. Figure 5–19 depicts data resulting from this variation of the alternating treatments design.

Graphic Display   When using either variation of the alternating treatments design—with or without a baseline phase—certain procedures should be followed. The alternating treatments can occur sequentially within one session, or from one session to the next as in the morning and afternoon of the same day, or on successive days. This scheduling should be counterbalanced so that, for example, the treatment employed on the morning of one day should be employed on the afternoon of the second and so on. (In research situations, similar counterbalancing should be undertaken for other possible confounding variables such as the person administering the treatment and location of treatment.) This counterbalancing should control for the possibility of carryover and/or sequencing effects (Barlow & Hayes, 1979).

In order to increase the likelihood of noticing any existing differences in the effectiveness of treatment, the stimulus signalling that a particular treatment will be in effect should immediately precede the treatment (Sulzer-Azaroff & Mayer, 1977). This discriminative stimulus should be consistent in form as well as in presentation. For example, the teacher may say, "This is treatment A" and "This is treatment B," or "Now we are going to use a Number Line," and "Now we are going to use Counting Chips." Or the teacher might color-code worksheets to indicate to the student the treatment condition that is in effect.

> Because confounding factors such as time of administration have been neutralized (presumably) by counterbalancing, and because the two treatments are readily discriminable by subjects through instructions or other discriminative stimuli, differences in the individual plots of behavior change corresponding with each treatment should be attributable to the treatment itself, allowing a direct comparison between two (or more) treatments. (Barlow & Hayes, 1979, p. 200)

This allows the teacher to infer a demonstrated experimental control and therefore a functional relationship between one (or more) of the independent variables and the dependent variable.

When using the alternating treatments design, the teacher may graph data in one of two ways. If a number of treatments are being compared, a baseline condition may be imposed prior to implementation of the treatments. This design consists of two phases, baseline and treatments, and may be represented as in Figure 5–20. Another application of the alternating treatments design incorporates the baseline condition (no treatment) simultaneously with the other treatments. This version of the design is reflected in Figure 5–21.

Whichever version of the alternating treatments design is selected, the data points for each variable should be differentiated by using symbols or colors. The lines resulting from connecting points should also be coded so that the data for each variable yield a clearly distinguishable curve.

Graphs should be analyzed to see if the data curve of one treatment is vertically separated from the other curves. This "fractionation" indicates that the associated treatment is most effective (Ulman & Sulzer-Azaroff, 1975). The conclusion that a significant difference exists among the independent variables can be drawn only if the data curves are separated from one another and do not cross at any point other than at the very beginning

FIGURE 5–20
Alternating
treatments design
with a baseline
condition

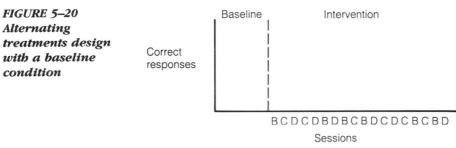

FIGURE 5–21
Alternating treatments design with a control condition

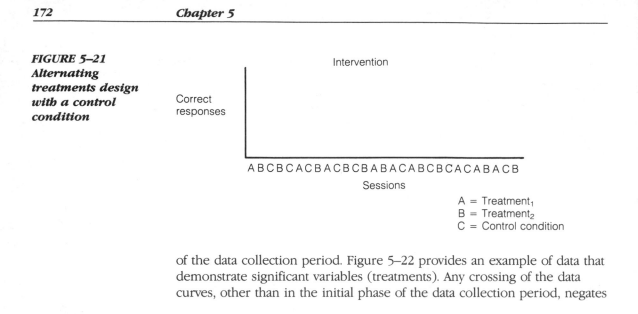

of the data collection period. Figure 5–22 provides an example of data that demonstrate significant variables (treatments). Any crossing of the data curves, other than in the initial phase of the data collection period, negates

FIGURE 5–22
Graph of data indicating the effectiveness of treatment A

FIGURE 5–23
Data indicating no difference among alternating treatments

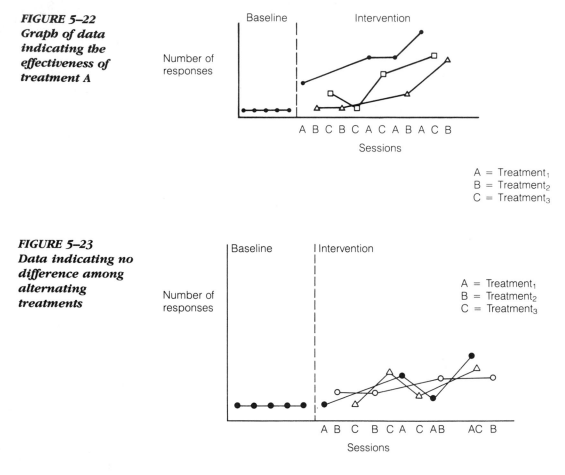

the possibility of a significant difference in effectiveness. An example of nonsignificant findings is illustrated in Figure 5–23.

Research Application

An alternating treatments design was used by Singh and Singh (1985) to compare the effectiveness of two error-correction procedures and a no-training condition on the number of oral reading errors made by four students. The two error-correction procedures were word-supply (the teacher told the student the correct word and the student pointed to and repeated the word), and word analysis (the teacher helped the student to sound out the word). Figure 5–24 displays the results for each of the four students in this study. Overall results indicated that both error-correction procedures greatly reduced the number of oral-reading errors of all students when compared to the no-training condition. The word-analysis method was significantly more effective than was the word-supply method of error-correction.

Teaching Application

For teachers, the alternating treatments design can provide crucial feed-back about the comparative effectiveness of teaching techniques, as the following example shows.

Marcia Learns Sight Vocabulary

Mr. Hagan was a resource teacher for mildly retarded elementary students. He wanted one of his students, Marcia, to learn basic sight vocabulary on a first-grade level. He chose 15 words and established that Marcia's baseline rate for reading them was zero. Mr. Hagan then divided the words into three groups of five. One set he printed on cards accompanied by an audio recording tape that Marcia could use to hear the words pronounced. He assigned a peer tutor to work with Marcia on the second set, and the teacher worked on the third set. Mr. Hagan recorded and graphed the number of words Marcia pronounced correctly each day for each set. Within a week, Marcia was pronouncing correctly the group of words learned with the peer tutor at a higher rate than either of the other sets. Mr. Hagan concluded that, for Marcia, peer tutoring was the most efficient way to learn sight vocabulary.

Advantages and Disadvantages

Using an alternating treatments design can help teachers individualize instruction.

The alternating treatments design is an efficient way for teachers to compare the effects of several intervention procedures at once. It may be extremely useful for determining instructional methods for individual students. Functional relationships established using this design are considered weak when compared to those established using reversal or multiple baseline designs. However, in many cases this disadvantage would be relatively unimportant to the teacher.

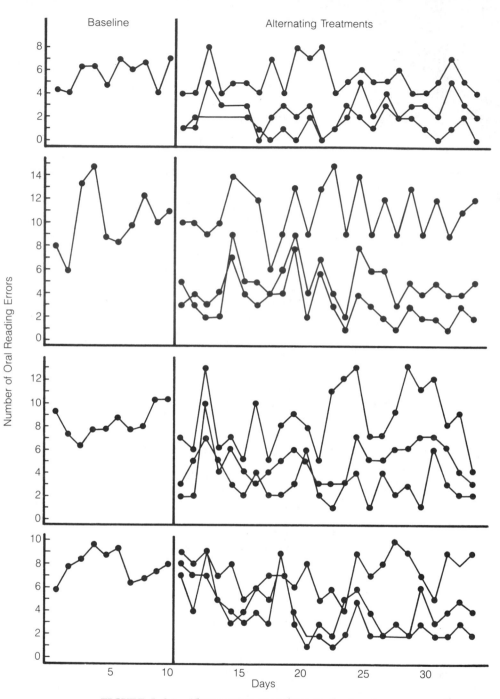

FIGURE 5–24 Alternating treatments design in a research application

Note: Data points are the means from three daily sessions in the baseline phase and are from each session in the alternating treatments phase. The same conditions prevailed in the three daily baseline sessions.

From "Comparison of Word-supply and word-analysis-error-correction procedures on oral reading by mentally retarded children" by J. Singh and N. Singh. American Journal of Mental Deficiency, 1985, 90, 64–70. Copyright 1985 by the American Association on Mental Deficiency.

The changing conditions design reflects reality— teachers keep trying different techniques until they find one that works.

CHANGING CONDITIONS DESIGN

A **changing conditions design** is used to investigate the effect of two or more treatments (independent variables) on the behavior of a student (dependent variable). Unlike the alternating treatments design, the treatments in a changing conditions design are introduced consecutively.

This design is ideal for the teacher who finds it necessary to try a number of interventions before finding one that is successful. The changing conditions design is also referred to as an *ABC design,* because each treatment phase is labeled with a different letter of the alphabet.

The data resulting from a changing conditions design cannot be used to determine the existence of a functional relationship between the dependent variable and any of the tested independent variables. As is the case with an AB design, the data can give only an indication of the effectiveness of a particular procedure.

The changing conditions design can, however, be altered to meet the qualifications of a research design by increasing the number of baselines. This modification calls for a return to baseline conditions before the initiation of a different intervention technique.

Implementation

The first step in implementing a changing conditions design is to collect data under baseline conditions in order to assess the student's present level of performance. Once a stable baseline is established, the teacher can introduce the selected treatment and measure its effectiveness through data collection.

If the data for the first intervention period do not demonstrate a change in the performance of the student, or if the change in performance is not in the desired direction, the teacher may design a second treatment. This second treatment can be either a complete change in strategy or a slight modification of the earlier treatment. A dashed line should be drawn on a graph of both sets of data to indicate the session on which the second treatment was initiated.

This process of redesigning treatment conditions is repeated until the desired effect on the student's behavior is achieved. Generally, the teacher should expect to see some evidence of a change in the student's behavior within five intervention sessions.

Graphic Display

The format for the changing conditions design is similar to the one used for the basic AB design. A baseline phase is followed by the intervention phases, with a dashed vertical line separating the data associated with each specific treatment. Figure 5–25 illustrates this basic format. Figure 5–26 illustrates the format for a changing conditions design with repeated baselines.

Research Application

Smith (1979) used a changing conditions design to measure the effect of a number of teaching conditions on a 12-year-old boy's oral reading. The dependent variable measured was the number of words read orally by the

FIGURE 5–25
A variation of the changing conditions design

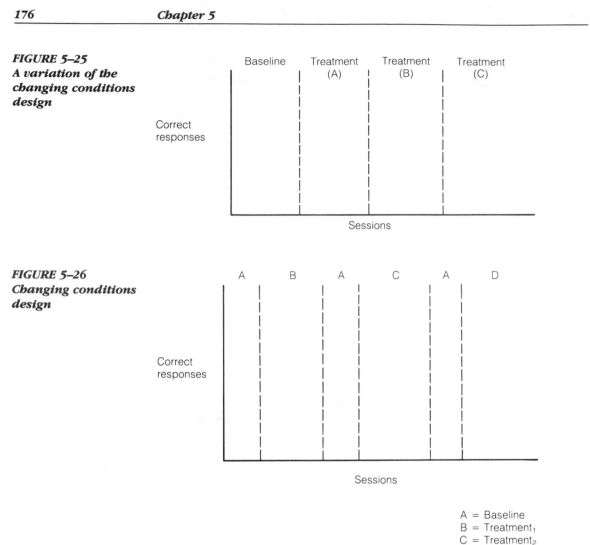

FIGURE 5–26
Changing conditions design

A = Baseline
B = Treatment₁
C = Treatment₂
D = Treatment₃

student (per minute) and the number of reading errors made. (Note: the numbers in the ovals represent the average number of words read per phase. The numbers in the rectangles represent the average number of errors per phase.) The following five conditions were included in the study:

1. Baseline—John was asked to read from his book.

2. Modeling—The teacher read the first page of a new story from the child's text. John was then asked to read orally.

3. Modeling plus correction—The previous condition was altered by adding a correction procedure. The teacher corrected John when he made an error and offered the correct word if he did not know it.

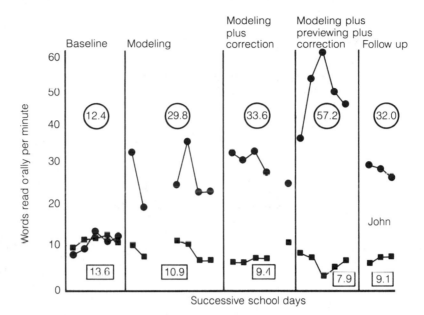

FIGURE 5–27   Research application of a changing conditions design

Note. From "The improvement of children's oral reading through the use of teacher modeling" by D.D. Smith. Journal of Learning Disabilities, 1979. *12*, 172–175. Copyright 1979 by Journal of Learning Disabilities. Reprinted by permission.

4. Modeling, previewing, and correction—After the teacher read, John reread the same passage and continued reading until the instructional time (5 minutes) elapsed. The correction procedure remained in effect.

5. Follow-up—Baseline conditions were reinstituted. Figure 5–27 presents the data that were recorded under each of the above conditions.

Turner, Hersen, and Bellack (1977) used a changing conditions design with repeated baselines to measure the effect of various interventions on the auditory hallucinations of a 33-year-old woman. The dependent variables measured were the percent of time of hallucinations, the duration of hallucinations, and the frequency of hallucinations. The following conditions (phases) were included in the study. Figure 5–28 presents the data that were recorded under each condition.

1. Baseline—the woman was seated in a chair and asked to raise her finger each time she heard voices and hold her finger up as long as the voices continued.

2. Social Disruption—the woman was engaged in conversation on unrelated topics throughout the session.

3. Baseline Condition.

4. Social Disruption.

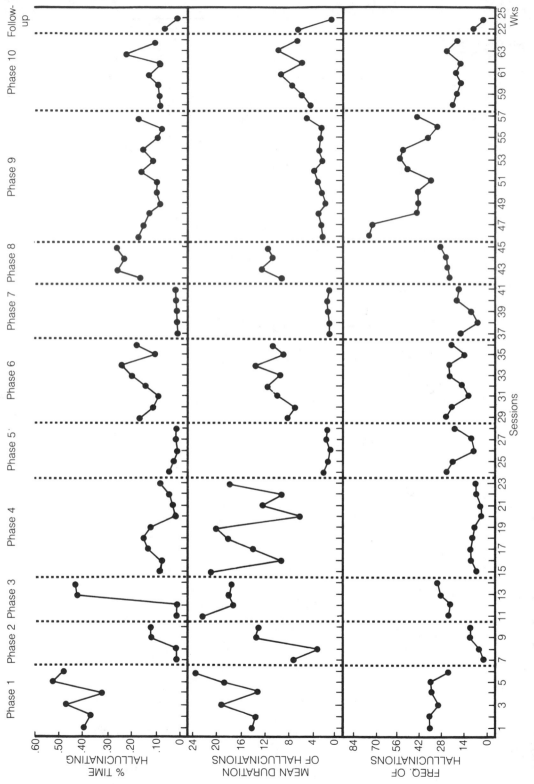

FIGURE 5–28 *Example of a changing conditions design*

Source: S. Turner, M. Hersen, & A. Bellack. "Effects of social disruption, stimulus interference, and aversive conditioning on auditory hallucinations," *Behavior Modification*, 1, p. 254. Copyright © 1977 by Sage Publications, Inc. Reprinted by permission of Sage Publications, Inc.

5. Stimulus Bell—an electrically operated bell was paired with the onset and duration of hallucinations.

6. Baseline Condition.

7. Stimulus Bell.

8. Baseline Condition.

9. Electrical Shock—electrical shock to the wrist was administered contingent on the signal and continued until the woman signaled that the voices ended.

10. Baseline Condition.

11. Follow-up—22 weeks and 25 weeks following discharge under baseline conditions.

It should be noted that with the design format presented in Figure 5–26, changing conditions with repeated baselines, the teacher may determine the existence of a "weak" functional relationship. To strengthen this determination, a replication phase of one or more of the interventions would have to be included as illustrated in Figure 5–28.

Teaching Application

Here's how a changing conditions design can be used in the classroom.

Roberta Learns to Shoot Baskets

Mr. Woods was recently hired to teach physical education at an elementary school. When he arrived at work, Mr. Woods was approached by the special education teacher, Ms. Jones. She was concerned about Roberta, a physically handicapped student who would be in Mr. Woods' gym class. Roberta, who was confined to a wheelchair, had difficulty with eye-hand coordination. Ms. Jones hoped that the student could learn to throw a basketball. Learning to play basketball would provide both coordination training and a valuable leisure skill for Roberta. Mr. Woods agreed that the basketball skill seemed appropriate.

Baseline.

Mr. Woods decided to use a systematic approach to instruction. He asked Roberta to throw the basketball 20 times to see how often she could place the ball through a lowered hoop. This procedure was followed for five gym periods with no additional instruction until a baseline performance rate was determined. Mr. Woods then decided to use a modeling technique; he showed Roberta how to throw the ball and then asked her to imitate him. Very little improvement was noted in five class periods. Mr. Woods met with the special education teacher to determine what could be done.

1st condition.

2nd condition.

Ms. Jones carefully reviewed all of the data and suggested a change in conditions. She explained that a change in intervention seemed necessary and that a modeling procedure be used in combination with keeping score on a chart. Mr. Woods agreed to try this. In two weeks, Roberta showed improvement but still missed more baskets than she hit. A final condition was implemented whereby modeling, scorekeeping, and a correction procedure were used together. Mr. Woods now showed Roberta how to throw, recorded her score, and showed her exactly what she did wrong when she missed. This

3rd condition

combination of procedures resulted in Roberta's being able to throw a basket-ball through a hoop 15 out of 20 times. A suggestion was made to Roberta's parents that a hoop be constructed at her home so that she could enjoy her new skill after school.

Advantages and Disadvantages

Changing conditions is a teaching design.

The changing conditions design with a single baseline allows the teacher to compare the effects of a number of interventions on student behavior. Although no functional relationship can be established, recording data in this format allows the teacher to monitor the effects of various procedures on student behavior. The teacher should be aware, however, that what she may be seeing is the cumulative effects of the interventions rather than the effects of any one intervention in isolation. Individual analysis of the effects of the interventions can be made with the changing conditions with re-peated baselines format. The teacher who records data systematically in a changing criterion design will have a record of the student's progress and a good indication of what procedures are effective with that student.

EVALUATING SINGLE-SUBJECT DESIGNS

Statistical Analysis

The purpose for using behavior modification and applied behavior analysis techniques in the classroom is to achieve, and verify, meaningful changes in a student's behavior. The effectiveness of an intervention is commonly judged against both an experimental criterion and a therapeutic criterion. The *experimental criterion* is the verification that an independent variable (an intervention) was responsible for the change in the dependent variable (a behavior). Single-subject designs demonstrating within-subject replica-tions of effect satisfy this criterion (Baer, Wolf, & Risley, 1968; Kazdin, 1977; Barlow & Hersen, 1984).

The *therapeutic criterion* is a judgment as to whether the results of the teacher's intervention are "important or of clinical or applied significance" (Kazdin, 1982). For example, the teacher should ask herself whether it is truly meaningful to increase a student's grade from a D − to a D (Baer et al., 1968) or to decrease a child's self-destructive behavior from 100 to 60 instances per hour (Kazdin, 1977) or to reduce a student's off-task behav-ior in a resource room while it remains at high in the regular classroom. Kazdin (1977) suggests a third evaluation criterion: *social validation,* the "social acceptability of an intervention program" (p. 430). Social validation is discussed at length in Chapter 2.

Intervention effects in applied behavior analysis are usually evaluated through *visual inspection* of the graph displaying the plotted data points of the various phases (conditions). Interpretation of data based on visual in-spection sounds unrefined and may certainly be subjective. It may there-fore be viewed by some as a weak form of evaluation. Evaluation resulting from visual inspection, however, reveals only strong intervention results, and that is what teachers are seeking.

The grossness and subjectivity of visual inspection are somewhat modified by common agreement that certain characteristics of the graphed data should be evaluated. These characteristics include the *means* of the data in the phases, the *levels* of performance in the phases, the *trends* in performance, and the *rapidity* of behavior change (Kazdin, 1982).

Evaluation of changes in means focuses on the change in the average rate of student performance across phases of a design. Within each phase the mean (average) of the data points is determined and may be indicated on the graph by drawing a horizontal dashed-line corresponding to the value on the ordinate scale. Visual inspection of the relationship of these means will help to determine if the intervention resulted in consistent and meaningful changes in the behavior in the desired direction of change. In Figure 5–29, Foxx and Shapiro (1978) have supplied such indicators of means. The viewer can easily see the relative position of the students' disruptive behavior across the various design phases.

Evaluation of the level of performance refers to the increase or decrease in student performance from the end of one phase to the beginning of the next phase. The teacher wants to evaluate the magnitude and direction of this change. "When a large change in level occurs immediately after the in-

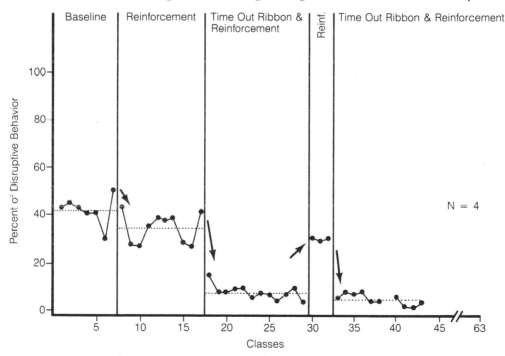

FIGURE 5–29 The mean percent of time spent in disruptive classroom behavior by four subjects

Note. The horizontal broken lines indicate the mean for each condition. The arrow marks a one-day probe (Day 39) during which the timeout contingency was suspended. A followup observation of the teacher-conducted program occurred on Day 63. *Source:* R. Foxx and S. Shapiro, "The Timeout Ribbon: A Nonexclusionary Timeout Procedure." *Journal of Applied Behavior Analysis, 11,* 1978, p. 131. Reproduced by permission.

troduction of a new condition, the level change is considered abrupt, which is indicative of a powerful or effective intervention" (Tawney & Gast, 1984, p. 162). Tawney and Gast (1984) suggest the following steps to determine and evaluate a change in level between two adjacent conditions: (1) identify the ordinate value of the last data point of the first condition and the first data point value of the second condition; (2) subtract the smallest value from the largest; and (3) note whether the change in level is in an improving or decaying direction (p. 162). In Figure 5–29 the arrows have been added to indicate level changes.

Evaluation of trend in performance focuses on systematic and consistent increases or decreases in performance. Trends in data are most often evaluated using a procedure known as the quarter-intersect method (White & Liberty, 1976). Evaluation of trends is based on lines of progress developed from the median value of the data points in each phase. Steps for computing lines of progress are illustrated in Figure 5–30. Trend lines can provide: (a) an indication of the direction of behavior change in the past; and (b) a prediction of the direction of behavior change in the future. This information can help the teacher to determine whether to change the intervention.

Taking this process one step further will yield a split-middle line of progress (White & Haring, 1980). This line of progress is drawn so that an equal number of data points fall on and above the line as fall on and below the line. As illustrated in Figure 5–31, if the data points do not naturally fall in such a pattern, the line is redrawn higher or lower, parallel to the original line, until the balance of data points is equal.

A fourth characteristic that may be evaluated through visual inspection is the rapidity of the behavior change. This refers to the length of time between the onset or termination of one phase and changes in performance. The sooner the change occurs after the intervention has been applied (or withdrawn), the clearer the intervention effect. It should be noted that "rapidity of change is a difficult notation to specify because it is a joint function of changes in level and slope (trend). . . . A marked change in level and in slope usually reflects a rapid change" (Kazdin, 1982, p. 316).

While visual inspection is useful, convenient, and basically reliable for identifying or verifying strong intervention effects, much of the current published research using single-subject designs has accompanied visual inspection with statistical verification of effects (e.g., t test, F test, R test, time series analysis). Kazdin (1976) offers three reasons to support the use of statistical techniques:

1. To assist in distinguishing subtle effects from chance occurrence.

2. To analyze the effects of a treatment procedure when a stable baseline cannot be established.

3. To assess the treatment effects in environments that lack control.

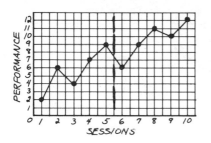

1. Divide the number of data points in half by drawing a vertical line down the graph.

In this example, there are ten data points, therefore the line is drawn between sessions 5 and 6. If there had been an odd number of data points, this line would have been drawn through a session point.

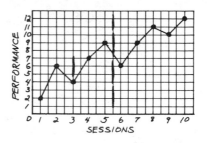

2. On the left half of the graph, find the mid-session and draw a vertical line.

In this example, there are five data points; therefore, the line is drawn at session 3. If there had been an even number of sessions, this line would have been drawn between two session points.

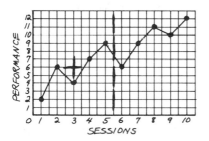

3. On the left half of the graph, find the mid-performance point and draw a horizontal line.

In this example, the data point at performance value 6 is the mid-performance point because there are two data points below it and two data points above it. If there had been an even number of data points, this line would have been drawn between the two middle points.

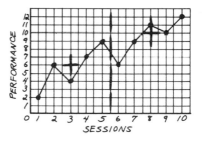

4. Repeat steps 2 and 3 on the right half of the graph.

In this example, session 8 is the mid-session, and the data point at performance value 10 is the mid-performance point.

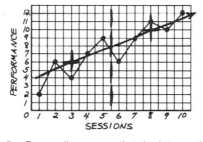

5. Draw a line connecting the intersections of both halves of the graph. This is the trend line for the data.

FIGURE 5–30   Steps for computing lines of progress

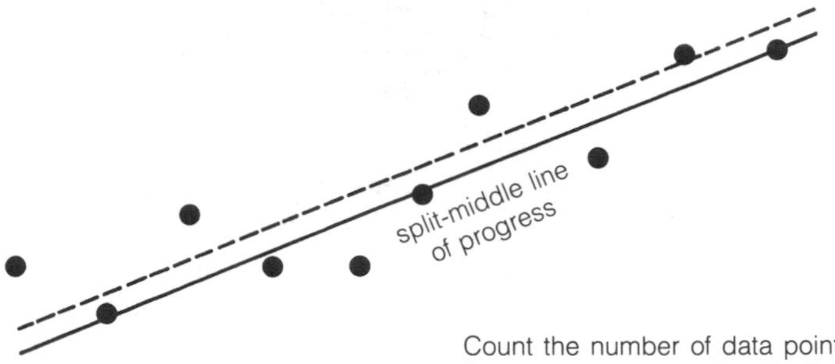

10. M. opens refrigerator.

1 min.

Count the number of data points which fall above and below the line drawn in [the procedure described in Figure 5-30]. There should be the same number of data points falling on and above the line as there are falling on and below the line. If not, move the line up or down (keeping it parallel to the original line) until a balance is achieved.

FIGURE 5–31   Split-middle line of progress

Source: O. White and N. Haring, *Exceptional Teaching.* Columbus, Ohio: Charles E. Merrill, 1980, p. 118. Reproduced by permission.

More information about advanced uses of visual inspection and especially about statistical evaluation in single-subject designs can be found in texts by Barlow and Hersen (1984), Kazdin (1982), and Tawney and Gast (1984).

Replication

A frequently asked question regarding the use of single-subject designs for research is whether the results can be generalized. If a study shows that a procedure is effective with a single subject, does this mean that it will be effective with others? Applied behavior analysts do not assume generalizability of research results based on a single successful intervention. Instead, they depend on *replication*—the repeated application of the same intervention with different subjects. The fact that systematic teacher praise increases one student's rate of doing math problems may not be a convincing argument for the use of praise. However, documentation that such praise has increased production of not only math problems but also many other academic and social behaviors with dozens of students at many different ages is convincing. By replication, applied behavior analysts gradually identify procedures and techniques effective with many students. These procedures and techniques can then be adopted by others with considerable confidence that they will work.

SUMMARY

In this chapter we have discussed a number of single-subject designs for research and teaching. We have described procedures for implementation, graphic representation, and data analysis for each design. It is the systematic analysis of data collected in a planned and organized form that distinguishes the applied behavior analyst from the mere behavior modifier. The teacher who is familiar with design can quickly evaluate the results of his or her own instruction and ultimately share these results with others. The continuing process of applied behavior analysis results in more effective teaching that is ultimately reflected in more adaptive academic and social behaviors of students.

REFERENCES

ALBERTO, P., JOBES, N., SIZEMORE, A., & DORAN, D. 1980. A comparison of individual and group instruction across response tasks. *The Journal of the Association for the Severely Handicapped, 5*, 285–293.

BAER, D., WOLF, M., & RISLEY, T. 1968. Some current dimensions of applied behavior analysis. *Journal of Applied Behavior Analysis, 1*, 91–97.

BARLOW, D., & HAYES, S. 1979. Alternating treatments design: One strategy for comparing the effects of two treatments in a single subject. *Journal of Applied Behavior Analysis, 12*, 199–210.

BARLOW, D., & HERSEN, M. 1984. *Single case experimental designs: Stragegies for studying behavior change.* New York: Pergamon Press.

CARR, E.G., NEWSOM, C.D., & BINKOFF, J.A. 1980. Escape as a factor in the aggressive behavior of two retarded children. *Journal of Applied Behavior Analysis, 13*, 101–117.

COOPER, J. 1981. *Measuring behavior.* Columbus, Ohio: Charles E. Merrill.

FOXX, R., & SHAPIRO, S. 1978. The timeout ribbon: A nonexclusionary timeout procedure. *Journal of Applied Behavior Analysis, 11*, 125–136.

GOLDSTEIN, R.S., MINKIN, B.L., MINKIN, N., & BAER, D.M. 1978. Finders, keepers?: An analysis and validation of a free-found-ad policy. *Journal of Applied Behavior Analysis, 11*, 465–473.

HALL, R.V., & FOX, R.G. 1977. Changing-criterion designs: An applied behavior analysis procedure. In B.C. Etzel, J.M. LeBlanc, & D.M. Baer (Eds.), *New developments in behavioral research: Theory, method and application.* Hillsdale, N.J.: Erlbaum. (In honor of Sidney W. Bijou)

HARTMANN, D.P., & HALL, R.V. 1976. The changing criterion design. *Journal of Applied Behavior Analysis, 9*, 527–532.

HERSEN, M., & BARLOW, D.H. 1976. *Single-case experimental designs: Strategies for studying behavior change.* New York: Pergamon Press.

HORNER, R.D., & BAER, D.M. 1978. Multiple-probe technique: A variation on the multiple baseline. *Journal of Applied Behavior Analysis, 11*, 189–196.

KAZDIN, A.E. 1976. Statistical analyses for single-case experimental designs. In M. Hersen & D. Barlow (Eds.), *Single-case experimental designs: Strategies for studying behavior change*, pp. 265–316. New York: Pergamon Press.

KAZDIN, A. 1977. Assessing the clinical or applied importance of behavior change through social validation. *Behavior Modification, 1*, 427–451.

KAZDIN, A. 1982. *Single-case research designs.* New York: Oxford University Press.

MURPHY, R., & BRYAN, A. 1980. Multiple-baseline and multiple-probe designs: Practical alternatives for special education assessment and evaluation. *The Journal of Special Education, 14*, 325–335.

NELSON, G.L., & CONE, J.D. 1979. Multiple-baseline analysis of a token economy for psychiatric inpatients. *Journal of Applied Behavior Analysis, 12*, 255–271.

REPP, A. 1983. *Teaching the mentally retarded.* Englewood Cliffs, N.J.: Prentice-Hall, Inc.

SIDMAN, M. 1960. *Tactics of scientific research.* New York: Basic Books.

SINGH, J., & SINGH, N. 1985. Comparison of word-supply and word-analysis error-correction procedures on oral reading by mentally retarded children. *American Journal of Mental Deficiency, 90*, 64–70.

SMITH, D. 1979. The improvement of children's oral reading through the use of teacher modeling. *Journal of Learning Disabilities, 12*, 172–175.

STOKES, T.F., FOWLER, S.A., & BAER, D.M. 1978. Training preschool children to recruit natural communities of reinforcement. *Journal of Applied Behavior Analysis, 11*, 285–303.

SULZER-AZAROFF, B., & MAYER, G.R. 1977. *Applying behavior-analysis procedures with children and youth.* New York: Holt, Rinehart, and Winston.

TAWNEY, J., & GAST, D. 1984. *Single subject research in special education.* Columbus, Ohio: Charles E. Merrill.

TURNER, S., HERSEN, M., & BELLACK, A. 1977. Effects of social disruption, stimulus interference, and aversive conditioning on auditory hallucinations. *Behavior Modification, 1*, 249–258.

ULMAN, J.D., & SULZER-AZAROFF, B. 1975. Multi-element baseline design in educational research. In E. Ramp & G. Semb (Eds.), *Behavior analysis: Areas of research and application,* pp. 377–391. Englewood Cliffs, N.J.: Prentice-Hall.

WHITE, O., & HARING, N. 1980. *Exceptional teaching.* Columbus, Ohio: Charles E. Merrill.

WHITE, O., & LIBERTY, K. 1976. Evaluation and Measurement. In N.G. Haring & R.L. Schiefelbusch (Eds.), *Teaching special children.* New York: McGraw-Hill.

Section Three
Applying Learning Principles

6
Arranging Consequences that Increase Behavior

Did you know that . . .

- Behaviorists did not invent positive reinforcement?
- Candy may not be a positive reinforcer?
- Sometimes it's better not to reinforce every instance of appropriate behavior?
- Negative reinforcement is not punishment?
- You don't need a lawyer to write a contract?

Because of recent widespread dissemination of information about behavioral principles, the term *reinforcement* has become part of the vocabulary of the general public. The word is widely used to describe pleasant events or rewards given to a person who complies with the requests of some behavior change agent. Thus, reinforcement has come to be associated with the stereotypic manipulative view of behavior modification. Reinforcement is sometimes conceptualized as an artificial tool created to make people engage in behaviors chosen by others. While many applied behavior analysts use the principles of reinforcement to change behavior, it is not true that reinforcement is an invention of behaviorists. Reinforcement is a naturally occurring phenomenon. Behaviorists have simply applied the effects of reinforcement in a thoughtful and systematic manner.

As Chapter 1 stated, reinforcement describes a relationship between two environmental events, some behavior (response) and some event or stimulus (consequence) that follows the response. The relationship is termed *reinforcement* only if the response increases or maintains its rate as a result of the consequence. Chapter 1 described two types of reinforcement: positive reinforcement—the *contingent presentation* of a consequence that increases behavior—and negative reinforcement—the *contingent removal* of some unpleasant stimulus.

If we were to examine the course of events in our daily lives, we would readily see that our continued performance of certain behaviors is due to the results or consequences of producing those behaviors. Every action we engage in results in some consequence. When our behavior results in a naturally occurring, desirable consequence, this experience serves as a motivating force for our continued performance. Consider these examples:

An office worker goes to work each day expecting to receive a check at the end of the week. If on Friday the check is delivered, and it is in an amount that the individual finds satisfying, it increases the probability that the person will go back to work on Monday.

A youngster on a little league team hits a double and receives cheers from fans and teammates. This result of her behavior will motivate the child to play again next Saturday.

When a baby coos at mother's approach, she cuddles him and spends more time playing with him. The mother's response will increase the frequency of the baby's cooing. If the baby continues to coo and giggle during this interaction, his behavior serves to reinforce mother's playing with him. This interaction will continue because it is reinforcing to both infant and mother.

A student spends 45 minutes each night during the week studying for a history exam on Friday. If the student makes an "A" on the exam, this consequence of studying will motivate her to study similarly before her next examination.

While many appropriate behaviors are maintained by naturally occurring reinforcers, this natural process may be insufficient to maintain all desirable behaviors. Teachers often find students for whom naturally occurring reinforcers fail to maintain appropriate behavior. Some students may see little immediate benefit from learning plane geometry or applied behavior analysis. Some students may be motivated by stronger competing reinforcers than those being offered by the teacher. These students may find the laughter of other students more reinforcing than the teacher's approval. Some students may not value the type of reinforcers the teacher offers. Grades, for example, may have little meaning to them. In such instances, the teacher must develop an interim program to systematically arrange opportunities for students to earn valued reinforcers. When naturally occurring reinforcers are not sufficiently powerful, the wise teacher looks for more powerful ones.

In this chapter, we shall describe procedures for the effective use of reinforcement procedures in changing classroom behavior. The majority of the chapter examines the use of positive reinforcement, while the final section describes classroom applications of negative reinforcement.

POSITIVE REINFORCEMENT

Positive reinforcement (S^{R+}) is the contingent presentation of a stimulus, immediately following a response, that increases the future rate and/or probability of the response. There are three operative words in this definition. The word *increases* makes it clear we are dealing with some form of reinforcement, for the stimulus will have the effect of increasing the probability that the response will occur again. The second operative word is *presentation*. When we use positive reinforcement, we intentionally present the student with a stimulus following the production of a response. The third operative word is *contingent*. The teacher will not present the consequence to the student unless and until the requested response is produced. If a teacher states the contingency, "Bill, when you finish all your math problems, you may play with the airplane models," the teacher is using positive reinforcement (if airplane models are reinforcing to Bill). The reinforcing stimulus (playing with airplane models) will be presented to the student contingent upon production of the requested behavior (completion of math problems). The examples in Table 6–1 illustrate the principle of positive reinforcement.

Whereas *positive reinforcement* refers to the relationship between a behavior and a consequence, the term *positive reinforcer* describes the consequent event itself. A positive reinforcer is a consequential stimulus (S^R) that

1. Increases or maintains the future rate and/or probability of occurrence of a behavior.

TABLE 6–1 *Examples of positive reinforcement*

	Stimulus	Response	S^{R+}	Effect
Example 1	Statement of the contingency and availability of appropriate materials	Bill completes math problems.	He is allowed to play with model planes.	Increased probability that Bill will complete next set of problems on time
Example 2		John sits upright in his seat.	Teacher presents him with a smile and words of praise.	Increased likelihood that John will continue to sit appropriately
Example 3		Sara brings in her homework each day this week.	She is appointed board monitor for next week.	Increased probability that Sara will continue to bring her homework everyday

2. Is administered contingently upon the production of a desired or requested behavior.

3. Is administered immediately following the production of the desired or requested behavior.

Choosing Effective Reinforcers

The reinforcing potential of an item or event depends on reinforcement history and deprivation.

Because a stimulus is defined as a positive reinforcer only as a result of its effect on behavior, no item or event can be termed a positive reinforcer until this relationship has been established. It therefore follows that a teacher cannot state with any real degree of certainty—in advance of evidence of such a relationship—what will or will not be a reinforcing consequence for any given student. What acts as a reinforcer for a particular student depends upon that person's *reinforcement history* (that is, what has motivated the student previously) and the *conditions of a deprivation state* (in other words, what does the student desire but not have or not get frequently). Therefore, what will serve as a reinforcer may be different for each student in the class. Preconceived notions of what should be reinforcing to a student are frequent reasons for the failure of intervention programs. When the desired behavior change does not occur, the teacher's first reaction is often to assume that reinforcement procedures have not worked, when in fact one of the fundamental notions of reinforcement has been violated: the *individualization of reinforcers.*

Ms. Ledbetter Wastes Silli-Circles

Ms. Ledbetter, who had been a second-grade teacher for 27 years, read about positive reinforcement in a newspaper article. The article described a program designed to teach handicapped children self-care skills such as dressing themselves. The researchers had used presweetened cereal as a positive reinforcer. Ms. Ledbetter was very excited about the prospect of motivating her students to do better in school. She therefore purchased a large economy-sized box of Silli-Circles, the newest sweetened oat cereal, and prepared to reinforce appropriate behavior.

Her lowest reading group included several students who lacked any reading skills whatever, as well as any apparent interest in acquiring such skills. When she called them to the reading circle, she announced, "Boys and girls, today when you answer a question correctly, you'll get a Silli-Circle. Won't that be nice?"

Ms. Ledbetter then proceeded with her lesson. She was disappointed to find that the students seemed no more interested in reading than before and that the few students who earned cereal either dropped it on the floor or attempted with little success to give it away.

"Well!" snorted Ms. Ledbetter, "I should have known. These newfangled ideas . . . That positive reinforcement stuff just doesn't work."

Ms. Ledbetter's conclusion that positive reinforcement doesn't work was based on her assumption that because cereal was a reinforcer for some students, it would be a reinforcer for all. Apparently, members of her reading group did not like Silli-Circles—or they had eaten a bowl for breakfast—or they couldn't read well enough to answer the questions. Therefore, Silli-Circles were not a positive reinforcer.

An aid in determining the potential effectiveness of reinforcers is reinforcer sampling. The method of sampling will vary depending upon the students' level of functioning. Students of higher capability can often simply be asked what items they would like to have as rewards for effort or achievement. Ms. Ledbetter could have saved a lot of grief if she had simply asked, "Do you like Silli-Circles?"

Another strategy for reinforcer sampling is using a prepared "reinforcer menu" like the one that appears in Figure 6–1. Each student would be asked to list the potential reinforcers in order of preference. A reinforcer menu should include a variety of potential reinforcers that teachers can

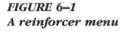

FIGURE 6–1
A reinforcer menu

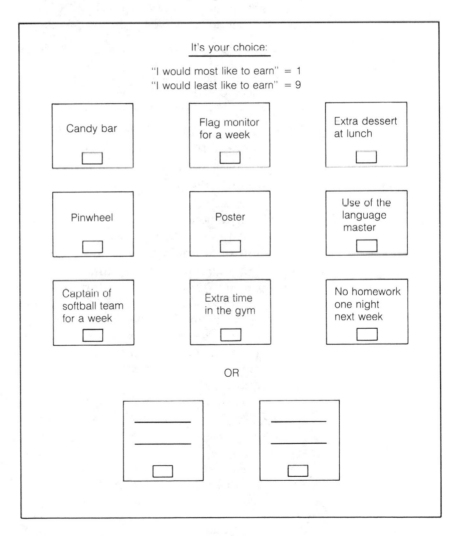

reasonably make available. The presentation of limited choices will prevent unrealistic selections (such as stereos and trips to Acapulco) that students might suggest if asked to select reinforcers. However, teachers may still wish to offer the opportunity for an open-ended response following some forced choices.

Reinforcer sampling for low-functioning students must be concrete.

To determine reinforcer preference with lower functioning students, it may be necessary to conduct the reinforcer sampling with actual objects or events. It is usually best to bring to a sampling session up to five items that the teacher thinks are potential reinforcers. These may include both edible items and toys. The student should be allowed to experience each of these items and then allowed to choose. For students with physical limitations microswitches may be used to enable them to access the array of items (Wacker, Berg, Wiggins, Muldoon, & Cavanaugh, 1985). The teacher then lists the items in order of preference based upon the number of times each was selected. As noted by Ayllon and Azrin (1968),

> the sampling of the event does not guarantee that the event will, in fact, be reinforcing to the individual; rather, the sampling allows any reinforcing properties to be exhibited if the event has any potential reinforcing properties. If the individual does not seek the event after it has been sampled, it will not be because of a lack of familiarity with it. (p. 92)

Hall and Hall (1980) suggest the following nine-step sequence for selecting potential reinforcers:

Step 1: "Consider the age, interests and appetites of the person whose behaviors you wish to strengthen." The teacher should select potential consequences that attempt to correspond to the chronological age and social background of the student. The offering of Fruit Loops or an opportunity to work with puzzles will probably have little motivational value to an adolescent.

Step 2: "Consider the behavior you wish to strengthen through reinforcement." The teacher should select potential consequences that attempt to correspond to the value, or effort required to produce the response. "If an employer offered to buy an employee a cup of coffee for working all weekend on a special job it would be unlikely that any worker would accept the offer." Similarly, offering a student 5 additional minutes of free time for completion of an entire day's written assignment is an opportunity the student will probably pass up.

Step 3: "List potential reinforcers considering what you know of the person, his or her age, interests, likes and dislikes and the specific behavior you have defined." This step allows the teacher to organize the potential reinforcers she is considering in an orderly and objective manner. Hall and Hall (1980) suggest that the potential reinforcers be organized according to categories, such as material reinforcers, activity reinforcers, and social reinforcers.

Step 4: "The Premack Principle." When selecting potential reinforcers, the teacher should consider watching the student and noting activities in which he likes to engage. The use of preferred activities as reinforcers was systematized by David Premack (1959). A discussion of the Premack Principle follows on page 202.

Step 5: "Consider asking the person." The teacher should remember that the best authority on the likes and dislikes of a student is that student. The mechanism most often used to determine a student's potential reinforcers is the previously mentioned Reinforcer Menu. For alternative forms of designing reinforcer menus the reader is referred to Raschke (1981).

Step 6: "Consider novel reinforcers." With this step Hall and Hall (1980) remind teachers that "varying the reinforcers is more effective than using the same reinforcers over and over." Repeated use of the same reinforcer can lead to boredom and satiation (see page 198), lessening the motivating effectiveness of a consequence.

Step 7: "Consider reinforcers that are natural." The authors suggest three advantages to the use of natural reinforcers. First, natural reinforcers such as recognition and privileges can be provided more easily and at lower cost than most edible and material consequences. Second, natural reinforcers are more likely to be available to the student after the behavior has been established. "In a natural situation, even though you discontinue to systematically reinforce the behavior you wanted to strengthen, the natural positive consequence you provided is more likely to be available on at least some occasion in the future." Third, natural reinforcers automatically occur on a contingent basis. Praise for a homework assignment well done will not naturally occur unless the behavior was performed.

Step 8: "Select the reinforcer or reinforcers you will use." Once the teacher has considered Steps 1–7, the authors suggest that the teacher now select the reinforcers that will most likely have the desired effect on the target behavior.

Step 9: "Make a record of the behavior." The teacher is reminded that the only way to confirm a consequence as a reinforcer is to observe its effect on the behavior. To objectively verify this effect the teacher should systematically document the change, if any, in the production of the behavior. Various methods for providing such documentation are presented in Chapter 4 of this text.

Making Reinforcers Contingent

If reinforcement is to be effective, the student must get the reinforcer only after performing the target behavior. There is a contingency in place: an "if . . ., then" requirement has been stated. Such a statement establishes a clear and explicit relationship between performing the behavior and receiving the reinforcer. If Clara finds that—regardless of whether she has

It may also be necessary to be sure that parents don't make potential reinforcers available noncontingently.

performed the target behavior—she can get a lollipop at the end of the day when the teacher is tired and has a few pops left over, Clara may decide that the teacher does not mean business: no contingency is actually in force. Implied in such a contingency and explicit in reinforcer delivery is that the teacher or some other specifically designated person is the source of the reinforcer. If the student can go to the classroom aide or some other adult during the day and get the same promised reinforcer without having performed the desired behavior, that student will quickly determine that there is no need to comply with the contingency.

Making Reinforcement Immediate

To be effective, a reinforcer should be delivered immediately after the target behavior is performed. This timing convinces the student of the veracity of the contingency and underlines the connection between a particular behavior and its consequence. Immediacy of delivery is also necessary to avoid the hazard of inadvertently reinforcing an intervening behavior. The longer the delay between the desired behavior and receipt of the reinforcer, the greater the possibility that the student may engage in a behavior not under the contingency or not desired. If the reinforcer is delivered following any intervening behavior, it is the latter whose probability will be increased.

Ms. Troutman Reinforces Chaos

Ms. Troutman was a teacher of students with severe social maladjustments. She taught in a self-contained special class. It was the first week of her first year in this setting, and she was determined to make it a success. She had taken a course in applied behavior analysis and decided to set the following contingency for her students:

If you complete at least 20 assignments during the week, *then* you may participate in a class party at 2:15 on Friday.

The students worked busily, and Ms. Troutman glowed with satisfaction as she wondered why people thought these students were difficult to teach. At 11:00 on Friday morning, the first student completed his 20th assignment. By noon, all seven students had fulfilled the contingency. The remaining hours before 2:15 were among the longest Ms. Troutman had ever spent. Even though the students yelled, fought, cursed, ran around, and generally created havoc, the party was held as scheduled. (Ms. Troutman at least had sense enough to know that if she failed to live up to her end of the contingency, the students would never believe her again.) Monday morning the class came in yelling, fighting, cursing, and running around. Ms. Troutman had reinforced chaos, and that's what she got.

Types of Reinforcers

Two major varieties of reinforcers are available to teachers: *primary reinforcers* and *secondary reinforcers*.

Primary Reinforcers

Primary reinforcers are stimuli that have biological importance to an individual. We can assume that they are innately motivating, because they are

Primary reinforcers can be powerful tools for changing behavior.

necessary to the perpetuation of life. Therefore, primary reinforcers are described as *natural, unlearned,* or *unconditioned* reinforcers. Given their biological importance, we may expect that they will be highly motivating to individual students. The major types of primary reinforcers include foods, liquids, sleep, shelter, and sex (the last reinforcer most commonly used in the form of entry to social activities that include both males and females). Obviously, the two most common and appropriate primary reinforcers for use in the classroom are food and liquids.

Edible reinforcers are used mainly with younger students and students of low functioning ability. Such reinforcers are usually used when teaching a new behavior. Because of their high motivational value, they quickly affect behavior.

Different strokes for different folks.

In spite of the almost mystical association between behavior modification and M&M candy, teachers rarely find it necessary to use edible reinforcers with older normal or mildly handicapped students. An imaginative teacher can choose from a great many other potential reinforcers. Reinforcing these students' behavior with candy or other treats is an example of behavioral overkill. Our advice to teachers is "Don't kill flies with an elephant gun." Besides being unnecessary, primary reinforcers may be perceived by students as insulting. A ninth-grader could hardly take seriously a teacher who said, "Great job on that algebra assignment, Casey. Here's your cookie." This is not to imply that teachers of older normal or mildly handicapped students should never use food as a reinforcer. An occasional treat may be very effective, when presented in an appropriate way. For example, a popcorn party for students who have met some contingency may be perfectly appropriate. It is amazing to observe how rapidly second- or third-graders complete assignments once the smell of the popping corn begins wafting about the room.

Deprivation is a condition that must be met if primary reinforcers are to be effective.

In order for primary reinforcers to be effective, the student whose behavior is to be reinforced must be in a state of deprivation in relation to that reinforcer. Using an edible reinforcer with a student who has just returned from lunch will have a diminished chance of effectiveness, because the student is not hungry. This by no means suggests that students should be starved so that food will be an effective reinforcer, but the necessity for a state of deprivation is a major drawback in the use of primary reinforcers. However, a student need not be hungry in order for limited amounts of special foods such as potato chips, raisins, ice cream, or candy to be effective reinforcers.

The opposite of deprivation is satiation. Satiation occurs when the deprivation state that existed at the beginning of an instructional session no longer exists, and the student's cooperation and attention have worn thin. A teacher of severely handicapped students who conducts a training session lasting perhaps 30 minutes may come to a point in the session when the primary reinforcer loses its effectiveness. The teacher will know this when the student's rate of correct responding slows down or—in the case of more assertive students—when the student spits the no-longer-reinforcer at the teacher. There are at least five ways a teacher may plan to prevent or delay satiation:

1. Assign a particular reinforcer to each instructional task. There is no need to use a single reinforcer all day long, in all subject areas, with any given student. To do so is to build in the potential danger of satiation. As a result of the reinforcer sampling that was conducted, an ordered list of several potential reinforcers is available so that more than one can be used during the day.

2. Shorten the instructional session in which the edible reinforcer is being used. Shorter sessions with fewer trials (controlled presentations) decrease the chances of satiation. Several short sessions may be held during the day.

3. If satiation occurs, try switching to an alternate reinforcer. Alternating salty foods and sips of liquid may be a very effective way of delaying satiation.

4. Decrease the size of the pieces of edible given for correct responses. Smaller portions also disappear more quickly, thus not creating artificially long periods between trials as the student continues to "savor" his raisin.

5. Have an array of edible reinforcers available for the student to select from following each correct response. In this way, the teacher may say, "Good, that is the correct answer. What would you like to eat as a reward?" Next to him or her is a plate with pieces of three or four different edibles from which to select. (Lovaas, 1977)

Carr, Binkoff, Kologinsky, and Eddy (1978) delayed the onset of satiation and tied their use of reinforcers directly to the concepts they were teaching to a group of autistic students. While teaching the students manual signs for the words *milk, apple, cookie, candy,* and *banana,* the teacher reinforced a correctly formed sign with a piece of the food indicated. By this method, the students received only a small amount of the reinforcer requested, and the reinforcer was varied according to the randomization of the trial presentations. A problem with satiation is illustrated by the following story.

Mr. Alberto Eats Ice Cream

Jeff was a severely retarded student whose behavioral repertoire seemed to consist mainly of throwing materials and chairs. He sometimes varied this by hitting teachers and other students. In an effort to control this behavior, Mr. Alberto had tried more than a dozen potential primary reinforcers—from potato chips to candy—with no success whatsoever. Appropriate behavior stayed at a low rate and inappropriate behavior at a high rate.

In desperation, Mr. Alberto asked Jeff's mother if there was *anything* Jeff liked. (It occurred to him that this might have been a good thing to have done in the first place.)

"Of course," said Mother, "Jeff loves butter-pecan ice cream."

One quick trip to the grocery store later, Mr. Alberto began to make progress with Jeff. By the end of a week, the behavior was under reasonable control, and Mr. Alberto thought his troubles were over. Before long, however, the inappropriate behavior was back in full force. Once again, Mr. Alberto questioned Jeff's mother. Could she explain why butter-pecan ice cream wasn't working any more?

"Well, maybe," said Mother, "it's because he gets so much at home. I learned long ago that it was the only way to make him behave. I give him a whole carton some days."

As Mr. Alberto solaced himself with the remaining pint of butter-pecan, he prepared once again to go in search of the elusive reinforcer.

Edibles provide the teacher with a wide range of potential reinforcers—for example, cereal (Alberto, Jobes, Sizemore, & Doran, 1980), ice cream and soda (Newsom & Simon, 1977), and diet chocolate (Olenick & Pear, 1980). However, certain precautions are necessary. A review of medical records or consultation with parents is advisable. A student may have allergies to certain foods, be on a special diet, have a lactose intolerance, or be susceptible to diabetic reactions.

Common sense also suggests some precautions in selecting reinforcers. A teacher conducting language training would not use peanut butter as a reinforcer, because it is difficult to imitate sounds with your tongue stuck to the roof of your mouth. Liquid reinforcers may increase the number of toileting breaks necessary, delaying the session unduly.

Teachers should also note that the strong motivational property of certain reinforcers, especially edible reinforcers, has the potential to encourage responses that are incompatible with the target response. Balsam and Bondy (1983) illustrate this point with the example of using ice cream as a reinforcer for a young child. They suggest that the ice cream itself might stimulate so much approach behavior (i e , staring, reaching) that it interferes with the child's attending to the relevant antecedent and response requirement of the contingency. Similarly, if a teacher tells the class that if the students are good they will be allowed a special treat at lunch time, the students may become increasingly fidgety and inattentive in anticipation of the reinforcement. (For a full discussion of the theoretical and operant negative side effects of reinforcement, the reader is referred to Balsam and Bondy, 1983.)

Secondary Reinforcers

No teacher wants to make students dependent upon primary reinforcers for working or behaving appropriately. Primary reinforcers, even for very young or severely handicapped students, are a temporary measure to enable rapid acquisition of appropriate behavior. The teacher cannot send a student back into a regular classroom expecting a Tootsie Roll each time he can identify the word *dog* or to work expecting a chocolate cake or side of beef at the end of the work week. Primary reinforcers, if used, should eventually be replaced by secondary reinforcers. **Secondary reinforcers** include social stimuli, such as words of praise or the opportunity to engage in preferred activities; and a symbolic representation, such as a token, exchangeable for another reinforcer. Unlike primary reinforcers, secondary reinforcers do not have biological importance to individuals. Rather, their value has been learned or conditioned. Thus, secondary reinforcers are often called *conditioned reinforcers*. Some students have not learned to value secondary reinforcers and must be taught to do so before secondary reinforcers will be effective.

Pairing.   Students for whom secondary reinforcers have no value often need primary reinforcement in order to acquire appropriate behavior. However, to avoid dependence upon primary reinforcers, their use should always be in conjunction with some form of secondary reinforcer. The combined use of primary and secondary reinforcers is known as **pairing**. (The primary reinforcer is paired with a secondary reinforcer.) For example, when Jake behaves appropriately, his teacher may give him a bite of food and simultaneously tell him what a good job he has done. Through pairing we condition or teach the student to be motivated by the secondary reinforcer alone. Once this association has been established, the secondary reinforcer may be as effective as the primary reinforcer. The teacher may then gradually withdraw the primary reinforcer. Some students, of course, have a reinforcement history including paired association, allowing for the use of secondary reinforcers without the need for primary reinforcers.

Pairing teaches students to value secondary reinforcers.

Social reinforcers.   A category of secondary reinforcers that teachers use almost unconsciously includes demonstrations of approval or attention. There are a wide variety of forms of verbal or physical interactions associated with a job well done. As seen in the list that follows, the range of potential social reinforcers includes various nonverbal expressions, teacher proximity to the student, physical contact between teacher and student, the granting of privileges that carry status for the student among peers, and words and phrases that convey pleasure and approval of the student's performance. Social reinforcers have proved to be effective not only for teachers in changing and maintaining student behavior (for example, Nutter & Reid, 1978), but also for students in changing and maintaining teacher behavior (see Polirstok & Greer, 1977).

Expressions

smiling, winking, laughing, nodding, clapping

Contact

hugging, touching, shaking hands, holding hands, patting head, back

Proximity

sitting next to the student at lunch, sitting next to the student on bus trips, placing the student's desk next to the teacher's, sitting on teacher's lap during story time, being teacher's partner in a game

Privileges

having good work displayed, being leader of an activity, being classroom monitor, being team captain

Words and Phrases

"I like the way you are sitting." "That is excellent work." "You should be proud of what you have done." "That is just what I wanted you to do." "You should show this to your parents."

Of these social reinforcers, words and phrases are the most often deliberately used by teachers. Words and phrases used by teachers may be affirmatively described as forms of teacher *praise*. O'Leary and O'Leary (1977) have stated that teacher praise must have certain qualities to function effectively as reinforcement:

1. Praise must be delivered contingent on performance of the behavior to be reinforced. Noncontingent delivery of praise violates one of the essentials of the operational definition of reinforcement. Noncontingent delivery of praise removes the dependent relationship between student performance and teacher's affirmative attention and therefore does not increase the future probability of the behavior.

2. Teacher praise should specify the behavior or particulars of the behavior being reinforced. There should be no confusion on the part of the student as to why he is receiving the teacher's affirmative attention.

3. The praise should sound sincere. This means that the teacher should avoid the use of stock phrases. The praise statement should vary in both content and tone according to the situation and the preference of the student being praised. Just as a student can satiate on the unrelenting delivery of primary reinforcers, students can satiate on the unrelenting delivery of certain phrases. Such stock phrases soon lose their reinforcing quality, deteriorate into teacher patter, and are soon ignored and/or resented by the student.

Clinton and Boyce (1975) differentiate the words and phrases used by teachers into two types. They refer to *affirmative reinforcers*, which are characterized by words such as *good* and *fine,* and *informative reinforcers*, which are characterized by words such as *right* and *correct*. Informative reinforcers would commonly be referred to as *constructive feedback*. When feedback is used as a reinforcer, it is almost always prefaced with an affirmative, followed by a specific praise statement that focuses upon the ac-

curacy or appropriateness of the behavior, as in "Good girl! You have hit
the peg right down with the hammer" (Parsonson & Baer, 1978). Verbal
feedback may also be used to reinforce an attempt or approximation (Hall,
Sheldon-Wildgen, & Sherman, 1980): for example, "Good try, you got two
out of three right. Now do this one the same way." This form of feedback
has been used in such diverse situations as a teaching strategy for multihandicapped students (Reid & Hurlbut, 1977), a coaching strategy in football, gymnastics, and tennis (Allison & Ayllon, 1980), and a supervisory
strategy in sheltered workshops (Martin, Pallotta-Cornick, Johnstone, &
Goyos, 1980). Table 6–2 offers examples of constructive verbal feedback.
Some forms of feedback provide the student with a more tangible measure
of accuracy, as in Cronin and Cuvo's (1979) use of various-colored stars
(red stars equal a performance better than last time; gold stars equal 100%
accuracy). Graphing has also been used as feedback to measure and reinforce production in a workshop (Jens & Shores, 1969). It must be remembered however, that as with any consequent stimulus, feedback may not be
a reinforcer for all students in all situations. Depending on a student's
prior experience with feedback and the delivery characteristics previously
listed, feedback may serve as a reinforcer, a punisher, or as a powerless antecedent that does not alter inappropriate behavior or result in the continuation of desired behavior (Peterson, 1982).

Activity reinforcers.   An activity is the secondary reinforcement perhaps
most often used by teachers. The systematic use of such activity reinforcers
was described by Premack (1959) and is referred to as the *Premack Principle.* The Premack Principle states that individuals engage in certain behaviors at low frequencies, so these behaviors therefore have a low probability of occurrence. Other behaviors are engaged in at high frequencies, and
therefore have a high probability of occurrence. When low frequency behaviors are followed by high frequency behaviors, the effect is to increase
the probability of the low-frequency behavior. In other words, any activity
that a student voluntarily performs frequently may be used as a reinforcer
for any activity that he seldom performs voluntarily. When a teacher tells a
student that she may work on her airplane model in the back of the room
when she has finished the math assignment, or when a mother tells her
child he may play outside when he has finished eating his brussel sprouts,

TABLE 6–2
Constructive verbal
feedback

Affirmative	Feedback
"Great!	(Description of correct response)
	You finished your work on time."
"Good try!	(Reinforcement of approximation)
	You almost got finished this time."
"Much better!	(Suggestion for modification)
	If you keep trying not to make careless mistakes, you'll finish all of them next time."

they are using the Premack Principle. (Table 6–3 offers suggestions for secondary reinforcers.)

Kazdin (1975) suggests some limitations to the use of activity reinforcers. First, access to some high-preference activities cannot always follow the low-preference behavior immediately, thereby reducing the effectiveness of the high preference behavior as a reinforcer. For example, scheduling problems might prevent students' using the gym right after doing their math. Second, an activity may often be an all-or-none enterprise. It is either earned or not earned. This may limit flexibility in administration of the reinforcer. For example, a student either earns the right to go on a field trip or make a Halloween mask or he doesn't earn it. Such activities cannot be proportionally awarded depending on the degree of acceptable performance. However, this limitation is not invariably true with activity reinforcers. Some activities may be earned in increments of time: for example, the consequence of 1 minute of shooting baskets in the gym for each correctly spelled word is easily administered.

A third limitation on activities as reinforcers is that many activities must be freely available to students without reference to their performance. Examples include lunch periods, physical education requirements, or art and music classes. Finally, the use of an activity reinforcer may cause an interruption in the continuous performance of the target behavior. For example, a teacher would not want to allow a student to go to the gym and

TABLE 6–3
Examples of privileges and activities to be used as secondary reinforcers

Manage the class store for a week or two
Decorate a bulletin board
Select and plan next field trip or holiday party
Lead a class activity (*making popcorn, singing songs*)
Put on a skit or direct next class play
Use of media equipment (*videotape machine, tape recorder*)
Set schedule of day's activities and lessons
Take part in a tutoring program
Conduct a class lesson on a topic of your choice
Driver education study
No homework on night of your choice
Being excused from a test
Membership in school patrol or special committees
Monitorships (*chalkboard, messages, pets, plants, paper, playground equipment, flag, milk, class roll*)
Class representative, president
Captain of sport team or reading group
Access to gymnasium, library, games, toys, or other privileges
Citations and badges
Time to use paints and easel, craft center
Earn extra recess or free time
Free time for self-selected project
Operating equipment (*slide, filmstrip, or movie projector, ditto machine*)

shoot a basket after each correctly spelled word. However, some students may not continue to perform the target behavior unless some form of reinforcement is available after each response. In cases where the effectiveness of activity reinforcers seems lessened by such factors, the use of generalized conditioned reinforcers may be considered.

Generalized conditioned reinforcers.   When a reinforcer has been associated with a variety of behaviors or with access to a variety of other primary or secondary reinforcers, it may be termed a **generalized conditioned reinforcer** or simply a *generalized reinforcer*. For example, a smile or words of praise are reinforcing following a variety of behaviors: praise from the boss after a particularly difficult work assignment or from a spouse for a delicious dinner, a smile from a teacher following a clever verbal response in class, or a hug from mother for picking up dirty clothes.

A second type of generalized reinforcer includes those that are exchangeable for something of value. Money is the most obvious example of this kind of reinforcer. Money, which has little or no intrinsic value, is associated with access to many types of reinforcers: food, shelter, or clothing, admission to a concert, a ticket to the Superbowl, or a Mercedes.

The use of generalized reinforcers has a number of advantages. Kazdin and Bootzin (1972) suggest that

1. As opposed to certain edible or activity reinforcers, generalized reinforcers permit the reinforcement of a response at any time and allow sequences of responses to be reinforced without interruption.

2. Generalized reinforcers may be used to maintain performance over extended periods of time and are less subject to satiation effects due to their reinforcing properties and their relative independence of deprivation states.

3. Generalized reinforcers provide the same reinforcement for individuals who have different preferences.

Token reinforcers.   Because the use of money is unrealistic in most school settings, a generalized reinforcer known as a *token reinforcer* has become widely used. Token reinforcers are simply symbolic representations exchangeable for some reinforcer of value to students. Originally, Ayllon and Azrin (1968) described procedures for using a token reinforcement system on a ward in a mental hospital. While token systems are still used in hospital settings (Magrab & Papadopoulou, 1977; Nelson & Cone, 1979), variations have been used in residential settings for disturbed and delinquent adolescents (Cohen & Filipczak, 1971; Wood & Flynn, 1978), sheltered workshops (Welch & Gist, 1974) and high school and college classes (Keller & Sherman, 1974; Kulik, Kulik, & Carmichael, 1974). Today token systems are used in most special education and many regular education classes. They are used by teachers, aides, and parents in teaching specific skills (Heward & Eachus, 1979), in tutoring situations (Schwartz, 1977), in managing the integration of handicapped students into regular classes (Russo & Koegel, 1977), and in coordinating behavior-change programs between school and home (Trovato & Bucher, 1980).

Use of tokens is analogous to the use of money in the general society. Token reinforcers are exchangeable for a wide variety of primary and other secondary reinforcers. They are used as a transition between primary reinforcers and the natural community of secondary reinforcers. A token system may be adapted for use with a single student and a single behavior, one student and several behaviors, groups of students and a single behavior, and groups of students and several of the same or different behaviors.

A system of token reinforcement requires two components: the token itself and a back-up reinforcer. The token itself is delivered immediately upon the desired response. This token can be an object, such as a poker chip, button, star, play money, paperclip, or metal washer. It can also be a symbol, such as check marks, a hole punched in a card, points, or the ubiquitous happy face. In general, tokens should be portable, durable, and easy to handle.

Tokens won't work unless they can be exchanged for something.

Both the teacher and the student should keep an accurate record of the number of tokens earned. When the token is in the form of objects such as chips, a token box or some other receptacle can be designated for storing the tokens in an assigned location or at the student's desk. With low-functioning or younger students, chaining necklaces with tokens or building towers will assist in control of token loss. A dot-to-dot representation of the back-up reinforcer may be drawn. In this system, as each response occurs, two dots are connected. When all dots are connected, the picture is complete and the student has earned the back-up reinforcer (Trant, 1977). When the token is in the form of points earned, stamps, or check marks, a chart in the front of the classroom or some recording card similar to those displayed in Figure 6–2 may be used.

Use of token systems requires precautions against counterfeiting or theft. Any student with 30 cents can buy 100 paper clips, thus debasing the value of paper clip tokens and thereby the effectiveness of the system. A simple preventive measure is to mark tokens, objects, or symbols with some kind of code that allows validation of their source or confirmation of their ownership by a particular student. If check marks on a card are used, the teacher can randomly use different colored markers on different days. The chances of a student's having a puce-colored marker in school on the day the teacher chooses that color are minimal.

Birnbrauer, Wolf, Kidder, and Tague (1965) point out that tokens in and of themselves are unlikely to have reinforcing power. They attain their reinforcing value by being exchangeable for items that are reinforcing. Therefore, it must be clearly understood by the students that they are working for these tokens in order, at some point, to exchange them for the second component of the token system, the back-up reinforcer.

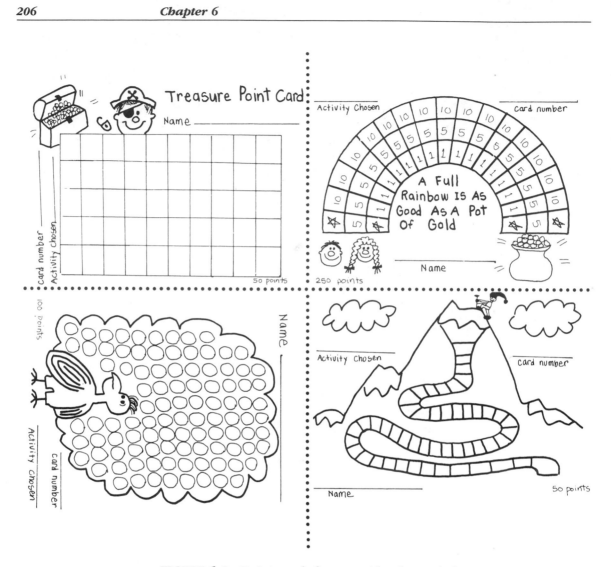

FIGURE 6–2   Point cards for use with token reinforcement systems

Note. From It's positively fun: Techniques for managing learning environments by P. Kaplan, J. Kohfeldt, & K. Sturla (Denver: Love Publishing, 1974). Copyright 1974 by Love Publishing. Reprinted by permission.

The selection of the back-up reinforcer is probably the most difficult aspect of the token system, especially if the system is being used with a group of students or an entire class. The teacher must select a wide enough variety of back-up reinforcers to provide a motivating item for each member of the class. Therefore, teachers should seek to include an assortment, such as edibles (cereals, crackers, cookies, juice), activities (going to the library, earning project time, listening to record player), objects (earning a game, notebook, or crayons), and privileges (being first in line, being collector of lunch money).

What students need to know about a token system.

When a teacher announces the initiation of a token system to a class or to an individual, students will want to know at least four things immediately. First, they will want to know what behaviors are required. As always, the contingency (if ..., then ...) should be clearly stated by the teacher and understood by the students. The descriptions of each behavior to be performed and the parameters of acceptability should be clearly stated or posted.

Second, the students will want to know what back-up reinforcers their tokens will buy. It is a good idea to keep representations of the back-up reinforcers, if not the items themselves, in full view in the classroom.

The third question might involve the cost of each back-up reinforcer in tokens. Based on their evaluation of the expense and desirability of the back-ups, the students will decide whether the reinforcer is worth the required behavior change. To get the process working initially, the teacher may price reinforcers to allow everyone quickly to acquire some. Students should learn at this very first exchange that if they earn a certain number of tokens, they may trade them in for a certain back-up reinforcer. Stainback, Payne, Stainback, and Payne (1973) suggest that prices of items, such as edibles, posters, and toy soldiers, should be in proportion to their actual monetary value. Pricing of activities and privileges is difficult to judge. Students should not earn back-ups too quickly nor should they be required to work inordinately hard. It will take a while for the teacher to judge the value of back-up reinforcers based upon their perceived desirability among class members.

Finally, the students will want to know when they can exchange the tokens for the back-up reinforcers. It is possible to allow exchange as soon as a student has enough tokens for a particular item. It is most common for a single exchange period to take place at the end of each day or week. In the early stages of a token system, especially with young or handicapped students, the period before the first exchange should be very short. It would be unwise to start a token system on Monday and schedule the first exchange period on Friday. Students need to see quickly how the exchange process works and that the teacher is indeed telling the truth. Therefore, we suggest that the first exchange period be either at lunch, at the end of school, or even at the morning break (using a cookie exchange, for example). Stainback et al. (1973) suggest that tokens be paid frequently in the early stages and that exchange times be held once or twice a day for the first 3 to 4 days and then gradually decreased in frequency until they are held only once a week by the third week.

Exchanging tokens for back-up reinforcers may take place in a variety of formats. The most common is the classroom store (Nelson & Cone, 1979; Stainback et al., 1973). In this format, the price-labeled back-up reinforcers are located on shelves in a corner of the classroom. During the designated exchange period, students may enter the store area and purchase any item they can afford. An interesting variation on the exchange is the class auction (Polloway & Polloway, 1979). In this format, the students are allowed to bid for each of the back-up reinforcers. Students may bid as high as they choose, up to the number of tokens they have earned.

One potential impediment to an effective token system occasionally arises as a result of exchange procedures. It is illustrated by the following case.

The Case of Charlie the Miser

Charlie was a student in Mr. Thomas' class for children with learning disabilities. He was a very intelligent boy with severe reading problems and many inappropriate behaviors. He, like Mr. Thomas' other students, responded well to a token reinforcement system. The students earned check marks on a card that was exchangeable for a variety of back-up reinforcers, including toys and privileges. The most expensive item cost three cards and many were available for one or two cards.

After several months Mr. Thomas noticed that both Charlie's behavior and his academic work had deteriorated drastically. This deterioration seemed to have happened overnight. Mr. Thomas could see no reason for it so he decided to do a very sensible thing. He asked Charlie what had happened. Charlie grinned and opened his work folder. "Look here," he chortled. "I got 11 cards saved up. I don't have to do anything for weeks, and I can still get anything I want from the store."

It is wise, as Mr. Thomas learned, to limit the number of tokens that a student can accumulate. If some students are saving for major events such as a class trip, or large items over a long period of time, tokens should be banked—with substantial penalty for early withdrawal.

See pages 54–57 for Professor Grundy's advice.

It is feasible to use token systems with an entire class or only with selected members of a class. If only some children earn tokens, others may question this. Professor Grundy's advice in the Chapter 2 vignette should help the teacher deal with this problem.

When using a system with an entire class, it is easiest to begin with some behavior that is a target for change for the entire class. For example, the teacher might initially give points for academic assignments finished or hands raised during class discussions. Once students have become familiar with the exchange system, the program can be individualized. While still distributing tokens for the initially targeted behavior, the teacher can integrate many types of academic tasks and social behaviors into the system. For example, Marty might earn tokens for neatness, Debbie for increased speed, and Sara for speaking loudly enough to be heard. Or, instead of this individual focus, the teacher might choose to expand the initial system for classroom behavior management to include additional behaviors that are appropriate for all class members. The following is an example of a teacher's criteria for classroom points (Schumaker, Hovell, & Sherman, 1977, p. 453).

When discussion held

4 pts. Student listens and contributes 3 times to discussion.

3 pts. Student listens and contributes twice to discussion.

2 pts. Student listens and contributes once to discussion.

1 pt. Student pays attention and listens to discussion.

0 pt. Student does not listen to discussion.

When in-class assignment given

4 pts. Student works all of class time on assignment.

3 pts. Student works ¾ of class time on assignment.

2 pts. Student works ½ of class time on assignment.

1 pt. Student starts work on assignment.

0 pt. Student does no work on assignment.

In classes where there is no opportunity for participation (reading on own, movie, lecture)

4 pts. Student is extremely attentive to subject (movie, teacher, guest lecturer) throughout class.

2 pts. Student is generally attentive to subject.

0 pts. Student does not attend to subject.

A token system can also be used for teaching complex academic tasks. For instance, if a teacher is trying to teach a class of students to write appropriate paragraphs, instead of awarding 20 points for writing a paper on "How I Spent My Summer Vacation," the teacher might use the token system by awarding points for the following behaviors:

1 point for bringing pen and paper to class

1 point for beginning the writing assignment on time

1 point for completing the writing assignment on time

1 point for each sentence that begins with a capital letter

1 point for each sentence that ends with a period

Once the students have begun to master these items, the teacher can then replace them with a point system for a more complex writing task. For example, by the fourth or fifth writing lesson, instead of awarding points for bringing paper and pen, the teacher might begin to award points for sentences using the appropriate plural suffix. Providing many alternative ways to earn a few points can be important in a token system. Some students find working toward several relatively simple objectives less frustrating and easier to undertake than attempting to earn many points for a long assignment that they may feel inadequate to complete. This approach also builds some measure of assured success into the assignment.

Token reinforcers can be very useful in management of a classroom. Ayllon and Azrin (1968) suggest that there are advantages to using tangible reinforcers (tokens) that may make them more effective than generalized social reinforcers (smiles and praise).

1. The number of tokens can bear a simple quantitative relation to the amount of reinforcement.

2. The tokens are portable and can be in the student's possession even when he or she is in a situation far removed from the classroom.

3. No maximum exists on the number of tokens a subject may possess.

4. Tokens can be used directly to operate devices for the automatic delivery of reinforcers.

5. Tokens are durable and can be continuously present during the period prior to exchange.

6. The physical characteristics of the tokens can be easily standardized.

7. The tokens can be made fairly indestructible so they will not deteriorate during the period prior to exchange.

8. The tokens can be made unique and nonduplicable so that the experimenter can be assured that they are received only in the authorized manner. (p. 77).

9. The use of tokens provides the student with a tangible means of continuous feedback. By having custody of the token objects or point card, the student can follow personal progress toward the criterion set in the contingency—whether it is for a behavior to be brought under control or for the acquisition of an academic objective.

10. The use of tokens enables more precise control by the teacher over administration of reinforcers. As noted by Kazdin (1977), a teacher's vocal tone will differ each time she or he says "Good working," as will the exact phrasing of "good," "pretty good," "very good," while in each instance meaning to convey an equal praise statement. Token reinforcement does not suffer from this subjective influence.

11. Tokens can be carried by the teacher and delivered unobtrusively. Thus, the administration of a token can be immediate without interfering with the student's performance of the target response or with other students' work.

12. A system of token reinforcement allows for differential valuing of performance. It does not require the delivery of reinforcement on an all-or-nothing basis. The student may initially be given a token for each correctly spelled word and later earn reinforcement for a 20-out-of-20 performance. The criterion for performance can be changed as performance improves.

13. A system of token reinforcement allows the student to become accustomed to delayed gratification of wants.

14. The use of a token system allows for greater versatility than is possible with other reinforcement systems. This versatility is related both to the wide variety of back-up reinforcers that may be selected and to the variety of behaviors that may be placed under a contingency for earning tokens.

15. The most important advantage a token system provides is its ease of generalizability. Unlike primary reinforcers or certain activity reinforcers, tokens can easily be used across settings (in other classrooms, in the cafeteria, on field trips) and with different behaviors simultaneously (in-seat behavior *and* correct spelling). They can also easily be administered by more than one teacher as well as parents.

16. Tokens can often maintain behavior at a higher level than other secondary reinforcers such as praise, approval, and feedback (Birnbrauer et al., 1965; Kazdin & Polster, 1973; O'Leary, Becker, Evans, & Saudargas, 1969).

Contracting

It can be difficult for a teacher to use a reinforcement system serving a number of students and having a variety of objectives for managing behavior and instruction. On a hectic day, a teacher might off-handedly state a contingency for a student without thinking it through. Later the teacher may not remember the details of what was said and thus not be in a position to enforce the contingency. To complicate this uncertainty, students tend to rewrite reality to suit themselves. "You said I could go outside if I finished my math. You didn't *say* I had to get them right." A simple way to systematize the use of reinforcement is contracting. **Contracting** is placing the contingency for reinforcement into a written document. In the contract, the teacher creates a permanent product that can be referred to if questions arise.

As with any contract, the classroom contract should be the product of reasonable negotiations between the parties involved—namely, the teacher and the student. Though the exact wording of a written contract will depend upon the sophistication of the student for whom it is designed, each will contain some form of the basic "If . . ., then . . ." statement, as seen in Figure 6–3. A written contract should always contain those elements minimally necessary for any reinforcement contingency: the behavior, the conditions, the criterion, the reinforcer.

The contract should contain precisely written statements describing the behavior required to avoid later disagreement about what was really meant. This description should include the parameters within which the behavior is to be performed and the criterion level for meeting the terms of the contract. After discussion of the criterion, the student should understand the method or instruments that will be used to evaluate performance. The contract should also include the type, amount, and method of delivery of the reinforcement.

In addition to these basic items, dates for an interim and final review should appear in the contract. The interim date reminds the teacher of the need to monitor progress and allows renegotiation if the behavior required is unrealistic or if there is an instructional component to be added. Listing the final review date sets the student's time limit for fulfilling the terms of the contract.

FIGURE 6–3 *Formats used in contracting*

Note. From *It's positively fun: Techniques for managing learning environments* by P. Kaplan, J. Kohfeldt, & K. Sturla (Denver: Love Publishing, 1974). Copyright 1974 by Love Publishing. Reprinted by permission.

Once the terms of the contract have been discussed and written down, the teacher should answer all questions the student may have. To ensure that the student understands the terms of the contract, he or she should read it back to the teacher and then restate the terms in different words. If this process results in a very different statement, the contract should be rewritten in easier language. Once a contract is finalized, the teacher and the student should both sign it, and each should have a copy.

Homme (1970, pp. 18–20) suggests basic rules for the use of reinforcers in contracting (numbers 1–5) as well as characteristics of proper contracting (6–10):

See Chapter 8 for more on reinforcement of small approximations.

1. "The contract payoff (reward) should be immediate." This rule follows what has been stated as one of the essential elements of an effective reinforcer: it must be administered immediately upon performance of the target behavior.

2. "Initial contracts should call for and reward small approximations." This form of *successive approximations*—that is, progressive steps toward the target behavior—is particularly useful for behaviors the student has never performed before, a criterion level set too high, or a behavior category which is too broad (such as "clean your room").

3. "Reward frequently with small amounts." Homme states that experience has shown that "it is far more effective to give frequent, small reinforcements than a few large ones." Frequent delivery of reinforcement allows for closer monitoring of the behavior change in progress by both the teacher and the student.

4. "The contract should call for and reward accomplishment rather than obedience." It is suggested by Homme that contracts which focus upon accomplishments lead to independence. Therefore, appropriate wording should be "If you accomplish such and such, you will be rewarded with such and such," as opposed to "If you do what I tell you to do, I will reward you with such and such."

5. "Reward the performance after it occurs." This rule restates an essential element of a reinforcer: it must be administered contingently. Inexperienced teachers sometimes state contingencies such as "If you get to go on the field trip today, you must do all your work next week." They are usually disappointed with the effects of such statements.

6. "The contract must be fair." The "weight" of the reinforcement should be in proportion to the amount of behavior required. The ratio set up in the contract should be fair to both the teacher and the student. Asking the student to finish 2 out of 20 problems correctly for 30 minutes free time is just as unfair as requiring 20 out of 20 problems correct for 2 minutes free time.

7. "The terms of the contract must be clear." Ambiguity causes disagreement. If the teacher and the student do not agree on the meaning of the contract, the teacher may decide that contracting is more trouble than it is worth. The student may decide that neither the teacher nor the system can be trusted.

8. "The contract must be honest." According to Homme, an honest contract is one which is (a.) "carried out immediately, and (b.) carried out according to the

terms specified in the contract." This can be assured if both the teacher and the student have freely engaged in the contract negotiation. Teachers should avoid imposing a "contract" on the student.

  9. "The contract must be positive."

appropriate: "I will do _____, if you do _____."

inappropriate: "I will not do _____, if you do _____."

                  "If you do not _____, then I will _____."

                  "If you do not _____, then I will not _____."

10. "Contracting as a method must be used systematically." As with any form of reinforcement strategy, if contracting is not done systematically and consistently, it becomes a guessing game: "Does she really mean it this time?"

Writing contracts brings added advantages to a reinforcement system:

1. A written contract is a permanent document that records the variables of the original contingency for consultation by both teacher and student.

2. The process of negotiation that leads to the contract enables students to see themselves as active participants in their learning as they each take part in settting their own expectations or limitations.

3. The writing of a contract emphasizes the individualization of instruction.

4. Contracting provides interim documents that state current objectives between IEP meetings. Such information may be shared with parents.

TABLE 6–4
Categories and examples of reinforcers available for classroom use

Class	Category	Examples
Primary Reinforcers	1. Edible reinforcers	Foods, liquids, e.g., pieces of cracker, sips of juice, pudding.
	2. Sensory reinforcers	Exposure to controlled visual, auditory, tactile, olfactory, or kinesthetic experience, e.g., face stroked with furry puppet, taped music through head-phones.
Secondary Reinforcers	3. Tangible (material) reinforcers	E.g., certificates, badges, stickers, rock star posters, balloons
	4. a) Privilege reinforcers	E.g., monitorships, team captain, excused from homework
	b) Activity reinforcers	E.g., play activities, special projects, access to media, extra math problems to do (yes there is such a student!)
	5. Generalized reinforcers	E.g., tokens, points, credits
	6. Social reinforcers	E.g., expressions, proximity, contact, words and phrases, feedback, seating arrangements.

Suggested categories and examples of potential reinforcers available for classroom use are summarized in Table 6–4. The teacher is cautioned that it would be erroneous to interpret such a table as a scheme for ordinal selection of reinforcement categories on a contrived to natural continuum. One might be tempted to identify, for example, the use of edible reinforcers as contrived or artificial in the classroom. However, such a designation depends on the target behavior, the setting, and the age of the student. Any category or specific stimulus could be described as a contrived or natural instance of reinforcer use. Food as reinforcer during feeding therapy or access to the water fountain as a reinforcer for properly lining up after a PE period may be considered natural consequences of the target behaviors. It may be helpful to distinguish between items or events ordinarily available in a given environment (i.e., natural) and those temporarily added to the environment for increased consequence intensity (i.e., contrived).

Variations in Administration of Reinforcers

A basic reinforcement system has the following design:

The teacher presents the antecedent discriminative stimulus.

The student performs the requested response.

The teacher presents the student with an appropriate reinforcer.

This basic scheme focuses upon the administration of an individually selected reinforcer designed for the particular student. Reinforcement is, however, a flexible strategy that can be adapted to a number of situations that arise in the management of a classroom. Based on the type of contingency and manner of administering consequences, Kazdin (1977) devised a matrix to represent these variations, shown in Figure 6–4. Although the matrix was originally proposed for variation in use of token systems, it is equally appropriate with any type of reinforcement system.

Type of Contingencies

	Individualized	Standardized	Group
Individual	1	2	3
Group	4	5	6

Manner of Administering Consequences

FIGURE 6–4   *Variations in administration of consequences*

Note. From *The token economy: A review and evaluation* by A.E. Kazdin (New York: Plenum Press, 1977). Reprinted by permission.

As Figure 6–4 shows, there are two options for administering consequences. First, reinforcement can be administered individually: the particular student who has performed the requested response is given the cereal,
free time, or the appropriate number of tokens. Second, reinforcement can
be administered to students as a group: given an acceptable performance
by the whole class, for example, all 30 students will earn extra time for
crafts on Wednesday afternoon.

There are three options for the type of contingency a teacher may establish for receipt of reinforcement. These are represented in Figure 6–4
along the top of the matrix. For the first one, *individualized contingencies,*
the behavior requested and the criterion of performance required are specific to the behavior or instructional needs of a particular student. Under
the second option *standardized contingencies,* the teacher sets a requirement that is applied equally to all members of a class or to several class
members. Using the third alternative, *group contingencies,* some behavior
is required of a group of students and reinforcement is based upon performance of the group as a whole.

Interaction among the two manners of administration and the three
types of contingencies yields the six-celled matrix in Figure 6–4.

Cell 1 shows a system in which both the contingency for reinforcement
and the manner of its delivery are individualized. The behavior and
criterion are specific to a particular student, and the reinforcement is
delivered only to that student.

a. Randy, if you complete 17 or your 20 arithmetic problems correctly,
then you may have 10 minutes extra use of the computer terminal.

b. Randy, for each arithmetic problem you solve correctly, you will
receive one token.

Cell 2 shows a system that provides the same contingency for
reinforcement for all class members (standard) but individualizes the
manner of delivery to each student.

a. Class, each of you who completes 17 of your 20 math problems
correctly may have an extra 10 minutes use of the computer terminal.

b. Class, each of you may earn one token each time you raise your hand
before asking a question.

Cell 3 shows a system that sets the same contingency for reinforcement for
a particular group of students, but individualizes the manner of reinforcer
delivery to each group member.

a. Math Group B, if you students can develop 10 original problems that
require multiplication for their solution, you all may have 10 extra minutes
at the computer terminal when you choose, to use whichever program you
wish.

b. Each boy in the class may earn a token for placing his tray on the cart
when he has finished eating his lunch.

Cell 4 shows a system that requires each member of a group to perform a specific behavior in order for the group as a unit to be reinforced.

a. Math Group B, you are to present a 15-minute program to the class concerning the use of multiplication. Randy, you are responsible for an explanation of the basic computation procedure; Carol, you are to explain the relationship between multiplication and addition; Nicholas, you are to demonstrate how to work problems from our math workbooks; and Sandy, you are to pose three original problems for the class to solve. At the end of the presentation, the four of you may use the computer terminal together for one of the game programs.

b. The following students may go to the gym to play basketball if they write an essay on basketball: Gary, your essay must have at least four sentences; Jamie, your essay must have at least six sentences, and Cory, your essay must have at least ten sentences.

Cell 5 shows a system that sets the same contingency for reinforcement for all class members and allows each individual who has met this standard criterion to become a member of a group to be jointly reinforced.

a. Class, for homework you are to develop a problem that requires multiplication for its solution. All students who have brought in an appropriate problem will be allowed to go to the math lab tomorrow morning from 10:00 to 10:30.

b. All students who receive a grade of 100 on the geography test today will be exempt from geography homework tonight.

Cell 6 shows a system that sets the same contingency for reinforcement and same manner of reinforcer delivery for a group of class members.

a. Math Group B, here are 20 problems. If you come up with the correct answer for all 20, you may go to the math lab for 30 minutes.

b. Redbirds, if you all remember to raise your hand before speaking during our reading lesson, you may select your own books to take home this weekend.

Following a review of studies in which contingencies were applied to more than one individual at a time, Litow and Pumroy (1975) delineated three administrative systems. They categorized these as dependent, independent, and interdependent group-oriented contingency systems.

In *dependent group-oriented contingency systems,* "the same response contingencies are simultaneously in effect for all group members, but are applied only to the performances of one or more selected group members. It is the performance of the selected group members that results in consequences for the whole group" (p. 342). This contingency system is illustrated by Litow and Pumroy (1975) as the teacher who makes reinforcement for the entire class contingent upon the performance of a particular student(s). The remaining class members are *dependent* on the targeted student's performance for the reinforcement.

a. The opportunity for the class to have an extra session of physical education depends on Robert and Caroline passing Friday's spelling test.

b. The opportunity for the class to have an extra session of physical education depends on Samuel and Lorelei completing 90% of their written assignments.

c. The opportunity for the class to have an extra session of physical education depends on William and Bernice having called-out without raising their hands no more than seven times.

In *independent group-oriented contingency systems,* "the same response contingencies are simultaneously in effect for all group members, but are applied to performances on an individual basis. . . . In this type of contingency system, each member's outcomes are not affected (are *independent* of) the performances of the other group members" (p. 342). This contingency system is illustrated by the teacher who makes reinforcement for each class member contingent upon that class member being able to meet the contingency criterion level of performance. Those who fail to achieve the performance criterion will not receive the reinforcer.

a. The opportunity for an extra session of physical education is available to each student who passes Friday's spelling test.

b. The opportunity for an extra session of physical education is available to all students who complete 90% of their written assignments.

c. The opportunity for an extra session of physical education is available to each student having no more than three call-outs.

In *interdependent group-oriented contingency systems,* "the same response contingencies are simultaneously in effect for all group members, but are applied to a level of group performance. . . . Consequently in this type of contingency system each member's outcome depends (are *interdependent*) upon a level of group performance" (p. 343).

Litow and Pumroy (1975) list three types of group performance levels:

1. The contingency is stated so that each group member must achieve a set criterion level. Failure to achieve this criterion level by the class results in no class member receiving the reinforcer. For example, the opportunity for an extra session of physical education is contingent upon each member of the class earning at least 90% on Friday's spelling test.

2. The contingency is stated such that each group member's performance meets a criterion *average* for the entire group. For example, the opportunity for an extra session of physical education is contingent upon a class average of 90% of written work completed.

3. The contingency is stated so that as a group the class must reach a single highest or lowest level of performance. For example, the opportunity for an extra session of physical education is contingent upon the class as a whole having engaged in no more than 12 call-outs.

Always use the simplest effective system.

By using these variations of the reinforcer delivery system, the teacher can tailor a reinforcement system to a particular classroom. Every classroom is different: some groups, even in regular classrooms, need a formal token system; other groups may have members who need contracting or individual systems; many regular classrooms can be managed using relatively informal arrangements of social and activity reinforcers. In general, teachers should use the simplest, most natural system that will be effective.

Group Contingencies and Peer Pressure

Group contingencies, as described above, can be an extremely effective means of managing some students' behavior. Adolescents, particularly, may find working as a group to be more reinforcing than individual contingencies. Peer pressure is a powerful tool in group contingencies. Indeed, it is so powerful that group contingencies should be used with caution to avoid the negative side effect of undue pressure on some members of a group (Balsam & Bondy, 1983). Consider the following example.

Ms. Montgomery Teaches Spelling

Ms. Montgomery, a fifth-grade teacher, was concerned about her students' grades on their weekly spelling test. Some students did very well, but others spelled only a few words correctly. Ms. Montgomery thought for some time and came up with what she felt was a brilliant plan. She divided the students into pairs—one good speller and one poor. She then announced, "The grade that I enter in my grade book on Friday will be an average of what you and your partner make on the test." She sat back and watched the students busily drilling one another and assumed that her troubles were over. (The astute reader will have noticed by now that teachers who assume their troubles are over usually find that they are just beginning.)

She first began to see a flaw in her plan when she observed LeeAnn chasing Barney around the playground during recess, slapping him with her speller and yelling, "Sit down, dummy, you've got to learn these words."

She knew she had blown it when she got a phone call at home that night from LeeAnn's mother protesting that LeeAnn was going to fail spelling because that dumb Barney couldn't learn the words, and one from Barney's mother asking if Ms. Montgomery had any idea what might have caused Barney to spend the entire afternoon crying in his room.

Teachers sometimes think that these procedures work only on American kids. Saigh & Umar (1983) report that this also works for Sudanese students.

Ms. Montgomery violated one of the most important rules about setting group contingencies: *Be absolutely sure that each member of the group is capable of performing the target behavior.* If this rule is violated, the teacher risks subjecting students to verbal and physical abuse by their peers.

Another important caution is being sure that some member does not find it reinforcing to sabotage the group's efforts. Barrish, Saunders, and Wolf (1969) arranged a group contingency to modify disruptive out-of-seat and talking-out behavior in a class of 24 fourth-grade students. The class was divided into two teams during reading and math periods. Each instance of out-of-seat or talking-out by an individual team member resulted in a mark against the entire team. The team with the greatest number of marks would lose certain privileges. While this procedure was successful, an important modification was implemented. Two members of one of the teams consistently gained marks for their team. During one session, one of the members "emphatically announced" that he was no longer going to play the game. The teacher and the children felt that the consistent behavior of one student should not further penalize the entire team. The saboteur was removed from the team (and the group contingency) and made into a one-man team, thus applying an individual-consequences procedure until his behavior was brought under control and he could then be returned to one of the class teams. The authors stated that it appeared the expected effects of peer pressure, instead of bringing individual behavior under group control, may have served as a social reinforcer for the student's disruptive behavior.

Finally, the system must minimize the possibility of some members' performing the target behavior for others. If these factors are taken into account, group contingencies can be a very useful management device.

Schedules of Reinforcement

Schedules of reinforcement refer to patterns of timing for delivery of reinforcers. Until now, we have described the delivery of reinforcers for each occurrence of the target behavior. Delivery of reinforcement on a continuous basis is referred to as a **continuous schedule of reinforcement (CRF)**. That is, each time the student produces the target response she or he immediately receives a reinforcer. This schedule may be seen as having a one-to-one ratio—Response:Reinforcement ($R:S^R$).

Because of this dense ratio of response to reinforcement, CRF schedules are most useful in teaching new behaviors (acquisition), especially to young and handicapped students. It is necessary to ensure that a student who is learning a new behavior will receive a reinforcer for each response that is closer to a correct response. The process of reinforcing such successive approximations to the target behavior is called *shaping* and will be discussed in Chapter 8. For now, it is sufficient simply to be aware that a CRF schedule is usually used in shaping procedures. A CRF schedule may also be used when the target behavior initially has a very low frequency. It may be most effective during the early stages of any reinforcement system. There are, however, certain problems in using a CRF schedule.

Problems with CRF schedules.

1. A student whose behavior is on a CRF schedule may become satiated on the reinforcer, especially if a primary reinforcer is being used. Once correct responding is frequent, the continuous receipt of an edible will reduce the deprivation state and thereby reduce motivation for correct responding.

2. Continuous delivery of reinforcers may lead to accusations that teachers are training students to expect some type of reinforcement every time they do as they are told.

3. CRF schedules are not the most efficient way to maintain behavior following its initial acquisition or control. First, once behavior has been acquired, or its frequency increased, by reinforcement on a CRF schedule, teachers may terminate the intervention program. The transfer from continuous reinforcement to no reinforcement results in rapid loss of the behavior. This rapid loss of behavior when reinforcement is withdrawn is called *extinction* and will be discussed in Chapter 7. Second, CRF schedules may interfere with classroom routine. How long could (or would) a teacher continuously reinforce 4, 6, 8, or 30 students for raising their hands before speaking or for making the letter *a* correctly?

The problems caused by continuing the use of CRF schedules beyond the point of effectiveness may be solved through use of a variety of less-than-continuous schedules.

Intermittent Schedules

In the real world, the ability to delay gratification is necessary.

In intermittent schedules, reinforcement follows some, but not all, correct or appropriate responses (Skinner, 1953). Because each occurrence of the behavior is no longer reinforced, intermittent schedules put off satiation effects. Behaviors maintained on intermittent schedules are also more resistant to extinction. Intermittent schedules require greater numbers of correct responses for reinforcement. The student learns to delay gratification and to maintain appropriate behavior over longer periods of time.

The two categories of simple intermittent schedules most often used to increase frequency of response are **ratio schedules** and **interval schedules** (Ferster & Skinner, 1957; Skinner, 1953). To increase the duration of a response, the teacher may use *response duration schedules* (Stevenson & Clayton, 1970). For a discussion of multiple, compound, and concurrent schedules, consult Reynolds (1975).

Ratio schedules. Under ratio schedules, the number of times a target behavior occurs determines the timing of reinforcer delivery. Under a *fixed-ratio schedule (FR)*, the student is reinforced on completion of a specified number of correct responses. A behavior on an FR3 schedule would be reinforced immediately following the occurrence of every third correct response, a contingent ratio of three correct responses to each reinforcer (R, R, R:S^R). A student who must complete eight math problems correctly to earn the right to work the puzzles or who must correctly point to the blue object on eight trials before getting a bite of pretzel, would be on an FR8 schedule of reinforcement.

Behaviors placed on FR schedules have particular characteristics. The student generally has a higher rate of responding than on CRF schedules because increases in rate result in increases in the frequency of reinforcement. Because the time it takes the student to perform the specified number of correct responses is not considered when delivering reinforcers, FR schedules may result in inappropriate fluencies for a given behavior. For example, a student may work math problems so rapidly in order to earn a reinforcer that he makes more mistakes and his handwriting deteriorates. In addition to inappropriate fluency, FR schedules may cause another type of problem. As the schedule ratio increases (from FR2 to FR10, for example) the student will often stop responding for a period of time following delivery of the reinforcer, taking what is termed a *postreinforcement pause*.

The problems of fluency and postreinforcement pause are eliminated by transition to a *variable-ratio schedule (VR)*. Under a VR schedule, the target response is reinforced on the *average* of a specified number of correct responses. A behavior on a VR10 schedule would be reinforced on the average of every tenth correct response. Therefore, in a teaching or observation session, the student may be reinforced following the 2nd, 4th, 8th, 9th, 12th, 16th, and 19th correct responses.

After the occurrence of a behavior on the FR schedule has been established at the criterion level (as stated in the behavioral objective), the VR schedule will maintain the desired level or rate of responding. The unpredictability of reinforcer delivery on a VR schedule causes the student's rate of responding to even out, with little or no postreinforcement pausing. "The probability of reinforcement at any moment remains essentially constant and the 'student' adjusts by holding to a constant rate" (Skinner, 1953, p. 104).

Interval schedules.   Under interval schedules, the occurrence of at least one correct or appropriate response plus the passage of a specific amount of time are the determinants for delivery of the reinforcer. Under a *fixed-interval schedule (FI)*, the student is reinforced the first time he or she

"No, Mrs. Cole, a VR10 schedule does not mean 10 cookies every time Ralph gets it right."

performs the target response following the elapse of a specified number of minutes. A behavior on an FI5 schedule may be reinforced 5 minutes after the last reinforced response. The first correct response that occurs after the 5 minutes has passed is reinforced. Following the delivery of that reinforcer, the next 5-minute cycle begins.

Behaviors placed on FI schedules also have particular characteristics. Because the only requirement for reinforcement on an FI schedule is that the response occur at least once following each specified interval, behaviors occur at a relatively low rate as compared with behaviors on ratio schedules. This is especially true if and when the student becomes aware of the length of the interval and therefore aware of when reinforcement is possible. This rate of responding will be affected by the length of the interval (Skinner, 1953). If reinforcement is available every minute, responding will be more rapid than if it is available every 10 minutes. Behaviors on FI schedules also have a characteristic that parallels the postreinforcement pause of FR schedules. A student eventually realizes that additional correct responses before the interval ends does not result in reinforcement. It also becomes apparent that responses immediately after reinforcement are never reinforced. The rate of responding eventually is noticeably lower (or ceases) for a short period of time after each reinforcement (the initial portion of the next interval). This decrease in correct responding is termed a *fixed-interval scallop*, because of the appearance of the data when plotted on a cumulative graph.

The effects on rate of responding resulting from FI schedules are eliminated by transition to a *variable-interval schedule (VI)*. Under a VI schedule, the intervals are of different lengths, while their average length is consistent. A behavior on a VI5 min schedule would have a reinforcer available for the target response on the average of every 5 minutes. As in use of VR schedules, the unpredictability levels out student performance. Behaviors under a VI schedule are performed at higher, steadier rates without the appearance of fixed-interval scallops, because the student can no longer predict the length of the interval following delivery of a reinforcer and therefore cannot predict which response will be reinforced.

A technique for increasing the rate of responding under an interval schedule is use of a **limited-hold (LH)** contingency. A limited-hold restricts the time the reinforcer is available following the interval. That is, when the interval has elapsed and the next correct response will be reinforced, the reinforcer will remain available for only a limited time. In this case, students must respond quickly in order to earn reinforcers, whereas under a simple interval schedule they may delay responding and still be reinforced. An FI5 min/LH5 sec schedule would make a reinforcer available for 5 seconds following each 5-minute interval. For example, when a student is being trained to ride a bus, he learns that the bus comes every 15 minutes and that he must step in quickly when the doors open (naturally occurring reinforcement) as they only stay open for 30 seconds (FI15 min/LH30 sec).

Response-duration Schedules.   Under response-duration schedules, the continuous amount of time of a target behavior is the determinant for delivery of the reinforcer. Under a *fixed-response-duration schedule (FRD)*, the student is reinforced following completion of a specified number of minutes (or seconds) of appropriate behavior. A behavior on a FRD10 schedule would be reinforced immediately at the end of each 10 minutes of continuous appropriate behavior. A student who the teacher wants to remain seated during reading period and who the teacher verbally praises for in-seat behavior every 10 minutes is under a FRD10 schedule. If the behavior stops occurring at any point during the time period, the timing is restarted.

As is the case under FR and FI schedules, a pause following reinforcement may be seen under FRD schedules. In this case, the pause appears to be related to the length of the required time period for appropriate behavior. The longer the time period, the longer the pause. It may be expected that if the time period is too long or is increased too rapidly, the behavior will either decrease or stop occurring altogether. These problems are minimized by varying the length of the time periods required for reinforcement, using a *variable-response-duration schedule (VRD)*. Under a VRD schedule, continuous appropriate behavior is reinforced on the average of a specified time period. A behavior on a VRD10 schedule would be reinforced on the average of every 10 minutes.

Thinning Schedules of Reinforcement

A formal classroom reinforcement system should be viewed as a temporary structure used to produce rapid behavior change. Most teachers eventually plan to bring the students' behavior under the control of more natural reinforcers. Schedule **thinning** helps decrease dependence on artificial reinforcers. In thinning, reinforcement gradually becomes available less often, or in other words, becomes contingent upon greater amounts of appropriate behavior.

In thinning of reinforcement schedules, the teacher moves from a dense schedule (continuous) to a sparse schedule (variable). The ratio between correct responding and reinforcement is systematically increased. The following examples illustrate this concept.

1. A student may be on a CRF schedule (1:1) for correctly identifying vocabulary words on flash cards. As the student approaches the criterion of 90% accuracy, the teacher may move the student to an FR3 schedule (R, R, R:S) and then to an FR6, to a VR8, and to a VR10. With each schedule shift, the teacher requires the student to perform more correct responses in order to receive a reinforcer.

2. A student may be on an FRD5 for sitting in her seat during the time she is to work in a workbook. Once the student has been able to meet this criterion, the teacher may move her to an FRD10, FRD20, FRD30 schedule. With each schedule shift, the teacher is requiring the student to maintain longer periods of appropriate behavior in order to earn reinforcement.

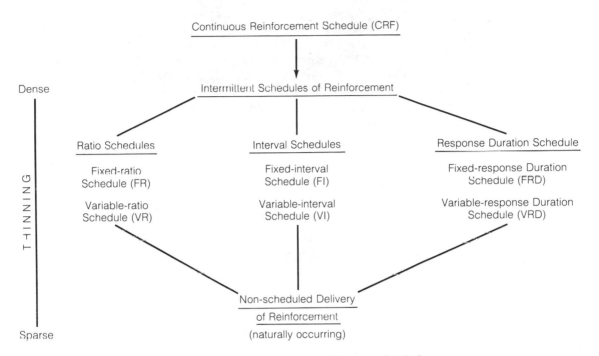

FIGURE 6–5   *Schedules for delivery of reinforcers*

Figure 6–5 presents an elemental scheme for schedule thinning. As schedule shifts are made from a continuous schedule to a fixed schedule to a variable schedule, a point is finally reached where predetermined timing of reinforcer delivery is no longer required. At this point the behavior is under the control of naturally occurring reinforcers.

Thinning schedules of reinforcement should result in the following:

1. Higher, steadier levels of responding as a result of moving to variable schedules

2. Decreasing expectation of reinforcement

3. Maintenance of the behavior over longer periods of time as the student becomes accustomed to delayed gratification

4. Removal of the teacher as a necessary behavior monitor

5. Transfer of control from the reinforcer to more traditional methods, such as teacher praise and attention (O'Leary & Becker, 1967), especially if schedule thinning is done in conjunction with pairing social reinforcers with tokens or primary reinforcers

6. An increase in persistence in responding toward working for goals (reinforcers) that require greater amounts of correct responding (work)

One caution should be considered when thinning schedules. Behaviors are subject to an effect known as **ratio strain**. Ratio strain occurs when the schedule has been thinned so quickly that the ratio for correct responding and reinforcement is too large. In such instances, the student doesn't earn reinforcement often enough to maintain his or her responding, and there is a marked decrease in rate. The student may stop responding altogether. If teachers see this effect occurring, they should return to the last schedule that resulted in an acceptable rate of responding and then thin again, but in smaller schedule shifts.

Professor Grundy Goes to Las Vegas

As Professor Grundy prepared to leave the house one morning, Mrs. Grundy handed him a stack of envelopes.

"Oliver, dear," she asked, "would you mind putting these in the mail for me?"

"More contests?" sneered Professor Grundy. "You've entered every contest, every sweepstakes, bought every raffle ticket ever offered. And how often have you won?"

"Well," answered Mrs. Grundy, "there was the pickle dish six years ago, the steak knives the year after that, and last year . . ."

"Minerva," scolded the Professor, "I've heard of resistance to extinction, but your behavior is being maintained on a VI3 *year* schedule. I'd think that was a little lean, even for you."

Imagine the Professor's surprise when he received a phone call later that morning. "Oliver," burbled Mrs. Grundy, "I've won a trip for two to Las Vegas! How's that for a reinforcer?"

The very next weekend, the Grundys flew off to claim their prize. As they walked into the lobby of their hotel, they passed a rank of slot machines.

"Humph! Just as I thought—primary and secondary reinforcers galore!"

"Just one moment, my dear," said Professor Grundy. "Let me try a couple of quarters in one of these. After all, when in Las Vegas . . ."

An hour later, Mrs. Grundy checked into their room. Three hours later, she ate a solitary dinner. At midnight, she returned to the lobby where the professor was still pulling the handle on the slot machine.

"Oliver," she insisted, "you must stop this."

"Just a few more minutes, Minerva," pleaded the professor, "I know that a few more quarters will do it."

Mrs. Grundy watched for a few minutes as Professor Grundy pulled the handle. Occasionally, the machine paid off with a few quarters.

"Oliver," snorted Mrs. Grundy, as she turned on her heel and stalked off, "I've heard of resistance to extinction, but your behavior is being maintained on a VR27 schedule of reinforcement. How utterly ridiculous!"

NEGATIVE REINFORCEMENT

Although positive reinforcement is most often used when the teacher's goal is to increase the rate or frequency of a target behavior, another procedure is available. *Negative reinforcement (S^R-)* is the contingent removal of an aversive stimulus immediately following a response that increases the future rate and/or probability of the response.

The first operative word in this definition is, once again, *increases*, which implies that some form of reinforcement is taking place. The second operative word is *removal*. Whereas in positive reinforcement a stimulus is *presented* to the student, in negative reinforcement something is removed from the student's environment. The third operative word is *contingent*. The teacher will not remove the aversive condition unless and until the requested response is produced. If a teacher states the contingency "Bill, you must stay in the room by yourself and finish all your math problems before you may join the rest of the class in the gym," that teacher is using negative reinforcement. The aversive condition of being left behind in the classroom while the rest of the class goes to the gym will be removed contingent upon completion of the math assignment that Bill should have completed earlier.

Negative reinforcement works because the student performs the behavior to *escape* the aversive stimulus. It is not necessary, however, for an aversive stimulus to be present in order for negative reinforcement to work. Negative reinforcement also works when a student performs some behavior in order to *avoid* an aversive stimulus. The disruptive behavior of a student who joins in to avoid the ridicule of his classmates is maintained by negative reinforcement.

Teachers often inadvertently use negative reinforcement. When a student cries over an assignment and the teacher removes it, that student learns that crying will result in the termination of an aversive stimulus (schoolwork); so the child cries more often. Parents and teachers who give in to children's temper tantrums or whining find that the child has tantrums or whines with increased frequency.

Aversive stimuli may lead to aggressive reactions.

The use of aversive stimuli in the classroom should be minimized. As we will discuss in greater detail in Chapter 7, such stimuli may lead to aggressive reactions. The child who is confined to his room until he gets "every last space ship *off* the floor" is likely to kick the unfortunate cat who comes to investigate the clattering and banging involved in the process. Escape and avoidance behavior may not be limited to the aversive stimulus but may result in a student escaping (running out of the room) or avoiding the entire school setting (playing truant). Mild aversive stimuli may sometimes be justified, but positive reinforcement is the procedure of choice for increasing or maintaining behavior.

Professor Grundy Teaches About Reinforcement

Professor Grundy's graduate class had turned in their observation assignments, muttering and grumbling all the while. After collecting the papers, Professor Grundy launched into his lecture on reinforcement (remarkably similar . . .). One of the students came up to the podium at the end of the lecture. Beaming, she said, "It's about time, Professor. I took this course to learn how to manage a classroom. For weeks, all we've talked about is history, theory, and all that technical junk. It was worth living through that just to hear tonight's lecture. I was going to drop the course, but now I won't miss a week."

"Why," asked Professor Grundy, lighting his pipe to hide the grin on his face, "do you think I wait to talk about reinforcement until *after* we're done with all that technical junk?"

SUMMARY

This chapter described procedures to increase or maintain appropriate academic or social behaviors. Positive reinforcement, the preferred approach, is the presentation of a stimulus contingent upon appropriate behavior. Positive reinforcers may be either primary or secondary. Negative reinforcement is the removal of an aversive stimulus contingent upon the performance of the target behavior. We suggested specific ways in which students' behavior can be changed using these procedures. We hope that we have also positively reinforced your reading behavior and that you are now prepared to continue.

REFERENCES

ALBERTO, P., JOBES, N., SIZEMORE, A., & DORAN, D. 1980. A comparison of individual and group instruction across response tasks. *The Journal of the Association for the Severely Handicapped, 5*, 285–293.

ALLISON, M.G., & AYLLON, T. 1968. Behavioral coaching in the development of skills in football, gymnastics, and tennis. *Journal of Applied Behavior Analysis, 13*, 297–314.

AYLLON, T., & AZRIN, N. 1968. *The token economy: A motivational system for therapy and rehabilitation.* New York: Appleton-Century-Crofts.

BALSOM, P.D., & BONDY, A.S. 1983. The negative side effects of reward. *Journal of Applied Behavior Analysis, 16*, 283–296.

BARRISH, H.H., SAUNDERS, M., & WOLF, M.M. 1969. Good behavior game: Effects of individual contingencies for group consequences on disruptive behavior in a classroom. *Journal of Applied Behavior Analysis, 2*, 119–124.

BIRNBRAUER, J.S., WOLF, M.M., KIDDER, J.D., & TAGUE, C.E. 1965. Classroom behavior of retarded pupils with token reinforcement. *Journal of Experimental Child Psychology, 2*, 219–235.

CARR, E.G., BINKOFF, J.A., KOLOGINSKY, E., & EDDY, M. 1978. Acquisition of sign language by autistic children. I: Expressive labelling. *Journal of Applied Behavior Analysis, 11*, 489–501.

CLINTON, L., & BOYCE, K. 1975. Acquisition of simple motor imitative behavior in mentally retarded and near mentally retarded children. *American Journal of Mental Deficiency, 79*, 695–700.

COHEN, H., & FILIPCZAK, J. 1971. *A new learning environment*. San Francisco: Jossey Bass.

CRONIN, K.A., & CUVO, A.J. 1979. Teaching mending skills to mentally retarded adolescents. *Journal of Applied Behavior Analysis, 12*, 401–406.

FERSTER, C.B., & SKINNER, B.F. 1957. *Schedules of reinforcement*. New York: Appleton-Century-Crofts.

HALL, C., SHELDON WILDGEN, J., & SHERMAN, J.A. 1980. Teaching job interview skills to retarded clients. *Journal of Applied Behavior Analysis, 13*, 433–442.

HALL, R.V., & HALL, M.C. 1980. *How to select reinforcers*. Lawrence, Kans.: H&H Enterprises, Inc.

HEWARD, W.L., & EACHUS, H.T. 1979. Acquisition of adjectives and adverbs in sentences written by hearing impaired and aphasic children. *Journal of Applied Behavior Analysis, 12*, 391–400.

JENS, K.G., & SHORES, R.E. 1969. Behavioral graphs as reinforcers for work behavior of mentally retarded adolescents. *Education and Training of the Mentally Retarded, 4*, 21–28.

KAPLAN, P., KOHFELDT, J., & STURLA, K. 1974. *It's positively fun: Techniques for managing learning environments*. Denver: Love Publishing.

KAZDIN, A.E. 1975. *Behavior modification in applied settings*. Homewood, Ill.: Dorsey Press.

KAZDIN, A.E. 1977. *The token economy: A review and evaluation*. New York: Plenum Press.

KAZDIN, A.E., & BOOTZIN, R.R. 1972. The token economy: An evaluative review. *Journal of Applied Behavior Analysis, 5*, 343–372.

KAZDIN, A.E., & POLSTER, R. 1973. Intermittent token reinforcement and response maintenance in extinction. *Behavior Therapy, 4*, 386–391.

KELLER, F., & SHERMAN, J. 1974. *The Keller plan handbook*. Menlo Park, Calif.: W.A. Benjamin.

KULIK, J., KULIK, C., & CARMICHAEL, K. 1974. The Keller plan in science teaching. *Science, 183*, 379–383.

LITOW, L., & PUMROY, D.K. 1975. A brief review of classroom group-oriented contingencies. *Journal of Applied Behavior Analysis, 8*, 341–347.

LOVAAS, O.I. 1977. *The autistic child: Language development through behavior modification*. New York: Irvington Publishing.

MAGRAB, P.R., & PAPADOPOULOU, Z.L. 1977. The effect of a token economy on dietary compliance for children on hemodialysis. *Journal of Applied Behavior Analysis, 10*, 573–578.

MARTIN, G., PALLOTTA-CORNICK, A., JOHNSTONE, G., & GOYOS, A.C. 1980. A supervisory strategy to improve work performance for lower functioning retarded clients in a sheltered workshop. *Journal of Applied Behavior Analysis, 13*, 183–190.

NELSON, G.L., & CONE, J.D. 1979. Multiple-baseline analysis of a token economy for psychiatric inpatients. *Journal of Applied Behavior Analysis, 12*, 255–271.

NEWSOM, C.D., & SIMON, K.M. 1977. A simultaneous discrimination procedure for the measurement of vision in nonverbal children. *Journal of Applied Behavior Analysis, 10*, 633–644.

NUTTER, D., & REID, D.H. 1978. Teaching retarded women a clothing selection skill using community norms. *Journal of Applied Behavior Analysis, 11*, 475–487.

O'LEARY, K.D., & BECKER, W.C. 1967. Behavior modification of an adjustment class. *Exceptional Children, 33*, 637–642.

O'LEARY, K.D., BECKER, W.C., EVANS, M.B., & SAUDARGAS, R.A. 1969. A token reinforcement program in a public school: A replication and systematic analysis. *Journal of Applied Behavior Analysis, 2*, 3–13.

O'LEARY, K.D., & O'LEARY, S.G. (Eds.). 1977. *Classroom management: The successful use of behavior modification*, 2nd ed. New York: Pergamon Press.

OLENICK, D.L., & PEAR, J.J. 1980. Differential reinforcement of correct responses to probes and prompts in picture-name training with severely retarded children. *Journal of Applied Behavior Analysis, 13*, 77–89.

PARSONSON, B.S., & BAER, D.M. 1978. Training generalized improvisation of tools by preschool children. *Journal of Applied Behavior Analysis, 11*, 363–380.

PETERSON, N. 1982. Feedback is not a new principle of behavior. *The Behavior Analyst, 5*, 101–102.

POLIRSTOK, S.R., & GREER, R.D. 1977. Remediation of mutually aversive interactions between a problem student and four teachers by training the student in reinforcement techniques. *Journal of Applied Behavior Analysis, 10*, 707–716.

POLLOWAY, E., & POLLOWAY, C. 1979. Auctions: Vitalizing the token economy. *Journal for Special Educators, 15*, 121–123.

PREMACK, D. 1959. Toward empirical behavior laws: I. Positive reinforcement. *Psychological Review, 66*, 219–233.

RASCHKE, D. 1981. Designing reinforcement surveys —Let the student choose the reward. *Teaching Exceptional Children, 14*, 92–96.

REID, D.H., & HURLBUT, B. 1977. Teaching nonvocal communication skills to multihandicapped retarded adults. *Journal of Applied Behavior Analysis, 10*, 591–603.

REYNOLDS, G.S. 1975. *A primer of operant conditioning*. Glenview, Ill.: Scott Foresman.

RUSSO, D.C., & KOEGEL, R.L. 1977. A method for integrating an autistic child into a normal public-school classroom. *Journal of Applied Behavior Analysis, 10*, 579–590.

SAIGH, P., & UMAR, A. 1983. The effects of a good behavior game on the disruptive behavior of Sudanese elementary school students. *Journal of Applied Behavior Analysis, 16*, 339–344.

SCHUMAKER, J.B., HOVELL, M.F., & SHERMAN, J.A. 1977. An analysis of daily report cards and parent-managed privileges in the improvement of adolescents' classroom performance. *Journal of Applied Behavior Analysis, 10*, 449–464.

SCHWARTZ, G.J. 1977. College students as contingency managers for adolescents in a program to develop reading skills. *Journal of Applied Behavior Analysis, 10*, 645–655.

SKINNER, B.F. 1953. *Science and human behavior*. New York: Macmillan.

STAINBACK, W., PAYNE, J., STAINBACK, S., & PAYNE, R. 1973. *Establishing a token economy in the classroom*. Columbus, Ohio: Charles E. Merrill.

STEVENSON, J., & CLAYTON, F. 1970. A response duration schedule: Effects of training, extinction, and deprivation. *Journal of the Experimental Analysis of Behavior, 13*, 359–367.

TRANT, L. 1977. Pictorial token card (communication). *Journal of Applied Behavior Analysis, 10*, 548.

TROVATO, J., & BUCHER, B. 1980. Peer tutoring with or without home-based reinforcement, for reading remediation. *Journal of Applied Behavior Analysis, 13*, 129–141.

WACKER, D., BERG, W., WIGGINS, B., MULDOON, M., & CAVANAUGH, J. 1985. Evaluation of reinforcer preferences for profoundly handicapped students. *Journal of Applied Behavior Analysis, 18*, 173–178.

WELCH, M.W., & GIST, J.W. 1974. *The open token economy system: A handbook for a behavioral approach to rehabilitation*. Springfield, Ill.: Charles C Thomas.

WOOD, R., & FLYNN, J.M. 1978. A self-evaluation token system *versus* an external evaluation token system alone in a residential setting with predelinquent youth. *Journal of Applied Behavior Analysis, 11*, 503–512.

7 Arranging Consequences that Decrease Behavior

Did you know that . . .

- You can use positive reinforcement to get students to stop doing inappropriate things?
- It is easier to give than to implement advice to "ignore it and it will go away"?
- Students who are sent out in the hall often enjoy themselves thoroughly?
- Spanking can be positive reinforcement?
- The teacher who told you to "chew that gum until you are sick of it" was using a form of behavior modification?

When teachers complain about a student's misbehavior, other teachers usually offer sympathy and advice for reducing or eliminating the inappropriate behaviors. Their suggestions most often emphasize punishment, in their terms defined as applying an aversive stimulus following the inappropriate behavior: for example, a rap on the back of the head, reading the riot act, a ruler across the knuckles, or a prolonged visit to the principal where the student "will learn how this school is run." The use of punishment can become a reflex, because in terms of simply and immediately stopping the occurrence of a behavior, it works! Using punishment is thus negatively reinforcing to teachers. However, this immediate reinforcement may make teachers lose sight of the side effects or reactions that may accompany the use of such procedures.

This chapter describes a broad range of behaviorally based alternatives to punishment that will have the same effect of reducing the occurrence of inappropriate behavior(s). These alternatives are presented as a sequenced hierarchy. The sequence moves from the most positive approaches used for behavior reduction (those that use reinforcement strategies) to the most aversive approaches. To be sure, the use of aversive consequences is included within this hierarchy. However, due to professional considerations and awareness of the undesirable side effects that aversive stimuli produce, these approaches are at the bottom of the hierarchy. Indeed, in our hierarchy three levels of options are presented before aversive consequences are even mentioned. These alternatives, with their individual constraints, are presented as viable alternatives to punishment, as they too have the desired effect of reducing the occurrence of inappropriate behavior.

PROCEDURAL ALTERNATIVES FOR BEHAVIOR REDUCTION

The hierarchy outlined in Figure 7–1 has four levels of options for reducing inappropriate behaviors. Level I is the first choice to consider, while Level IV is, in most instances, the choice of last resort.

Figure 7–1
Hierarchy of procedural alternatives for behavior reduction

Procedure of choice

Level I	Strategies of differential reinforcement
	a. Differential reinforcement of low rates of behavior (DRL)
	b. Differential reinforcement of other behavior(s) (DRO)
	c. Differential reinforcement of incompatible behavior (DRI)
Level II	Extinction (terminating reinforcement)
Level III	Removal of desirable stimuli
	a. Response cost
	b. Time-out procedures
Level IV	Presentation of aversive stimuli
	a. Unconditioned aversive stimuli
	b. Conditioned aversive stimuli
	c. Overcorrection procedures

Level I offers three strategies using differential reinforcement: differential reinforcement of low rates of behavior, differential reinforcement of other behaviors, and differential reinforcement of incompatible behaviors. These are options of first choice because, by selecting them, the teacher is employing a positive (reinforcement) approach to behavior reduction.

Level II refers to extinction procedures as an option. Using extinction means withdrawing or no longer delivering the reinforcers that maintain some behavior.

Level III contains the first set of options using what will be defined as a punishing consequence. However, these options—such as response cost and time-out procedures—still do not require the application of an aversive stimulus. Rather, the administration of these options may be seen as a mirror image of negative reinforcement. In the use of negative reinforcement, an aversive stimulus is contingently removed in order to increase a behavior. Level-III options require removal of a desirable stimulus in order to decrease a behavior.

The options in Level IV of the hierarchy are to be selected after unsuccessful attempts at the first three levels have been documented or when their continued performance presents an imminent danger to the student or to others in the class. The options in this level include the application of unconditioned or conditioned aversive stimuli or the use of an overcorrection procedure. Administration of these options may be seen as the mirror image of a reinforcement procedure.

Positive reinforcement:

Stimulus is contingently *presented* to *increase* a behavior.

Presentation of Aversive Stimuli:

Aversive stimulus is contingently *presented* to *decrease* a behavior.

The following sections discuss procedures at each level in greater detail.

LEVEL I: STRATEGIES OF DIFFERENTIAL REINFORCEMENT

Differential Reinforcement of Lower Rates of Behavior

DRL schedules may be used when shaping behavior (see Chapter 8).

Differential reinforcement of lower rates of behavior (DRL) is the application of a specific schedule of reinforcement, used to decrease the rate of behaviors that, while tolerable or even desirable in low rates, are inappropriate when they occur too often or too rapidly. For example, contributing to a class discussion is a desirable behavior; dominating a class discussion is not. Doing math problems is appropriate; doing them so rapidly that careless errors occur is not. Burping occasionally, while hardly elegant, is tolerable; burping 25 times an hour is neither. For behaviors whose elimination is desirable, DRL schedules may be used to decrease the rate of the behavior gradually.

In the initial laboratory version of this technique, DRL provided for a reinforcer to be delivered contingent upon a response, separated from a preceding response by a minimum amount of *time*. This format of DRL is referred to as *interresponse-time DRL*. The DRL procedures discussed in this chapter are modifications of the initial laboratory format. The DRL format most commonly used in the classroom provides for reinforcement delivery "when the number of responses in a specified period of time is less than, or equal to, a prescribed limit" (Deitz & Repp, 1973).

The latter DRL modification based upon the number of responses emitted has two variations: *full-session DRL* and *interval DRL*. The full-session DRL compares the total number of responses in an entire session (for instance, 30 minutes) with a preset criterion. A reinforcer is delivered if responding is at or below that criterion. Such an approach may be seen in studies by Edelson and Sprague (1974), Hall et al., (1971), and Harris and

Sherman (1973). Interval DRL involves dividing a session into intervals (for example, dividing the 30-minute session into six 5-minute intervals) and delivering reinforcement at the end of each interval in which responding is below or equal to a specified limit (for an example, see Deitz & Repp, 1974).

DRL procedures have been used with individual students and with groups of students in order to reinforce lower numbers of responses during class sessions. If baseline levels of the target behavior are high, the teacher may choose successively lower DRL limits in order to bring the rate into an acceptable range. This kind of numerically successive approach is analogous to using a changing criterion design. If baseline levels of the target behavior are low to moderate, the teacher may choose to set a single criterion in an acceptable range. If the acceptable limit is zero, that too may be set as the criterion.

Recall the changing criterion design discussed in Chapter 5.

In their initial experiments in the use of DRL, Deitz and Repp (1973) employed both criterion-setting strategies. In the first, an 11-year-old moderately retarded boy had a baseline of talkouts within a 50-minute session that averaged 5.7, with a range of 4–10 talkouts per session. He was told that if he talked out 3 or fewer times in a 50-minute session, he would be allowed 5 minutes of free play time at the end of the day. During this intervention, he averaged 0.93 talkouts per session with a range of 0–2.

In the second experiment, 10 moderately retarded students had an average of 32.7 talkouts with a range of 10–45. The students were told that if the group talked out 5 or fewer times in a session, each person would get two pieces of candy. This intervention yielded an average of 3.13 talkouts per session with a range of 1–6. In the third experiment, 15 high school girls in a regular class demonstrated a baseline level of 6.6 instances of inappropriate social discussion during a class period of 50 minutes. The intervention was planned in four phases: 6 or fewer inappropriate discussions, 3 or fewer, 2 or fewer, and then a zero rate in order to earn a free period on Friday.

As a guide in the use of DRL scheduling, Repp and Deitz (1979) suggest the following:

1. Baseline must be recorded in order to determine the average number of responses per full session or session intervals. This average occurrence may then serve as the initial DRL limit.

2. Reasonably spaced criteria should be established when using successively decreasing DRL limits so as to avoid both too frequent reinforcement and ratio strain, and so that the program can be faded out.

3. A decision must be made as to whether or not to provide feedback to the student(s) concerning the cumulative number of responses during the session. (pp. 223–224)

The primary advantage of DRL scheduling is its peculiar ability to reduce the occurrence of the behavior through delivery of reinforcement. It therefore offers the same advantages of reinforcement in general. In addi-

See Chapter 6 for discussion of the merits of reinforcement strategies.

tion, the approach is progressive in nature, because it allows the student to adjust, in reasonable increments, to successively lower rates rather than making a drastic behavioral change. The limits chosen should be within the abilities of the student and acceptable to the teacher. DRL is not a rapid means of changing behavior and therefore is inappropriate for use with violent or dangerous behavior.

Ms. Sinclair Teaches Stacy to Be Self-Confident

Stacy was a student in Ms. Sinclair's second-grade class. Stacy had excellent academic skills but raised her hand constantly to ask "Is this right?" or to say "I can't do this." If Ms. Sinclair had not been a behaviorist, she would have said that Stacy lacked self-confidence.

One morning, Ms. Sinclair called Stacy to her desk. She remembered that Stacy always volunteered to clean the blackboards after lunch. She told Stacy that she wanted her to learn to do her work by herself.

"If you really need help," she assured Stacy, "I'll help you. But I think three times in one morning is enough. If you raise your hand for help three times or less this morning, you may clean the blackboards when we get back from lunch."

Stacy agreed to try. Within a few days, she was raising her hand only once or twice during the morning. Ms. Sinclair praised her enthusiastically for being so independent. The teacher noticed that Stacy often made comments like "I did this all by myself, Ms. Sinclair. I didn't need help once." If Ms. Sinclair had not been a behaviorist, she would have said that Stacy was developing self-confidence.

Differential Reinforcement of Other Behaviors

Under the procedure called **differential reinforcement of other behaviors (DRO)**, a reinforcing stimulus is delivered when the target behavior is not emitted for a specified period of time (Reynolds, 1961). Whereas DRL reinforces gradual behavior reduction, DRO reinforces only zero occurrence. In fact, DRO is sometimes referred to as *differential reinforcement of zero rates of behavior.* In the previous chapter, reinforcement was seen as the delivery of consequent stimuli contingent upon the *occurrence* of a desired behavior. DRO involves the presentation of a consequent stimulus contingent upon the *nonoccurrence* of a behavior.

DRO may have at least three administrative variations, similar to those used with DRL procedures:

1. Reinforcement contingent upon a behavior's not occurring throughout an entire time period (the same as a DRL zero). For example, reinforcement is delivered only if talking out does not occur for an entire 30-minute period (DRO 30 mins.). Each time the student talks out, a new cycle is begun. This is a *full-session DRO.*

2. Reinforcement contingent upon the nonoccurrence of a behavior within a session that has been divided into intervals. Reinforcement is delivered at the end of each of these intervals in which the behavior has not

occurred. For example, a 30-minute session may be divided into 5-minute intervals and reinforcement delivered at the end of each of the 5-minute intervals in which the student has not talked out. The intervals may be of equal length or they may be of different length (that is, on the average of every 5 minutes, as is done in variable interval scheduling). The intervals may be gradually lengthened until the equivalent of a full-session DRO is reached.

3. DRO may be used with permanent-product data. For example, a happy face may be drawn on every paper that does not contain doodles.

The teacher who uses DRO procedures should be aware that the DRO schedule requires reinforcement of all behaviors except one (Ferster & Perrott, 1968). Therefore, in practice a student will be reinforced for performing any behaviors other than the target behavior. This procedure should always be paired with reinforcement of appropriate behaviors.

DRO has been used with a variety of behaviors, such as self-injurious behavior (Myers, 1975; Repp & Deitz, 1974; Rose, 1979), aggressive behavior (Repp & Deitz, 1974), exhibitionism (Lutzker, 1974), disruptions (Deitz, Repp, & Dietz, 1976; Repp, Deitz, & Deitz, 1976; Schmidt & Ulrich, 1969), stereotypic behavior (Harris & Wolchik, 1979; Repp, Deitz, & Speir, 1974), and the training of compliance behaviors (Goetz, Holmberg, & LeBlanc, 1975). Repp et al. (1974) used DRO procedures with three severely retarded individuals. The individuals were a 12-year-old girl whose target behavior was flapping her lips with her fingers, a 22-year-old woman whose target behavior was rocking, and a 23-year-old man whose target behavior was waving a hand in front of his eyes. A kitchen timer was set to a prescheduled number of minutes (the DRO interval). The student who did not emit the target stereotypic response during the interval was hugged and praised by the teacher when the bell rang. If the target stereotypic behavior occurred, the teacher said "no" and reset the timer. Very short intervals (40 seconds) were used initially. As the rates of behavior decreased, the intervals were lengthened. All the subjects showed greatly decreased rates of stereotypic behaviors.

As a guide to the use of DRO scheduling, Repp and Deitz (1979) suggest:

1. Baseline must be recorded not only to measure the inappropriate behavior, but also to schedule the DRO procedure properly. Because the size of the initial DRO interval can be crucial, it should be based on data rather than set arbitrarily. From the baseline an average interresponse time (time between responses) should be determined, and a slightly smaller interval should be designated as the initial DRO interval.

2. Criteria must be established for increasing the length of the DRO interval. The basic idea is

   a. to start at a small enough interval that the student can earn more reinforcers for not responding than he could earn for responding, and

b. to lengthen that interval over time. The decision to lengthen should be based on the success of the student at each interval length.

3. Possible occurrence of the undesirable behavior necessitates two additional decisions:

a. whether to reset the DRO interval following a response occurrence or merely to wait for the next scheduled interval, and

b. whether to consequate a response occurrence in any other way or just to ignore it.

4. Reinforcement should not be delivered immediately following a grossly inappropriate behavior even if the DRO interval has expired without the target response having occurred. (pp. 222–223)

Clarence Learns Not to Hit People

Clarence was a student in Mr. Byrd's resource class. He often hit the other students in the class, usually because someone had touched some possession of his. Having observed that Clarence hit someone an average of 12 times during the 1½ hour resource period, for an average interresponse time of 7.5 minutes, Mr. Byrd decided on an interval of 7 minutes. He told Clarence that he could earn a card worth 5 minutes to work on an art project for each 7 minutes that elapsed without hitting. When Clarence hit someone, Mr. Byrd reset the timer. He did this rather than simply not delivering the reinforcer at the end of the interval, because he was afraid that once Clarence had "blown it," he would engage in a veritable orgy of hitting until the end of the interval.

Within a few days, Clarence's rate of hitting was much lower, so Mr. Byrd lengthened the intervals to 8 minutes, then 10, then 15. Soon he was able to reinforce the absence of hitting at the end of the period and still maintain a zero rate.

Differential Reinforcement of Incompatible Behaviors

Procedures for **differential reinforcement of incompatible behaviors (DRI)** involve reinforcing a response that is incompatible in its structure or other characteristics with the behavior targeted for reduction. Mutually exclusive responses are chosen so that an appropriate response makes it physically impossible for the student to engage in the inappropriate behavior. For example, a student cannot simultaneously make stereotypic hand movements and play appropriately with toys (Favell, 1973). It is equally impossible to make such hand movements while imitating hand gestures modeled by a teacher (Weisberg, Passman, & Russell, 1973). A student cannot slap his face and play with a ball at the same time (Tarpley & Schroeder, 1979). Similarly, engaging in an art project makes it impossible to wander around the room (Patterson, 1965).

DRI procedures have been employed to modify a variety of behaviors: inappropriate classroom behavior (Ayllon, Layman, & Burke, 1972; Ayllon & Roberts, 1974; Deitz et al., 1976), hyperactivity (Patterson, 1965; Twardosz & Sajwaj, 1972), inappropriate speech (Barton, 1970), stereotypic behaviors (Baumeister & Forehand, 1971; Favell, 1973; Weisberg et al., 1973), aggressive behavior (Vukelich & Hake, 1971), and self-injurious behavior (Tarpley & Schroeder, 1979).

Ayllon and Roberts (1974) used DRI to bring the disruptive (out-of-seat, talking, hitting) behavior of 5 fifth-grade boys under control by reinforcing their academic performance. Awarding points for daily and weekly exchange was made contingent upon the percentage of correct answers in their reading workbooks during daily 15-minute sessions. During baseline, the mean percentages for both disruptive behavior and academic accuracy were in the 40% to 50% range. After intervention, disruptive behavior had decreased to an average of 5% of intervals and average academic accuracy had increased to 85%.

As a guide to the use of DRI scheduling, Repp and Deitz (1979) suggest:

1. A behavior that is incompatible to the undesirable behavior must be chosen. If there is no appropriate behavior that is opposite to the inappropriate behavior, then a behavior that is beneficial to the student should be selected and should be reinforced.

2. Baseline should be recorded to determine

   a. how often the inappropriate behavior occurs, and

   b. how often the chosen incompatible behavior occurs.

3. The schedule of reinforcement must be determined. In addition, a program for carefully thinning the schedule should be written so that the program can be phased out and the student's behaviors can come under control of natural contingencies in the environment. (p. 224)

LEVEL II: EXTINCTION

In contrast to Level I, which focuses on providing reinforcement, Level II, extinction, reduces behavior by abruptly withdrawing or terminating the positive reinforcer maintaining an inappropriate target behavior. As Chapter 6 mentioned, this abrupt withdrawal results in the cessation or extinction of behavior. When the behavior being maintained is an appropriate one, preventing extinction is the goal. Many inappropriate behaviors, however, are also maintained by positive reinforcement. A parent who gives children cookies or candy when they cry may be positively reinforcing crying. If the cookies are withdrawn, crying should diminish.

Extinction is most often used in the classroom to decrease behaviors that are being maintained by teacher attention. Teachers often pay attention to students who are behaving inappropriately, and many students find such attention positively reinforcing. This may be true even if the attention takes the form of criticism, correction, or threats. Some students' behavior may be positively reinforced by even such extreme measures as yelling and spanking.

It is often difficult for teachers to determine when their attention is positively reinforcing inappropriate behavior. Thus, a teacher may find it helpful to have someone else observe teacher-student interaction. Once the relationship between the teacher's attention and the student's behavior is verified by this method, extinction in the classroom most often takes the form of ignoring inappropriate behavior. The teacher withholds the previously given positive reinforcer (attention), and the inappropriate behavior extinguishes or dies out.

Extinction procedures have been used with a variety of behaviors: for example, disruptive classroom behavior (Zimmerman & Zimmerman, 1962), aggressive behavior (Brown & Elliott, 1965), tantrums (Carlson, Arnold, Becker, & Madsen, 1968), nonstudy behavior (Hall, Lund, & Jackson, 1968), and delusional speech (Liberman, Teigen, Patterson, & Baker, 1973).

Extinction is most often used in conjunction with reinforcing other more appropriate behaviors. Combining procedures this way appears to speed extinction (Scott, Burton, & Yarrow, 1967; Wilson & Hopkins, 1973). Used independently, "there is little or no evidence of constructive learnings. What is learned is that a certain behavior no longer provides an expected reward; the net effect is a reduction in the repertoire of behavior" (Gilbert, 1975, p. 28). If attention is given to appropriate behavior, this indicates to the student that the teacher's attention (S^{R+}) is still available, but that it is selectively available. It is not the student who is being ignored, just an inappropriate behavior.

"Just ignore it and it will go away. He's only doing it for attention." This statement is one of the most common suggestions given to teachers. In truth, extinction is much easier to discuss than to implement. It *will* go away, all right, but not necessarily rapidly or smoothly. The teacher who decides to implement an extinction procedure should give careful consideration to the following points.

Delayed Reaction

Problems with extinction.

The effects of extinction are not usually immediate. The extinction procedure may take a considerable amount of time to produce reduction in behavior. Once reinforcement is withheld, behavior continues for an indeterminate amount of time (Skinner, 1953). This characteristic, known as *resistance to extinction*, is particularly marked when behaviors have been maintained on intermittent schedules of reinforcement. The student continues to seek the reinforcer that a history of reinforcement has taught her or him will be the result. In an initial extinction phase to reduce aggressive behavior toward peers in a preschooler, Pinkston, Reese, LeBlanc, and Baer (1973) found it took 8 days to reduce the rate of behavior from 28% of total peer interactions to 6% of interactions. In a study of the effects of self-injurious behavior, Lovaas and Simmons (1969) report that "John hit himself almost 9000 times before he quit" (p. 146).

Increased Rate

The teacher should expect an increase in the rate and/or intensity of the behavior before significant reduction occurs (Watson, 1967). It's going to get worse before it gets better. In comments on one subject, Lovaas and Simmons (1969) say, "Rick eventually did stop hitting himself under this arrangement (i.e., extinction) but the reduction in self-destruction was not immediate, and even took a turn for the worse when the extinction was first initiated" (p. 146). On John and Gregg, two other subjects, these authors acknowledge that "the self-destructive behavior showed a very gradual drop over time, being particularly vicious in the early stages of extinction" (p. 147). Figure 7–2 displays this phenomenon in graphic data from Lovaas and Simmons (1969) and Pinkston et al. (1973).

A common pattern is that of the teacher who decides to ignore some inappropriate behavior such as calling out. When a student finds that a previously reinforced response is no longer effective, the student then begins to call out louder and faster. If, after a period of time, the teacher says, "Oh, all right, Ward, what do you want?" the teacher has reinforced the behavior at its new level of intensity and may find that it remains at this level. Once an extinction procedure has been implemented, the teacher absolutely must continue ignoring whatever escalation of the behavior occurs.

Controlling Attention

It is ridiculous to say to a student, "Can't you see I'm ignoring you?" Of course, what the student *can* see is that the teacher is *not* ignoring her. Even nonverbal indications that the teacher is aware of the misbehavior may be sufficient to prevent extinction. The teacher who stands rigidly with teeth and fists clenched is communicating continued attention to the student's behavior. It takes a great deal of practice to hit just the right note. We have found that it helps to have something else to do.

1. Become *very* involved with another student, perhaps praising the absence of the target behavior in her—"I like the way you raised your hand, Lou; that's the *right* way to get my attention."

2. Read something or write busily.

3. Recite epic poetry subvocally.

4. Carry a worry rock or beads.

5. Stand outside the classroom door and kick the wall for a minute.

Extinction-induced Aggression

The last suggestion in the preceding section is related to another phenomenon that may occur: extinction-induced aggression by the student in early stages of extinction procedures (Azrin, Hutchinson, & Hake, 1966). In search of the previously available reinforcer, the student says, in effect, "You only think you can ignore me. Watch this trick." The pattern of escalation and aggression that occurs in the early stages of extinction is illustrated by a typical interaction between a hungry customer and a defective vending machine. The customer puts a quarter in the machine (a previously reinforced response) and pushes the appropriate button. When no reinforcer is forthcoming, the customer then pushes the button again . . .

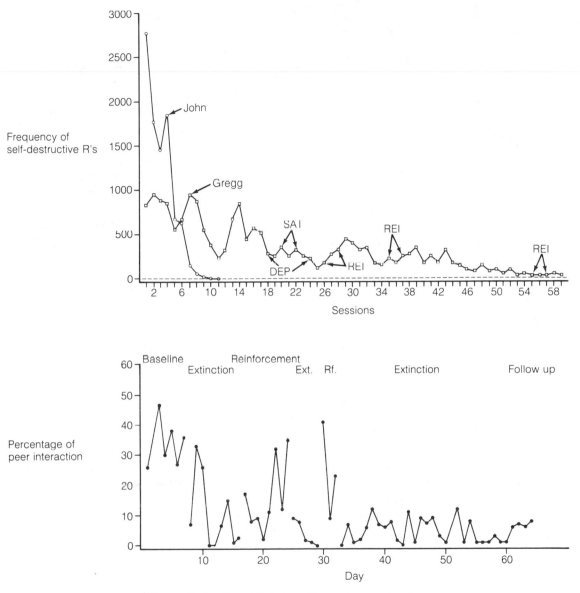

Figure 7–2   Data from studies using extinction procedures for behavior reduction

and again . . . and faster . . . and harder. Before her response is extinguished, she will very likely deliver a sharp rap or swift kick to the unreinforcing machine.

Spontaneous Recovery

The teacher can also expect the possible reappearance of an extinguished behavior. This phenomenon is known as *spontaneous recovery* (Skinner,

1953) and may occur after the behavior has been extinguished for some time. The student tries once again to see if the extinction rule is still in effect. Such reemergence of the behavior can be quickly terminated by ignoring it. However, failure to ignore it may result in rapid relearning on the part of the student.

Imitation or Reinforcement by Others

The behavior the teacher is ignoring may spread to other class members. If other students see that a particular student is getting away with a misbehavior, that he is not being punished for it, they may imitate the behavior (Bandura, 1965). This may serve to reinforce the behavior. It may also result in a number of students' performing a misbehavior instead of just one, making the behavior that much harder to ignore. The use of an extinction procedure relies upon the teacher's ability to terminate the reinforcing stimulus for the inappropriate behavior. This is one of the hardest aspects of conducting an extinction procedure. In a classroom setting, the best bet is that the behavior is being reinforced by attention from the teacher (yelling) or from the classmates (laughing). In order to determine the reinforcing stimulus, the teacher may have to test several suspicions systematically, attempting to eliminate one potential reinforcer at a time.

It is frequently difficult to control the reinforcing consequences delivered by peers. Successful approaches to this problem have been used by Patterson (1965), who reinforced peers for withholding attention during when the target student was out-of-seat, talked or hit others; by Solomon and Wahler (1973), who selected 5 high-status peers and trained them in the use of extinction and reinforcement of appropriate behavior; and by Pinkston et al. (1973), who attended to the peer being aggressed against while the aggressor was ignored.

Limited Generalizability

Although extinction is effective, it appears to have limited generalizabilty. That is, the behavior may occur just as frequently as ever in settings where extinction is not in effect. Liberman et al. (1973) reported no generalization of treatment to routine interchanges with staff on the ward. Lovaas and Simmons (1969) reported that behavior in other settings is unaffected when extinction is used only in one setting. Extinction may be required in all necessary environmental settings.

Benoit and Mayer (1974) prepared a flow chart that summarizes factors to be considered before selecting extinction as an approach to eliminating undesirable behavior. As seen in Figure 7–3, their chart suggests six considerations before making a decision to use extinction, here stated as questions to guide teachers' decision making:

Questions to ask yourself before implementing an extinction procedure.

1. Can the behavior be tolerated temporarily based on its topography (e.g., Is it aggressive?) and on its current rate of occurrence?

2. Can an increase in the behavior be tolerated?

3. Is the behavior likely to be imitated?

4. Are the reinforcers known?

5. Can reinforcement be withheld?

6. Have alternative behaviors been identified for reinforcement?

A social consequence such as teacher attention is not always the maintaining consequence of a behavior. "Some persons do things not for attention or praise, but simply because it feels good or is fun to do" (Rincover, 1981, p. 1). In such instances sensory consequences rather than teacher consequences may be maintaining the behavior. This seems to be particularly true of certain stereotypic or self-injurious behaviors. A student's stereotypic hand-flapping behavior may be maintained by the visual input resulting from the behavior. A student's self-injurious, self-scratching behavior may be maintained by the tactile input resulting from the behavior. When sensory consequences can be identified as the reinforcer of a behavior, the form of extinction known as *sensory extinction* may be employed

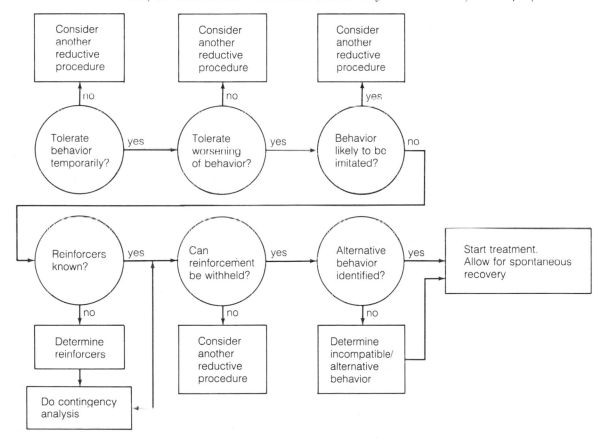

Figure 7–3 Flow chart indicating sequence of considerations in use of extinction

(Rincover, 1981). Sensory extinction attempts to remove the naturally occurring sensory consequence of the behavior rather than directly preventing the behavior itself. The hand-flapping may be reduced by placing weights on the student's arm thereby making hand-flapping more work, reducing its frequency, and fading the reinforcer (Rincover, 1981). Self-scratching may be reduced by covering the area with heavy petroleum jelly, thereby eliminating the tactile consequence of the behavior. Difficulties arise in using sensory extinction with precison due to difficulty in identifying the sensory consequence (if there is one) and the difficulty in eliminating "all the sensory consequences inherent in many commonly occurring stereotypic responses, such as rocking or clapping" (Aiken & Salzberg, 1984, p. 298).

Mr. Medlock Extinguishes Arguing

Judy was a student in Mr. Medlock's fourth-grade class. Whenever Mr. Medlock told Judy to do something, she argued with him. Mr. Medlock had found himself having conversations like this:

"Judy, get to work."

"I am working, Mr. Medlock."

"No you're not, Judy. You're wasting time."

"I'm getting ready to work."

"I don't want you to get ready. I want you to do it."

"How do you expect me to work if I don't get ready?"

He realized one day that he was having childish arguments with a 9-year-old and that his behavior was reinforcing Judy's arguing. He decided to put this behavior on extinction. The next day he said, "Judy, get to work." When Judy began to argue that she was working, he walked away.

Judy muttered to herself for a while and then said loudly, "I ain't gonna do this dumb work, and you can't make me." Mr. Medlock held on and continued to ignore her comments.

Emily raised her hand. "Mr. Medlock," she simpered, "Judy says she's not going to do her work."

"Emily," said Mr. Medlock quietly, "take care of yourself."

"But, Mr. Medlock, she said you can't make her," countered Emily.

Mr. Medlock realized that his only hope was to ignore Emily's behavior too. He got up and began to walk around the room, praising students who were working and reminding them of the math game they were going to play when work was finished. Soon Emily went back to her work. Judy, however, began to tap her pencil ostentatiously on her desk. Mr. Medlock continued to interact with other students. Judy finally shrugged and began to do her assignment. When she had been working for several minutes, Mr. Medlock walked casually over to her and said, "Good job, Judy. You've already got the first two right. Keep it up."

It occurred to Mr. Medlock that Judy's delay in starting to do her work was probably reinforced by his nagging and that if he ignored this procrastination as well, Judy would probably begin to work more quickly.

PUNISHMENT

The two remaining levels of the hierarchy, *Levels III* and *IV*, contain options for behavior reduction that may be termed *punishment*. As is the case with the term *reinforcer*, we use a functional definition of the term *punisher*. A punisher is a consequent stimulus (S^P) that

1. decreases the future rate and/or probability of occurrence of behavior.

2. is administered contingently upon the production of an undesired or inappropriate behavior.

3. is administered immediately following the production of the undesired or inappropriate behavior.

It must be clearly understood that the terms *punishment* and *punisher* as used in this context are defined functionally. Any stimulus can be labeled a punisher if its contingent application results in a reduction of the target behavior. A punisher, like a reinforcer, can be identified only by its effect upon behavior—not on the nature of the consequent stimulus. For example, if a father paddles his son for throwing toys, and the son stops throwing toys, then the paddling was a punisher. If the son instead continues to throw toys, then the paddling was not a punisher. If each time a student calls out, her teacher reduces her free play time by 1 minute or takes tokens away and this results in the reduction or cessation of the calling out, then the consequence was a punisher. If the behavior continues, the consequence was not a punisher. Again, this is a definition of the term *punisher* from a functional perspective.

Punishers, like reinforcers, may also be naturally occurring phenomena. Punishers are not simply techniques devised by malevolent behaviorists to work their will upon students. Consider the examples and label the punishers:

Jeannie toddles into the kitchen while her father is cooking dinner. As father's back is turned, Jeannie reaches up and touches the saucepan on the front burner. She jerks her hand away crying and thereafter avoids touching the stove.

Theresa has finished her math assignment quickly. Proudly, she raises her hand and announces this fact to her teacher. The teacher assigns her 10 additional problems to work. The next day Theresa works more slowly and fails to finish her assignment before the end of math class.

Gary is a retarded youngster who is mainstreamed. On the first day that he attends Mr. Johnston's fourth-grade reading class, he stumbles through the oral reading passage. The other students ridicule him and he subsequently refuses to leave the special class to attend the fourth grade.

Mrs. Brice, a first-year teacher, decides to use praise with her junior-high social studies class. She greets each student who arrives on time with effusive compliments and a happy face sticker. The next day no students arrive on time.

LEVEL III: RESPONSE COST

Response cost attempts to reduce behavior through removal of a reinforcer. The procedure itself may be defined as the withdrawal of specific amounts of reinforcer contingent upon inappropriate behavior. As implied by this definition, "some level of positive reinforcement must be made available in order to provide the opportunity for ... withdrawing that reinforcement" (Azrin & Holz, 1966, p. 392). If the use of a response cost procedure empirically results in the desired behavior reduction, withdrawal of the reinforcement functions as a punisher.

Response cost may be seen as a system of leveling fines, a familiar event. A city government has, as a means of behavior control and fund raising, a whole system of fines for predefined, inappropriate behavior. We, the citizens, have possession of the pool of reinforcers—the dollar bills we have earned working. The city administration withdraws specific amounts of these reinforcing stimuli contingent upon such inappropriate behaviors as littering, staying too long at a parking meter, and speeding. Similarly, McSweeny (1978) reported that the number of directory assistance phone calls made in Cincinnati decreased significantly when a charge for these calls was instituted. Marholin and Gray (1976) found that when cash shortages were subtracted from employees' salaries, the size of shortages was sharply reduced.

A token system can incorporate response cost procedures. If a teacher informs students that they will earn one token for each of ten math problems they solve correctly, that teacher is employing a token reinforcement system. If, on the other hand, the teacher hands out ten tokens to each student and informs them that, for each problem they solve incorrectly, one token will be "repossesed," that teacher is employing a response cost procedure. In practice, a response cost procedure is most often (Kazdin & Bootzin, 1972) and most effectively (Phillips, Phillips, Fixsen, & Wolf, 1971) used in combination with a token reinforcement system. In such a combined format, the students concurrently earn the pool of reinforcers and lose the reinforcers as fines for misbehavior. The students have ongoing access to future reinforcers.

In classrooms, response cost procedures have the versatility to be used in modifying social behaviors (rule violation, off-task) and academic performance, such as completion of math problems (Iwata & Bailey, 1974), as well as in managing individual students or groups of students without the undesirable side effects usually associated with punishment (Kazdin, 1972).

There are a number of practical cautions in the use of a response cost procedure. First, the teacher must have the ability to withdraw the reinforcer once given. It is probably unwise to attempt to use response cost pro-

Problems with response cost.

cedures with edible primary reinforcers. The student who has at his desk a cup of candies that are to be contingently withdrawn is apt immediately to eat all the candy as his first inappropriate behavior. The slight young teacher in the secondary classroom who walks over to the football tackle and announces that he is to return five tokens may find that the student answers, "In a pig's eye" (or something to that effect). It is best in such an instance to use points, which can be withdrawn without being physically repossessed.

Careful consideration must also be given to the magnitude of the penalties—the number of tokens or points being withdrawn. Research has produced mixed recommendations. For example, Burchard and Barrera (1972) used severe fines, whereas Siegel, Lenske, and Broen (1969) used mild fines; both got good results. An important point to remember is that exacting large fines may make tokens worthless. If students learn that an entire day's work can be wiped out by a fine, they are unlikely to work very hard.

Another problem may occur when all the reinforcers have been withdrawn. Then what happens? Consider, for example, the substitute teacher who is assigned to the ninth-grade remedial class for the day. One of the thoughts uppermost in her mind, in addition to the educational welfare of her temporary charges, is making it in one piece to lunchtime. When the students enter the room, she announces that if they cooperate and work hard during the morning hours, she will permit them 30 minutes of free time. If she fines them 5 minutes whenever anyone misbehaves, by 10 o'clock the students may have very little free time left. Once the reinforcement system has become debased to this extent, the student energy involved in being good far outweighs the remaining amount of the reinforcer.

When using response cost, as with all management systems, students must clearly understand the rules of behavior and the penalties for infractions. Clear understanding will avoid lengthy conversations at the time of misbehavior, when the teacher should simply have to describe the infraction and exact the fine.

Before selecting a response cost procedure, the teacher should answer the following questions:

1. Have more positive procedures, such as differential reinforcement strategies, been considered?

Guide to setting up a response cost procedure.

2. Does the student currently have, or have access to, a pool of reinforcers?

3. Have the rules of appropriate behavior and the results (fines) for infractions been clearly explained and understood?

4. Has the ratio of the size of the fine to each instance of misbehavior been thought out?

5. Can the reinforcers be retrieved?

6. Will appropriate behavior be reinforced in conjunction with the use of response cost?

Time-out Procedures

Time-out procedures serve as punishment by denying a student, for a fixed period of time, the opportunity to receive reinforcement contingent upon a response. *Time-out* is a shortened form of the term *time out from positive reinforcement*. Before using time-out procedures, the teacher must be sure that reinforcing consequences for appropriate behavior are available in the classroom. These consequences might be a formal reinforcement system or simply activities that the students enjoy. Identification of available reinforcers may be relatively simple when the students in a classroom are working for a specific item, such as tokens. The specific reinforcers in many classrooms where this is not the case are more difficult to identify. In some classrooms, reinforcers are infrequent, thus limiting the effectiveness of time-out procedures.

Time-out procedures may be categorized according to the method of denying access to reinforcers. While the student is still within the classroom, the teacher may make an environmental rearrangement. These procedures may be termed *nonseclusionary time-out*. For more disruptive behaviors, the student may be denied access to reinforcers by being isolated outside the classroom. These procedures are called *seclusionary time-out*.

Nonseclusionary Time-out Procedures

In **nonseclusionary time-out** procedures, the student is not removed from the instructional setting; rather, the teacher denies the student access to reinforcers through a temporary manipulation of the environment. Teachers use this procedure in its most common form when faced with a generalized, minor disturbance. They may tell the students to put their heads on their desks, or they may turn out the room lights (Baron, Kaufman, & Rakavskas, 1967) to eliminate the mutual reinforcers of talking together and laughing at one another. Similarly, time-out may be accomplished by turning off a record player that is being played too loudly.

During a group lesson, Foxx and Shapiro (1978) used a different approach. They had students wear a ribbon-tie while behaving in a socially appropriate manner. A student's ribbon was removed for any instance of misbehavior. This removal signaled an end to teacher attention and an end to the student's participation in activities and access to edible reinforcers for 3 minutes. In a study by Adams, Martin, and Popelka (1971), a misbehavior prompted the teacher to turn on a recorded tone. While the tone was audible, access to reinforcers was denied to the student. In classrooms where token systems are in effect, simply taking away a student's point card for a specified number of minutes provides a period of time-out.

Another form of nonseclusionary time-out is *contingent observation* (May et al., 1974). Contingent observation involves moving a student to the edge of an activity so that she or he can still observe other students' behavior being reinforced. In one study (Porterfield, Herbert-Jackson, & Risley, 1976), the teacher told the student what he should have done, moved him to the periphery of the activity, and told him to watch how the other students performed a particular behavior. After 1 minute, he was asked if he had learned how to emit the same behavior appropriately. If he said "yes," he was returned to the group activity. If he said "no," he remained in time-out for an additional minute.

A type of nonseclusionary time-out that has been used with more severe behavior problems is known as facial or visual screening. Visual screening procedures have been successfully used with self-striking (Lutzker, 1978; Singh, Beale, & Dawson, 1981; Zegiob, Alford, & House, 1978), hand clapping and vocalization (Zegiob, Jenkins, Becker, & Bristow, 1976), and thumb biting to the point of severe flesh damage (Singh, 1980). This procedure involves occluding the student's eyes with a terry-cloth bib or the teacher's hand usually for a period of 3 seconds (though data by Singh et al., 1981 suggest that a 1 minute duration is the most effective) following cessation of the behavior without removing him from the instructional setting. Once the bib or hand is removed, the ongoing lesson is immediately resumed. McGonigle, Duncan, Cordisco, and Barrett (1982) used visual screening with four students engaged in various stereotypic and self-injurious behaviors. Their procedure involved, "Teacher placing one hand over the child's eyes so as to preclude any source of visual input, while holding the back of the child's head with the other hand. Duration of the visual screening treatment was a minimum of 5 seconds for each child. Criterion for release from visual screening was contingent upon nondisruptive behavior following expiration of the minimum time requirement" (p. 463).

Exclusionary time-out involves removing the student from an activity as a means of denying access to reinforcement. It is not always necessary to remove the student completely from the classroom. Exclusion (May et al., 1974) may be accomplished simply by removing the student from the immediate activity area to another part of the room. Observation and subsequent modeling of reinforced behavior (as with contingent observation) are not components of this procedure. The student is removed to a chair facing a corner or placed in a screened-off area of the room. A variation of such exclusion was employed by Baer, Rowbury, and Baer (1973). When a student misbehaved during an activity for which she was earning tokens, she was placed in the middle of the room out of reach of the work activity, thereby being denied access to earning tokens.

Seclusionary Time-out Procedures

In most instances, time-out is a procedure associated with the use of a time-out room (Birnbrauer, Wolf, Kidder, & Tague, 1965; Lahey, McNees, & McNees, 1973; MacPherson, Candee, & Hohman, 1974). This involves removing the student from the classroom itself to a room identified for total

"I don't understand it, professor—time-out just isn't working anymore."

social isolation, contingent upon a misbehavior. In such a room, access is denied to all potential reinforcers resulting from the teacher, classmates, or the classroom. Such a procedure is sometimes termed **seclusionary time-out**.

Unfortunately, time-out rooms have been seriously misused or their use mismanaged in many instances. Therefore, this procedure has been the subject of negative publicity and even considered and regulated in litigation (*Wyatt v. Stickney*, 1972). Various states and school districts now have procedural regulations for use of time-out rooms (May et al., 1974). Concern has arisen basically because of two aspects of the use of time-out rooms—duration in the room and physical features of the room. Gast and Nelson (1977b) address both these concerns:

> The duration of each time-out period should be brief. One to five minutes generally is sufficient. It is doubtful that time-out periods exceeding 15 minutes serve the purpose for which they are intended. (p. 463) If a seclusion time-out is to be employed, the time-out room should:
>
> 1. Be at least 6 by 6 feet in size.
> 2. Be properly lighted (preferably recessed, with the switch outside the room).
> 3. Be properly ventilated.
> 4. Be free of objects and fixtures with which children could harm themselves.
> 5. Provide the means by which an adult could continuously monitor, visually and auditorily, the student's behavior.
> 6. Not be locked. A latch on the door should be used only as needed, and only with careful monitoring. (p. 463)

A teacher should work through the following sequence of steps in using a time-out room or any of the forms of this procedure:

*Guidelines for
seclusionary time-out.*

1. Before beginning to use time-out as a management procedure, identify the behavior(s) that will result in use of a time-out procedure. Be sure the students understand the behavior. Explain the behavior expected of students while they are in time-out. Tell how long the time-out period will last.

2. When the misbehavior occurs, reidentify it. Tell the student in a calm manner, "That is fighting. Go to time-out for __ minutes." No other conversation should ensue. Ignore any statements the student may make as an excuse for misbehavior or relating to feelings about time-out. If necessary, lead the student to the time-out area. If you encounter resistance from the student during this procedure, Hall and Hall (1980) suggest that the teacher

  a. gently but firmly lead the student to time-out.

  b. be prepared to add time to time-out if the student refuses to go or yells, screams, kicks, or turns over furniture.

  c. require the student to clean up any mess resulting from resistance to time-out before the student may return to classroom activities.

  d. be prepared to use a back-up consequence for students who refuse time-out.

3. Once a student enters the time-out area, the time begins. Check your watch or set a timer. Gast and Nelson (1977a) review three formats for contingent release from time-out room procedures:

  a. release contingent upon a specified period (for example, 2 minutes) of appropriate behavior.

  b. release contingent upon a minimum duration of time-out, with an extension until all inappropriate behavior has terminated.

  c. release contingent upon a minimum duration of time-out, with a specified extension (such as 15 seconds) during which no inappropriate responses are exhibited.

4. Once the interval of time has ended, return the student to the previous appropriate activity. Do not comment on how well the student behaved while in time-out. A student should be returned to the activity he or she was engaged in before time-out to avoid negatively reinforcing an escape from that activity.

In order to monitor the effects of time-out and to substantiate proper and ethical use of the procedure, records should be kept of each occasion when time-out is implemented, especially when a time-out room is used. Records should include at least the following information (Gast & Nelson, 1977b):

Monitoring the use of time-out.

1. the student's name

2. the episode resulting in the student's placement in time-out (behavior, activity, other students involved, staff person and so on)

3. the time of day the student was placed in time-out

4. the time of the day the student was released from time-out

5. the total time in time-out

6. the type of time-out (contingent observation, exclusion, or seclusion)

7. the student's behavior in time-out

Prior to selecting a time-out procedure, the teacher should consider the following questions concerning its use:

1. Have more positive procedures—for example, differential reinforcement strategies—been considered?

Questions to ask before using time-out.

2. Have both nonseclusionary and seclusionary time-out procedures been considered?

3. Can time-out be implemented with minimal student resistance? Can the teacher handle the possible resistance?

4. Have the rules of appropriate behavior and the results for misbehavior been clearly explained and understood? Have the rules of behavior while in time-out been clearly explained and understood?

5. Have district regulations concerning the use of time-out procedures been reviewed and complied with?

6. Will appropriate behavior be reinforced in conjunction with the use of time-out?

The following vignette is an example of what might happen should a teacher fail to consider questions like these.

Ms. Sutton Tries Time-Out

Ms. Sutton, a second-grade teacher, read about time-out. She decided that she would use it to teach Aaron not to hit other students. She did not have access to a time-out room, but concluded that putting Aaron outside the classroom in the hall would be just as good.

The next time Aaron hit someone, Ms. Sutton told him, "Aaron, you hit somebody. You have to go to time-out."

She sat him in a chair in the hall and went back to teaching reading. At the end of the period, about an hour later, she went to get him. He came back into the room and hit Elaine before he even got to his seat. Once again, he went to the hall. This pattern was repeated throughout the morning.

Aaron spent most of the time in the hall and the rest of the time hitting people. Ms. Sutton concluded that time-out was an ineffective procedure. Later that day, she heard Aaron say to Elaine, "Hey, I got this all figured out. If I hit you, I get to go sit in the hall. I don't have to do my work, and I get to talk to all

the people in the hall. The principal even came by and asked me what I did. Boy, did I tell him!"

Factors that may make time-out ineffective.

Time-out will not be effective if positive reinforcement is not available in the classroom, if students escape tasks while in time-out, or if reinforcing consequences are available during time-out.

LEVEL IV: PRESENTATION OF AVERSIVE STIMULI

The presentation of an aversive stimulus as a consequence of an inappropriate behavior is, in popular usage, identified with the term *punishment*. As indicated earlier in this chapter, teachers turn to this form of punishment almost by reflex. Perhaps, because many people have been disciplined at home and at school by being yelled at or hit, they therefore learn that way to handle inappropriate behavior on the part of others, at least on the part of those who are physically smaller. From a more functional perspective, this form of punishment is often used because it has three powerful advantages. First and foremost, the use of an aversive stimulus rapidly stops the occurrence of a behavior and has some long-term effects (Azrin, 1960). A child throwing a tantrum who is suddenly hit across the backside will probably stop immediately; a pair of students gossiping in the back of the room will stop when the teacher screams at them. Secondly, the use of aversive stimuli facilitates learning by providing a clear discrimination between acceptable and nonacceptable behavior or between safe and dangerous behavior (Marshall, 1965). The student slapped for spitting, shocked for self-injurious behavior, or hit by a car while dashing across the street clearly and immediately sees the inappropriateness of the behavior. Thirdly, the aversive consequence following a student's inappropriate behavior vividly illustrates to other students the results of engaging in that behavior and, therefore, tends to lessen the probability that those others will engage in the behavior (Bandura, 1965).

What makes the presentation of aversive consequences so appealing to teachers and parents?

In listing these advantages, *we are not recommending the use of aversive consequences* (especially involving physical contact) as a routine management procedure in the classroom or institution. We merely acknowledge the behavioral effects resulting from their use. Physical or other strong aversive consequences are justified only under the most extreme instances of inappropriate behavior. They are appropriate only when safety is jeopardized or in instances of long-standing serious behavior problems. Aversive consequences should be used only after considering appropriate safety and procedural guidelines. It is suggested that guidelines should minimally include

Reread the story on pages 43–44 of Chapter 2. Was the presentation of an aversive stimulus justified in this instance?

1. *demonstrated* and *documented* failure of alternative nonaversive procedures to modify the target behavior.

2. informed written consent by the student's parents or legal guardians, through due process procedures and assurance of their right to withdraw such consent at any time (Rimm & Masters, 1979).

3. the decision to implement an aversive procedure made by a designated body of qualified professionals.

4. a prearranged timetable for review of the effectiveness of the procedure, and discontinuance of the procedure as soon as possible.

5. periodic observation to ensure consistent and reliable administration of the procedure by the staff member.

6. documentation of the effectiveness of the procedure as well as evidence of increased accessibility to instruction.

7. having only designated staff member(s) administer the procedure. (The staff member should have prior instruction in the procedure and have reviewed published studies in the use of the procedure to be familiar with procedure specific guidelines and possible negative effects.)

8. positive reinforcement of incompatible behavior, whenever possible, as a part of any program using aversive stimuli.

Krasner (1976) has pointed out an important distinction between effectiveness and acceptability. It is not the effectiveness of aversive procedures that is in question, but their acceptability to parents, the public, and many professionals. Techniques involving aversive consequences understandably cause concern to many people. While their use may be acceptable in cases such as self-injurious behavior if proper safeguards are implemented, it is doubtful that such drastic measures could or even should become accepted as routine classroom management procedures.

Aversive stimuli may be categorized into two groups: unconditioned aversive stimuli and conditioned aversive stimuli. **Unconditioned aversive stimuli** result in physical pain or discomfort to the student. Examples of such stimuli sometimes used to punish children are slapping, biting, or pinching. This class of stimuli includes anything that causes pain—naturally occurring consequences, such as contact with a hot stove, or contrived consequences, such as the application of electric shock. Because these stimuli immediately produce a behavior change without the need for any previous experience, they may also be termed *universal, natural*, or *unlearned punishers*.

Unconditioned aversive stimuli also include consequences that may be described as *mild* aversives (McGonigle et al., 1982; Singh, 1979; Zegiob et al., 1976). Rather than causing pain, these noxious stimuli result in annoyance, discomfort, or irritation (Dorsey, Iwata, Ong, & McSween, 1980; Singh, 1980; Zegiob et al., 1978). These mild aversive consequences include the administration of substances and the use of physical control.

Substances used as aversive consequences of inappropriate behavior include water, citric acid, and aromatic ammonia. The directing of a fine mist of water towards a subject's face contingent upon the occurrence of a tar-

geted behavior has been employed with behaviors ranging from mouthing of objects (resulting in vomiting or hand-biting for some subjects) to head-banging and tearing of flesh from lips and forearms (Dorsey et al., 1980; Gross, Berler, & Drabman, 1982). Putting citric acid (lemon juice) into the subject's mouth has been used to suppress rumination (Apolito & Sulzer-Azaroff, 1981; Sajwaj, Libet, & Agras, 1974), self-injurious behavior (Mayhew & Harris, 1979), and public masturbation (Cook, Altman, Shaw, & Blaylock, 1978). The strongest of the substance consequences has been the use of aromatic ammonia. Aromatic ammonia (smelling salts) in the form of capsules crushed and held under the subject's nose has been used to reduce self-injurious behavior (Altman & Haavik, 1978; Baumeister & Baumeister, 1978; Singh, 1979; Tanner & Zeiler, 1975) and outward aggression (Doke, Wolery, & Sumberg, 1983).

Physical control refers to procedures requiring direct, physical intervention in order to suppress a targeted behavior. Two such procedures are immobilization or physical restraint and contingent exercise. Immobilization has been used primarily with stereotypic (Reid, Tombaugh, & Van den Heuvel, 1981) and self-injurious behaviors (Hamad, Isley, & Lowry, 1983), and pica (Winton & Singh, 1983). The application of this procedure involves an adult's holding the child's arms immobile, at his side, for a predetermined amount of time.

Contingent exercise requires the student to engage in unrelated physical activity such as push-ups or deep knee bends as a consequence for a targeted behavior. This procedure has been used with self-injurious, stereotypic, and self-stimulatory behaviors (DeCatanzaro & Baldwin, 1978; Kern, Koegel, & Dunlap, 1984; Kern, Koegel, Dyer, Blew, & Fenton, 1982), as well as with Marine Corps recruits who annoy drill sergeants.

Anyone intending to use any of these aversive consequences *needs to review the original studies* for procedure-specific guidelines such as duration and schedule of administration of the consequence and, of special concern, the potential negative effects, both behavioral and physical, resulting from their use.

While these consequences are included within the category of unconditioned aversive stimuli, they are not necessarily universal punishers. Some students may find the procedure reinforcing rather than punishing. A student may indeed like the taste of lemon juice, may find it fun to have water sprayed in his or her face, or enjoy the physical contact with the teacher. In certain instances for example, physical restraint was found to be a reinforcing consequence, and therefore access to restraint contingent upon not performing a target behavior had the effect of behavior reduction (Favell, McGimsey, & Jones, 1978; Favell, McGimsey, Jones, & Cannon, 1981).

Conditioned aversive stimuli are stimuli a person learns to experience as aversive as a result of pairing with an unconditioned aversive stimulus. This class includes consequences such as words and warnings, vocal tones, or gestures. A child, for example, may have experienced being yelled at, paired with being spanked. Yelling may thus have become a conditioned aversive stimulus, as experience has proven to the student that yelling is associated with pain. The pain associated with a conditioned aversive stimulus may also be psychological or social pain or discomfort, usually in the form of embarrassment or ridicule from peers.

Verbal reprimands (shouting or scolding) are the most common form of conditioned aversive stimuli used in the classroom (Heller & White, 1975; Thomas, Presland, Grant, & Glynn, 1978; White, 1975). A series of studies was conducted to identify factors that influence the effectiveness of reprimands (Van Houten, et al., 1982). Three such factors were identified: (1) verbal reprimands were more effective delivered in conjunction with "nonverbal aspects of reprimands such as eye contact and a firm grasp"; (2) "reprimands were more effective when delivered from nearby the student than when delivered from across the room"; and (3) "the delivery of reprimands to one student reduced the disruptive behavior of an adjacent peer" (p. 81).

If unconditioned or conditioned aversive stimuli are to be used as consequences in a behavior-reduction program, they should be used as effectively as possible. As indicated in the functional definition, the teacher must be consistent and immediate in applying the consequences (Azrin, Holz, & Hake, 1963). The rules of behavior must be clearly associated with a contingency that has been previously stated: the "if . . ., then . . ." statement of cause and effect. The student must understand that the aversive consequence is not being arbitrarily applied. The immediacy of application convinces the student of the veracity of the contingency and underlines the connection between a particular behavior and its consequence.

In addition to ensuring consistency and immediacy, the teacher should avoid extended episodes of punishment. Consequences should be quick and directly to the point. If words are a punisher for a student, a few words—such as "Don't run in the hall"—may be far more effective than a 15-minute lecture, most of which the student tunes out.

Punishment is far less effective when the intensity of the aversive stimulus is increased gradually instead of being initially introduced at its full intensity (Azrin & Holz, 1966). With a gradual increase in intensity, the student has the opportunity to become habituated or desensitized to the intensity of the previous application. Such gradual habituation may eventually lead the teacher to administer an intensity level far above what was originally considered necessary to terminate the student's misbehavior.

A desire or actual attempt to escape from an aversive condition is a natural response. If punishment is to be effective in changing undesirable student behavior, the teacher will have to arrange the environment to prevent students' escaping punishment (Azrin, Hake, Holz, & Hutchinson, 1965).

The most important element of any program that includes punishment of inappropriate behavior is to be sure that any punishment is always used in *association with reinforcement* for appropriate behavior. Punishment involves very little learning on the part of the student. In effect, all the student learns is what behavior should not be engaged in. Reinforcing appropriate behavior instructs the student on appropriate or expected behaviors and provides an opportunity for successful or reinforced experiences.

Disadvantages of Using Aversive Stimuli

The disadvantages of using aversive consequences far outweigh the advantage of their immediate effect. The following limitations of such procedures should make teachers stop and consider very carefully before choosing to use aversives:

1. In the face of aggressive punishers, the student sees three behavior options:

   a. The student may become aggressive and strike back (for example, yell back at the teacher or even become physically aggressive). A reaction may be triggered that will result in escalation of the situation.

   b. The student may become withdrawn. He or she may stand there, completely tune out the punisher and remain tuned out for the remainder of the day, thereby preventing any learning.

   c. The student may engage in some form of escape and avoidance behavior. Once a person has run out of the room, a punisher in the classroom can have no immediate effect.

2. We know that a most basic and powerful form of instruction and learning takes place through modeling or imitation. Because a teacher is a figure of respect and authority, his or her behavior is closely observed by the students. The teacher becomes a model of adult behavior in reacting to various types of situations. The teacher who yells or hits is, in effect, saying to the students that this is how an adult reacts and copes with undesirable behaviors in the environment. Students may, through such a model, learn an inappropriate form of behavior.

3. Unless students are guided to make a clear distinction about what behavior is being punished, they may come to fear and avoid the teacher or the entire setting in which the punishment occurred. Consider

   a. The fourth-grade teacher whose students flinch when she walks up and down between the rows of desks. She is a slapper.

   b. The little girl in an institution who suddenly would not sleep in her room at night. She stayed up crying and screaming until she was taken out in the hall. It was discovered that an attendant had been taking her into her room to be paddled for bad behaviors during the day.

4. Many interactions that teachers consider punishers function as positive reinforcers instead. A child may find making an adult lose control and look ridiculous very reinforcing.

A final disadvantage to the use of punishment is illustrated by the following vignette.

Professor Grundy Teaches Dennis a Thing or Two

Professor Grundy's 5-year-old nephew, Dennis, was spending the week at the professor's house. One of Dennis' more unpleasant habits was jumping on the bed. Mrs. Grundy had asked Dennis not to do it, but this had been completely ineffective.

Professor Grundy was sitting in his easy chair, smoking his pipe, and reading a professional journal. He heard the unmistakable "crunch, crunch, crunch" of the bedsprings directly overhead in the guest bedroom. "Minerva," he said, "the time has come for drastic action. Where is that flyswatter?"

"Oliver, you're not going to beat the child, are you?" asked Minerva.

"Certainly not, my dear," replied the professor. "I shall merely apply an unconditioned aversive stimulus, contingently, at a maximum intensity."

The professor took the flyswatter in hand and tiptoed up the carpeted steps in his stockinged feet. He continued to tiptoe into the bedroom, where he observed Dennis happily jumping on the bed. Dennis did not observe the professor; his back was to the door. Grundy applied the immediate, contingent, and intense aversive stimulus saying, firmly, "Do *not* jump on the bed." Dennis howled.

"That, I think," said the Professor to Mrs. Grundy, "will teach Dennis a thing or two." In fact, Dennis did not jump on the bed all weekend. Professor Grundy was at home all the time and knew he would have heard the springs.

On Monday, when the Professor arrived home from the university, Minerva met him at the door. "Oliver," she said, "I don't know which thing or two you planned to teach Dennis, but the one he learned was not to jump on the bed when you were home. He's been at it all day."

"Oliver, let's first try offering him a treat if he doesn't jump on the bed."

Like Dennis, what students learn most often from punishment using aversive stimuli is not to perform the behavior when the person who applied the punishment is present. They learn not to get caught! They don't learn to behave appropriately.

Overcorrection

Overcorrection was developed as a behavior-reduction procedure that includes training in appropriate behaviors. Overcorrection procedures are therefore considered educative (Azrin & Foxx, 1971). The purpose of overcorrection is to teach students to take responsibility for their inappropriate acts and to teach them appropriate behaviors. Correct behavior is taught through an exaggeration of experience. The exaggeration of experience characteristic of overcorrection contrasts with a simple correction procedure in which a student rectifies an error of behavior but is not necessarily required to follow that with exaggerated or extended practice of the appropriate behavior.

There are two basic types of overcorrection procedures. *Restitutional overcorrection* is used when a setting has been disturbed by a student's misbehavior. The student must overcorrect the setting she or he has disturbed. *Positive-practice overcorrection* is used when the form of a behavior is inappropriate. In this procedure, the student practices an exaggerated correct form of an appropriate behavior (Foxx & Azrin, 1973).

In general, overcorrection procedures should have the following characteristics (Foxx & Azrin, 1972):

1. They should be directly related to the misbehavior of concern so as not to become arbitrary and punitive.

2. The overcorrection should be immediately contingent upon the disruptive action.

3. Performance of overcorrection acts should require very active participation and effort, without pausing.

4. The duration of overcorrection should be extended.

Successful use of overcorrection has been demonstrated using various durations. Overcorrection periods may last anywhere from 30 seconds (Azrin, Kaplan, & Foxx, 1973) or 1 minute (Freeman, Graham, & Ritvo, 1975) to as long as 2 hours (Foxx, 1976), depending on the behavior involved. Successful reduction of stereotypic hand movements has been reported after requiring a student to engage in an overcorrection procedure for 2.5 minutes (Epstein, et al., 1974) as well as 5 minutes (Foxx & Azrin, 1973).

Restitutional Overcorrection

A restitutional overcorrection procedure requires that the student restore or correct an environment that he or she has disturbed not only to its original condition, but beyond that. For example, when the teacher catches a student throwing spit-balls, she is employing simple correction when she says, "Michael pick that up and throw it in the trash." She is employing restitutional overcorrection when she says, "Michael pick that up and throw it in the trash, and now pick up all the other papers on the floor."

This form of environmental restoration was used by Azrin & Foxx (1971) as part of their toilet-training program. When a child had an accident, she or he had to undress, wash the clothes, hang them to dry, shower, obtain clean clothing, dress, and then clean the soiled area of the lavatory. A variation was employed by Azrin and Wesolowski (1974) when, in order to eliminate stealing, they required the thief to return not only a stolen item but also an additional identical item to the person who had been robbed.

A review of studies by Rusch and Close (1976) shows restitutional overcorrection techniques used to reduce various classes of disruptive behavior:

1. In cases where objects were disturbed or rearranged, all objects (such as furniture) within the immediate area where a disruption occurred were straightened, not merely the originally disturbed objects.

2. In cases where someone annoyed or frightened others, all persons present were to be apologized to, not just those frightened by social misbehavior such as hitting or cursing.

3. In cases of self-inflicted oral infection, a thorough cleansing of the mouth with an oral antiseptic followed unhygienic oral contacts such as biting people or chewing objects.

4. In cases where medical attention is required due to injuries caused by acts of physical aggression, the aggressor had to cleanse and medicate such wounds.

5. In cases of agitation, a period of absolutely quiet was imposed following commotions such as shrieking and screaming.

Positive-practice Overcorrection

With positive-practice overcorrection, the student, having engaged in an inappropriate behavior, is required to engage in exaggerated or overly correct practice of the appropriate way of behaving. If, for example, a class runs to line up for recess, the teacher who makes them sit back down and line up again is using simple correction. The teacher who makes them sit back down, then practice getting in line several times while reciting the rules for doing it the right way is using positive-practice overcorrection.

Positive practice overcorrection of autistic-like behavior is sometimes called autism reversal.

To ensure the educative nature intended by this procedure, the practice should be of an alternative appropriate behavior topographically similar to the inappropriate behavior. Foxx and Azrin (1973) and Azrin et al. (1973) used this procedure to reverse the stereotypic behavior of autistic children. As seen in Figure 7–4, the researchers identified and then required the students to practice exaggerations of appropriate postures for extended periods of time. Azrin and Foxx (1971) used positive-practice procedures in

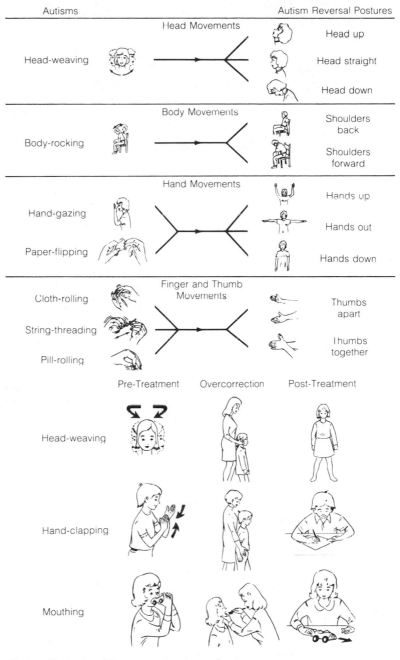

Figure 7–4   Positive practice in reducing stereotype behaviors

Note (top) From "The elimination of autistic self-stimulatory behavior by overcorrection by R. Foxx & N. Azrin, *Journal of Applied Behavior Analysis,* 1973. 1–14. Copyright 1973 by Society for the Experimental Analysis of Behavior, Inc. Reprinted by permission.

Note (bottom) From "Autism reversal Eliminating stereotyped self-stimulation of retarded individuals" by N. Azrin, S. Kaplan & R. Foxx. *American Journal on Mental Deficiency,* 1973, 70, 241–248. Copyright 1973 by Society for the Experimental Analysis of Behavior, Inc. Reprinted by permission.

their toilet-training program by artificially increasing the frequency of urinations through offering the students large quantities of appealing liquids. This technique increased the opportunity for practice and reinforcement. Azrin and Wesolowski (1975) eliminated floor sprawling by requiring the student to practice sitting on several chairs (one at a time, of course) for an extended time period. Classroom disturbances have been overcorrected by requiring students to practice reciting correct classroom procedures to the teacher during time periods such as recess (Azrin, Azrin, & Armstrong, 1977; Azrin & Powers, 1975; Bornstein, Hamilton, & Quevilon, 1977).

Overcorrection procedures should not themselves be allowed to become positively reinforcing. Indeed, a quality of aversiveness is involved in their use. Restitutional or positive-practice overcorrection procedures usually include the following components (Epstein et al., 1974; Rusch & Close, 1976):

Guidelines for using overcorrection.

1. telling the student she or he behaved inappropriately
2. stopping the student's ongoing activity
3. providing systematic verbal instructions for the overcorrection activity in which the student is to engage
4. forcing the practice of correctional behavior (manually guiding the desired movements, using as much bodily pressure as necessary, but reducing such pressure immediately as the person begins to perform the movement with verbal instruction alone)
5. returning the student to the ongoing activity

Before using an overcorrection procedure, teachers should consider the following management concerns:

1. Implementation of overcorrection requires the full attention of the teacher. She must be physically close to the student to ensure that he complies with the overcorrection instruction and ready to intervene with physical guidance if necessary.

2. Overcorrection procedures tend to be time consuming, sometimes lasting 5 to 15 minutes and possibly longer (Foxx & Azrin, 1973; Ollendick & Matson, 1976; Sumner, Meuser, Hsu, & Morales, 1974). However, recent studies suggest that short-duration implementation may be at least as effective as longer durations in facilitating behavior changes, especially when appropriate alternative behavior is being taught emphasizing the educative rather than punishment potential of the procedure (Carey & Bucher, 1983; Conley & Wolery, 1980).

3. Because physical contact with the student is involved in the use of overcorrection, the teacher should be aware of the possibility of aggression on the part of the student (Carey & Bucher, 1983; Rollings, Baumeister, & Baumeister, 1977) or attempts to escape and avoid the aversive situation.

4. During long periods of overcorrection the student may become so disruptive that the teacher cannot guide him through the overcorrection procedure (Matson & Stephens, 1977).

5. Because overcorrection often involves physical guidance for extended periods, it may be a very aversive procedure to the adults implementing it (Repp, 1983).

Two procedures that result in behavior reduction at times may be misidentified as a form of overcorrection or confused with those procedures. These other approaches are *negative practice* and *stimulus satiation*. The confusion may occur because, as in overcorrection, these procedures involve providing an exaggeration of experience.

Negative practice (Dunlap, 1928, 1930, 1932) may be confused with positive practice. Negative practice requires that the student repeatedly perform the *inappropriate* behavior. There is no pretense that the procedure is educative. The procedure is based upon the assumption that repeated performance will result in response fatigue or satiation. For example, if a certain student's inappropriate behavior is getting up and running around the room during a lesson, positive practice might involve extended time periods sitting in various chairs, whereas negative practice might involve requiring the student to run and run and run and run.

Negative practice has been used to reduce small motor behaviors (Dunlap, 1928, 1930). A limited number of studies report the use of this procedure. Those that have been published have dealt with the reduction of tics (Walton, 1961; Yates, 1958), smoking (De Lahunt & Curran, 1976), the elimination of grimacing and unusual body movements made by cerebral palsied children when speaking (Rutherford, 1940), and self-injurious behavior (Mogel & Schiff, 1967).

Negative practice depends on response fatigue or satiation. **Stimulus satiation**, on the other hand, depends on the student's becoming satiated with the antecedent to the behavior. Ayllon (1963) used stimulus satiation with an individual who hoarded large numbers of towels and stored them in her room in a psychiatric hospital. In order to reduce this hoarding behavior, nurses took towels to the woman when she was in her room and simply handed them to her without any comment. The first week she was given an average of 7 towels daily; by the third week, the number was increased to 60 towels. When the number of towels kept in her room reached 625, she started to take a few out. Thereafter, no more towels were given to her. During the next 12 months, the average number of towels found in her room was 1–5 per week, compared with the baseline range of 13–29.

Overcorrection may provide an alternative to aversive consequences in the classroom. It is important to remember that overcorrection procedures, although they have some aversive features, are to be used not as retaliative but as educative tools. The teacher's tone and manner make a difference in the way the procedures will be received by students. A teacher who uses an angry or haranguing tone of voice or unnecessary force when guiding students through overcorrection procedures may increase the probability of resistance. Firmness without aggression is the aim here.

SUMMARY

This chapter reviewed a number of procedures to decrease or eliminate inappropriate or maladaptive behaviors: differential reinforcement, reinforcement of incompatible behaviors, extinction, punishment, and overcorrection. These procedures are most usefully and constructively viewed as a hierarchy of approaches, from those emphasizing reinforcement to those having aversive features. (See again Figure 7–1.)

We stressed throughout the chapter that procedures to decrease behaviors should be chosen only when the behaviors in question are clearly interfering with a student's ability to learn or are presenting a danger to the student or to other people. Positive reinforcement of appropriate behavior should always be combined with any procedure to decrease or eliminate behavior.

REFERENCES

ADAMS, N., MARTIN, R., & POPELKA, G. 1971. The influence of timeout on stutterers and their dysfluency. *Behavior Therapy, 2*, 334–339.

AIKEN, J., & SALZBERG, C. 1984. The effects of a sensory extinction procedure on stereotypic sounds of two autistic children. *Journal of Autism and Developmental Disorders, 14*, 291–299.

ALTMAN, K., & HAAVIK, S. 1978. Punishment of self-injurious behavior in natural settings using contingent aromatic ammonia. *Behavior Research and Therapy, 16*, 85–96.

APOLITO, P., & SULZER-AZAROFF, B. 1981. Lemon-juice therapy: The control of chronic vomiting in a twelve-year-old profoundly retarded female. *Education and Treatment of Children, 4*, 339–347.

AYLLON, T. 1963. Intensive treatment of psychotic behavior by stimulus satiation and food reinforcement. *Behavior Research and Therapy, 1*, 53–61.

AYLLON, T., LAYMAN, D., & BURKE, S. 1972. Disruptive behavior and reinforcement of academic performance. *Psychological Record, 22*, 315–323.

AYLLON, T., & ROBERTS, M.D. 1974. Eliminating discipline problems by strengthening academic

performance. *Journal of Applied Behavior Analysis, 7*, 71–76.

AZRIN, N.H. 1960. Effects of punishment intensity during variable-interval reinforcement. *Journal of the Experimental Analysis of Behavior, 3*, 128–142.

AZRIN, N.H., & FOXX, R.M. 1971. A rapid method of toilet training the institutionalized retarded. *Journal of Applied Behavior Analysis, 4*, 89–99.

AZRIN, N.H., HAKE, D.G., HOLZ, W.C., & HUTCHINSON R.R. 1965. Motivational aspects of escape from punishment. *Journal of the Experimental Analysis of Behavior, 8*, 31–44.

AZRIN, N.H., & HOLZ, W.C. 1966. Punishment. In W.A. Honig (Ed.), *Operant behavior: Areas of research and application*. New York: Appleton-Century-Crofts.

AZRIN, N.H., HOLZ, W.C., & HAKE, D.F. 1963. Fixed-ratio punishment. *Journal of the Experimental Analysis of Behavior, 6*, 141–148.

AZRIN, N.H., HUTCHINSON, R.R., & HAKE, D.J. 1966. Extinction-induced aggression. *Journal of the Experimental Analysis of Behavior, 9*, 191–204.

AZRIN, N.H., KAPLAN, S.J., & FOXX, R.M. 1973. Autism reversal: Eliminating stereotyped self-stimulation of retarded individuals. *American Journal of Mental Deficiency, 78*, 241–248.

AZRIN, N.H., & POWERS, M. 1975. Eliminating classroom disturbances of emotionally disturbed children by positive practice procedures. *Behavior Therapy, 6*, 525–534.

AZRIN, N.H., & WESOLOWSKI, M.D. 1974. Theft reversal: An overcorrection procedure for eliminating stealing by retarded persons. *Journal of Applied Behavior Analysis, 7*, 577–581.

AZRIN, N.H., & WESOLOWSKI, M.D. 1975. The use of positive practice to eliminate persistent floor sprawling by profoundly retarded persons. *Behavior Therapy, 6*, 627–631.

AZRIN, V., AZRIN, N.H., & ARMSTRONG, P. 1977. The student-oriented classroom: A method of improving student conduct and satisfaction. *Behavior Therapy, 8*, 193–204.

BAER, A.M., ROWBURY, T., & BAER, D.M. 1973. The development of instructional control over classroom activities of deviant preschool children. *Journal of Applied Behavior Analysis, 6*, 289–298.

BANDURA, A. 1965. Influence of models' reinforcement contingencies on the acquisition of imitative responses. *Journal of Personality and Social Psychology, 1*, 589–595.

BARON, A., KAUFMAN, A., & RAKAVSKAS, I. 1967. Ineffectiveness of "time out" punishment in suppressing human operant behavior. *Psychonomic Science, 8*, 329–330.

BARTON, E.S. 1970. Inappropriate speech in a severely retarded child: A case study in language conditioning and generalization. *Journal of Applied Behavior Analysis, 3*, 299–307.

BAUMEISTER, A., & BAUMEISTER, A. 1978. Suppression of repetitive self-injurious behavior by contingent inhalation of aromatic ammonia. *Journal of Autism and Childhood Schizophrenia, 8*, 71–77.

BAUMEISTER, A.A., & FOREHAND, R. 1971. Effects of extinction of an instrumental response on stereotyped body rocking in severe retardates. *Psychological Record, 21*, 235–240.

BENOIT, R.B., & MAYER, G.R. 1974. Extinction: Guidelines for its selection and use. *The Personnel and Guidance Journal, 52*, 290–295.

BIRNBRAUER, J.S., WOLF, M.M., KIDDER, J.D., & TAGUE, C.E. 1965. Classroom behavior of retarded pupils with token reinforcement. *Journal of Experimental Child Psychology, 2*, 219–235.

BORNSTEIN, P., HAMILTON, S., & QUEVILLON, R. 1977. Behavior modification by long-distance: Demonstration of functional control over disruptive behavior in a rural classroom setting. *Behavior Modification, 1*, 369–380.

BROWN, P., & ELLIOTT, R. 1965. The control of aggression in a nursery school class. *Journal of Experimental Child Psychology, 2*, 102–107.

BURCHARD, J.D., & BARRERA, F. 1972. An analysis of timeout and response cost in a programmed environment. *Journal of Applied Behavior Analysis, 5*, 271–282.

CAREY, R., & BUCHER, B. 1983. Positive practice overcorrection: The effects of duration of positive practice on acquisition and response duration. *Journal of Applied Behavior Analysis, 16*, 101–109.

CARLSON, C.S., ARNOLD, D.R., BECKER, W.C., & MADSEN, G.H. 1968. The elimination of tantrum behavior of a child in an elementary classroom. *Behavior Research and Therapy, 6*, 117–120.

CONLEY, O., & WOLERY, M. 1980. Treatment by overcorrection of self-injurious eye gouging in preschool blind children. *Journal of Behavior Therapy and Experimental Psychiatry, 11*, 121–125.

COOK, J., ALTMAN, K., SHAW, J., & BLAYLOCK, M. 1978. Use of contingent lemon juice to eliminate public masturbation by a severely retarded boy. *Behavior Research and Therapy, 16*, 131–134.

DECATANZARO, D., & BALDWIN, G. 1978. Effective treatment of self-injurious behavior through a forced arm exercise. *Journal of Applied Behavior Analysis, 11*, 433–439.

DEITZ, S.M., & REPP, A.C. 1973. Decreasing classroom misbehavior through the use of DRL schedules of reinforcement. *Journal of Applied Behavior Analysis, 6*, 457–463.

DEITZ, S.M., & REPP, A.C. 1974. Differentially reinforcing low rates of misbehavior with normal elementary school children. *Journal of Applied Behavior Analysis, 7*, 622.

DEITZ, S.M., REPP, A.C., & DEITZ, D.E.D. 1976. Reducing inappropriate classroom behavior of retarded students through three procedures of differential reinforcement. *Journal of Mental Deficiency Research, 20*, 155–170.

DE LAHUNT, J., & CURRAN, J.P. 1976. Effectiveness of negative practice and self-control techniques in the reduction of smoking behavior. *Journal of Consulting and Clinical Psychology, 44*, 1002–1007.

DOKE, L., WOLERY, M., & SUMBERG, C. 1983. Treating chronic aggression. *Behavior Modification, 7*, 531–556.

DORSEY, M.F., IWATA, B.A., ONG, P., & MCSWEEN, T.E. 1980. Treatment of self-injurious behavior using a water mist: Initial response suppression and generalization. *Journal of Applied Behavior Analysis, 13*, 343–353.

DUNLAP, K. 1928. A revision of the fundamental law of habit formation. *Science, 67*, 360–362.

DUNLAP, K. 1930. Repetition in breaking of habits. *The Scientific Monthly, 30*, 66–70.

DUNLAP, K. 1932. *Habits, their making and unmaking*. New York: Liveright.

EDELSON, R., & SPRAGUE, R. 1974. Conditioning of activity level in a classroom with institutionalized retarded boys. *American Journal of Mental Deficiency, 78*, 384–388.

EPSTEIN, L.H., DOKE, L.A., SAJWAJ, T.E., SORRELL, S., & RIMMER, B. 1974. Generality and side effects of overcorrection. *Journal of Applied Behavior Analysis, 7*, 385–390.

FAVELL, J., 1973. Reduction of stereotypes by reinforcement of toy play. *Mental Retardation, 11*, 21–23.

FAVELL, J.E., MCGIMSEY, J.F., & JONES, M.L. 1978. The use of physical restraint in the treatment of self-injury and as positive reinforcement. *Journal of Applied Behavior Analysis, 11*, 225–241.

FAVELL, J., MCGIMSEY, J., JONES, M., & CANNON, P. 1981. Physical restraint as positive reinforcement. *American Journal of Mental Deficiency, 85*, 425–432.

FERSTER, C.B., & PERROTT, M.C. 1968. *Behavior principles*. New York: Appleton-Century-Crofts.

FOXX, R.M. 1976. Increasing a mildly retarded woman's attendance at self-help classes by overcorrection and instruction. *Behavior Therapy, 7*, 390–396.

FOXX, R.M., & AZRIN, N.H. 1972. Restitution: A method of eliminating aggressive-disruptive behavior of retarded and brain damaged patients. *Behaviour Research and Therapy, 10*, 15–27.

FOXX, R.M., & AZRIN, N.H. 1973. The elimination of autistic self-stimulatory behavior by overcorrection. *Journal of Applied Behavior Analysis, 6*, 1–14.

FOXX, R.M., & SHAPIRO, S.T. 1978. The timeout ribbon: A nonseclusionary timeout procedure. *Journal of Applied Behavior Analysis, 11*, 125–136.

FREEMAN, B., GRAHAM, V., & RITVO, E. 1975. Reduction of self-destructive behavior by overcorrection. *Psychological Reports, 37*, 446.

GAST, D., & NELSON, C.M. 1977a. Legal and ethical considerations for the use of timeout in special education settings. *The Journal of Special Education, 11*, 457–467.

GAST, D., & NELSON, C.M. 1977b. Time out in the classroom: Implications for special education. *Exceptional Children, 43*, 461–464.

GILBERT, G. 1975. Extinction procedures: Proceed with caution. *Mental Retardation, 13*, 25–29.

GOETZ, E.M., HOLMBERG, M.C., & LEBLANC, J.M. 1975. Differential reinforcement of other behavior and noncontingent reinforcement as control procedures during the modification of a preschooler's compliance. *Journal of Applied Behavior Analysis, 8*, 77–82.

GROSS, A., BERLER, E., & DRABMAN, R. 1982. Reduction of aggressive behavior in a retarded boy using a water squirt. *Journal of Behaviour Therapy and Experimental Psychiatry, 13*, 95–98.

HALL, R.V., FOX, R., WILLARD, D., GOLDSMITH, L., EMERSON, M., OWEN, M., DAVIS, F., & PORCIA, E. 1971. The teacher as observer and experimenter in the modification of disputing and talking-out behaviors. *Journal of Applied Behavior Analysis, 4*, 141–149.

HALL, R.V., & HALL, M. 1980. *How to use time out*. Lawrence, Kan.: H & H Enterprises.

HALL, R.V., LUND, D., & JACKSON, D. 1968. Effects of teacher attention on study behavior. *Journal of Applied Behavior Analysis, 1*, 1–12.

HAMAD, C., ISLEY, E., & LOWRY, M. 1983. The use of mechanical restraint and response incompatibility to modify self-injurious behavior: A case study. *Mental Retardation, 21*, 213–217.

HARRIS, S.L., & WOLCHIK, S.A. 1979. Suppression of self-stimulation: Three alternative strategies. *Journal of Applied Behavior Analysis, 12*, 185–198.

HARRIS, V.W., & SHERMAN, J.A. 1973. Use and analysis of the "Good Behavior Game" to reduce disruptive classroom behavior. *Journal of Applied Behavior Analysis, 6*, 405–417.

HELLER, M., & WHITE, M. 1975. Rates of teacher approval and disapproval to higher and lower ability classes. *Journal of Educational Psychology, 67*, 796–800.

IWATA, B.A., & BAILEY, J.S. 1974. Reward versus cost token systems: An analysis of the effects on students and teacher. *Journal of Applied Behavior Analysis, 7*, 567–576.

KAZDIN, A.E. 1972. Response cost: The removal of conditioned reinforcers for therapeutic change. *Behavior Therapy, 3*, 533–546.

KAZDIN, A.E., & BOOTZIN, R.R. 1972. The token economy: An evaluative review. *Journal of Applied Behavior Analysis, 5*, 343–372.

KERN, L., KOEGEL, R., & DUNLAP, G. 1984. The influence of vigorous versus mild exercise on autistic stereotyped behaviors. *Journal of Autism and Developmental Disorders, 14*, 57–67.

KERN, L., KOEGEL, R., DYER, K., BLEW, P., & FENTON, L. 1982. The effects of physical exercise on self-stimulation and appropriate responding in autistic children. *Journal of Autism and Developmental Disorders, 12*, 399–419.

KRASNER, L. 1976. Behavioral modification: Ethical issues and future trends. In H. Leitenberg (Ed.), *Handbook of behavior modification and behavior therapy*, pp. 627–649. Englewood Cliffs, N.J.: Prentice-Hall.

LAHEY, B.B., MCNEES, M.P., & MCNEES, M.C. 1973. Control of an obscene "verbal tic" through timeout in an elementary school classroom. *Journal of Applied Behavior Analysis, 6*, 101–104.

LIBERMAN, R.P., TEIGEN, J., PATTERSON, R., & BAKER, V. 1973. Reducing delusional speech in chronic, paranoid schizophrenics. *Journal of Applied Behavior Analysis, 6*, 57–64.

LOVAAS, O.I., & SIMMONS, J.Q. 1969. Manipulation of self-destruction in three retarded children. *Journal of Applied Behavior Analysis, 2*, 143–157.

LUTZKER, J. 1974. Social reinforcement control of exhibitionism in a profoundly retarded adult. *Mental Retardation, 12*, 46–47.

MACPHERSON, E.M., CANDEE, B.L., & HOHMAN, R.J. 1974. A comparison of three methods for eliminating disruptive lunchroom behavior. *Journal of Applied Behavior Analysis, 7*, 287–297.

MARHOLIN, D., II, & GRAY, D. 1976. Effects of group response-cost procedures on cash shortages in a small business. *Journal of Applied Behavior Analysis, 9*, 25–30.

MARSHALL, H. 1965. The effect of punishment on children. A review of the literature and a suggested hypothesis. *Journal of Genetic Psychology, 106*, 23–33.

MATSON, J., & STEPHENS, R. 1977. Overcorrection of aggressive behavior in a chronic psychiatric patient. *Behavior Modification, 1*, 559–564.

MAY, J., MCALLISTER, J., RISLEY, T., TWARDOSZ, S., & COX, C. 1974. *Florida guidelines for the use of behavioral procedures in state programs for the retarded.* Florida Division of Retardation, Tallahassee.

MAYHEW, G., & HARRIS, F. 1979. Decreasing self-injurious behavior: Punishment with citric acid and reinforcement of alternative behaviors. *Behavior Modification, 3*, 322–336.

MCGONIGLE, J., DUNCAN, D., CORDISCO, L., & BARRETT, R. 1982. Visual screening: An alternative method for reducing stereotypic behaviors. *Journal of Applied Behavior Analysis, 15*, 461–467.

MCSWEENY, A.J. 1978. Effects of response cost on the behavior of a million persons: Charging for directory assistance in Cincinnati. *Journal of Applied Behavior Analysis, 11*, 47–51.

MOGEL, S., & SCHIFF, W. 1967. Extinction of a head-bumping symptom of eight years' duration in two minutes: A case report. *Behavior Research and Therapy, 5*, 131–132.

MYERS, D.V. 1975. Extinction, DRO, and response-cost procedures for eliminating self-injurious behavior: A case study. *Behavior Research and Therapy, 13*, 189–192.

OLLENDICK, T., & MATSON, J. 1976. An initial investigation into the parameters of overcorrection. *Psychological Reports, 39*, 1139–1142.

PATTERSON, G.R. 1965. An application of conditioning techniques to the control of a hyperactive child. In L. P. Ullmann & L. Krasner (Eds.), *Case studies in behavior modification* (pp. 370–375). New York: Holt, Rinehart and Winston.

PHILLIPS, E.L., PHILLIPS, E.A., FIXSEN, D.L., & WOLF, M.M. 1971. Achievement place: Modification of the behaviors of pre-delinquent boys within a token economy. *Journal of Applied Behavior Analysis, 4*, 45–59.

PINKSTON, E.M., REESE, N.M., LEBLANC, J.M., & BAER, D.M. 1973. Independent control of a preschool child's aggression and peer interaction by contingent teacher attention. *Journal of Applied Behavior Analysis, 6*, 115–124.

PORTERFIELD, J.K., HERBERT-JACKSON, E., & RISLEY, T.R. 1976. Contingent observation: An effective and acceptable procedure for reducing disruptive behavior of young children in a group setting. *Journal of Applied Behavior Analysis, 9*, 55–64.

REID, J., TOMBAUGH, T., & VAN DEN HEUVEL, K. 1981. Application of contingent physical restraint to suppress stereotyped body rocking of profoundly retarded persons. *American Journal of Mental Deficiency, 86*, 78–85.

REPP, A. 1983. *Teaching the mentally retarded*. Englewood Cliffs, N.J.: Prentice-Hall.

REPP, A.C., & DEITZ, D.E.D. 1979. Reinforcement-based reductive procedures: Training and monitoring performance of institutional staff. *Mental Retardation, 17*, 221–226.

REPP, A.C., & DEITZ, S.M. 1974. Reducing aggressive and self-injurious behavior of institutionalized retarded children through reinforcement of other behaviors. *Journal of Applied Behavior Analysis, 7*, 313–325.

REPP, A.C., DIETZ, S.M., & DIETZ, D.E.D. 1976. Reducing inappropriate behaviors in classrooms and in individual sessions through DRO schedules of reinforcement. *Mental Retardation, 14*, 11–15.

REPP, A.C., DEITZ, S.M., & SPEIR, N.C. 1974. Reducing stereotypic responding of retarded persons by the differential reinforcement of other behavior. *American Journal of Mental Deficiency, 79*, 279–284.

REYNOLDS, G.S. 1961. Behavioral contrast. *Journal of the Experimental Analysis of Behavior, 4*, 57–71.

RIMM, D.C., & MASTERS, J.C. 1979. *Behavior therapy: Techniques and empirical findings*. New York: Academic Press.

RINCOVER, A. 1981. *How to use sensory extinction*. Lawrence, Kans: H & H Enterprises.

ROLLINGS, J., BAUMEISTER, A., & BAUMEISTER, A. 1977. The use of overcorrection procedures to eliminate the stereotyped behaviors of retarded individuals: An analysis of collateral behaviors and generalization of suppressive effects. *Behavior Modification, 1*, 29–46.

ROSE, T. 1979. Reducing self-injurious behavior by differentially reinforcing other behaviors. *AAESPH Review, 4*, 179–186.

RUSCH, F., & CLOSE, D. 1976. Overcorrection: A procedural evaluation. *AAESPH Review, 1*, 32–45.

RUTHERFORD, B. 1940. The use of negative practice in speech therapy with children handicapped by cerebral palsy, athetoid type. *Journal of Speech Disorders, 5*, 259–264.

SAJWAJ, T., LIBET, J., & AGRAS, S. 1974. Lemon-juice therapy: The control of life-threatening rumination in a six-month-old infant. *Journal of Applied Behavior Analysis, 7*, 557–563.

SCHMIDT, G.W., & ULRICH, R.E. 1969. Effects of group contingent events upon classroom noise. *Journal of Applied Behavior Analysis, 2*, 171–179.

SCOTT, P.M., BURTON, R.V., & YARROW, M.R. 1967. Social reinforcement under natural conditions. *Child Development, 38*, 53–63.

SIEGEL, G.M., LENSKE, J., & BROEN, P. 1969. Suppression of normal speech disfluencies through response cost. *Journal of Applied Behavior Analysis, 2*, 265–276.

SINGH, N. 1979. Aversive control of breath-holding. *Journal of Behavior Therapy and Experimental Psychiatry, 10*, 147–149.

SINGH, N. 1980. The effects of facial screening on infant self-injury. *Journal of Behaviour Therapy and Experimental Psychiatry, 11*, 131–134.

SINGH, N., BEALE, I., & DAWSON, M. 1981. Duration of facial screening and suppression of self-injurious behaviour: Analysis using an alternative treatments design. *Behavioral Assessment, 3*, 411–420.

SKINNER, B.F. 1953. *Science and human behavior.* New York: Macmillan.

SOLOMON, R.W., & WAHLER, R.G. 1973. Peer reinforcement control of classroom problem behavior. *Journal of Applied Behavior Analysis, 6,* 49–56.

SUMNER, J., MEUSER, S., HSU, L., & MORALES, R. 1974. Overcorrection treatment of radical reduction of aggressive-disruptive behavior in institutionalized mental patients. *Psychological Reports, 35,* 655–662.

TANNER, B.A., & ZEILER, M. 1975. Punishment of self-injurious behavior using aromatic ammonia as the aversive stimulus. *Journal of Applied Behavior Analysis, 8,* 53–57.

TARPLEY, H., & SCHROEDER, S. 1979. Comparison of DRO and DRI on rate of suppression of self-injurious behavior. *American Journal of Mental Deficiency, 84,* 188–194.

THOMAS, J.D., PRESLAND, I.E., GRANT, M.D., & GLYNN, T.L. 1978. Natural rates of teacher approval and disapproval in grade-7 classrooms. *Journal of Applied Behavior Analysis, 11,* 91–94.

TWARDOSZ, S., & SAJWAJ, T. 1972. Multiple effects of a procedure to increase sitting in a hyperactive, retarded boy. *Journal of Applied Behavior Analysis, 5,* 73–78.

VAN HOUTEN, R., NAU, P., MACKENZIE-KEATING, S., SAMEOTO, D., & COLAVECCHIA, B. 1982. An analysis of some variables influencing the effectiveness of reprimands. *Journal of Applied Behavior Analysis, 15,* 65–83.

VUKELICH, R., & HAKE, D.F. 1971. Reduction of dangerously aggressive behavior in a severely retarded resident through a combination of positive reinforcement procedures. *Journal of Applied Behavior Analysis, 4,* 215–225.

WALTON, D. 1961. Experimental psychology and the treatment of a tiqueur. *Journal of Child Psychology and Psychiatry, 2,* 148–155.

WATSON, L.S. 1967. Application of operant conditioning techniques to institutionalized severely and profoundly retarded children. *Mental Retardation Abstracts, 4,* 1–18.

WEISBERG, P., PASSMAN, R.H., & RUSSELL, J.E. 1973. Development of verbal control over bizarre gestures of retardates through imitative and nonimitative reinforcement procedures. *Journal of Applied Behavior Analysis, 6,* 487–495.

WHITE, M.A. 1975. Natural rates of teacher approval and disapproval in the classroom. *Journal of Applied Behavior Analysis, 8,* 367–372.

WILSON, C.W., & HOPKINS, B.L. 1973. The effects of contingent music on the intensity of noise in junior high home economics classes. *Journal of Applied Behavior Analysis, 6,* 269–275.

WINTON, A., & SINGH, N. 1983. Suppression of pica using brief-duration physical restraint. *Journal of Mental Deficiency Research, 27,* 93–103.

Wyatt v. Stickney, 344F. Supp. 373, 344F. Supp. 387 (M.D. Ala. 1972) affirmed sub nom. *Wyatt v. Aderholt,* 503 F. 2nd 1305 (5th Cir. 1974).

YATES, A. J. 1958. Symptoms and symptom substitution. *Psychological Review, 65,* 371–374.

ZEIGOB, L., ALFORD, G., & HOUSE, A. 1978. Response suppressive and generalization effects of facial screening on multiple self-injurious behavior in a retarded boy. *Behavior Therapy, 9,* 688.

ZEIGOB, L., JENKINS, J., BECKER, J., & BRISTOW, A. 1976. Facial screening: Effects on appropriate and inappropriate behaviors. *Journal of Behavior Therapy and Experimental Psychiatry, 7,* 355–357.

ZIMMERMAN, E.H., & ZIMMERMAN, J. 1962. The alteration of behavior in a special classroom situation. *Journal of the Experimental Analysis of Behavior, 5,* 59–60.

8 Differential Reinforcement: Stimulus Control and Shaping

Did you know that . . .

- Not all discrimination is the concern of the Equal Employment Opportunity Commission?
- Not all prompters work in the theater?
- You can teach anybody anything?
- Applied behavior analysts shape up behavior rather than shipping out students?

Chapters 6 and 7 discussed the arrangement of consequences that increase or decrease behavior. Two additional dimensions of behavior may concern teachers. The first is the circumstances under which students perform behaviors—do they perform them at the right time or in the right place? Many behaviors are defined as appropriate or inappropriate depending upon such factors. It is appropriate to say the word *went* when the teacher writes it on the board and asks, "What is this word?" Answering "went" when the teacher writes the word *go* is not appropriate. It is perfectly correct to run as rapidly as possible when competing in a track meet but not when moving down the hall to a locker. Yelling is an admirable behavior on the football field but not in the classroom. These behaviors do not need to be eliminated; students merely need to learn to perform them under the right circumstances. Applied behavior analysts call bringing behavior under the control of time, place, and circumstances *stimulus control*.

A second dimension of behavior change teachers are concerned with is developing behaviors that do not exist in the repertoires of their students. How can a teacher positively reinforce talking if a student never talks, or sitting if a student never sits, or reading if a student can't read?

The process of teaching students behaviors not currently in their repertoires is called *shaping*. A behavior that the student can perform is gradually guided, or shaped, into the target behavior.

Stimulus control and shaping are often used in conjunction to teach students academic and social behaviors. For that reason, and because both make use of differential reinforcement procedures, they are both described in this chapter. The first portion of the chapter describes in detail the phenomenon of stimulus control and stimulus control procedures for use in the classroom.

DIFFERENTIAL REINFORCEMENT FOR STIMULUS CONTROL

When we discussed in Chapters 6 and 7 the arrangement of consequences that increase or decrease behavior, we were concerned with what happens *after* a student performs a behavior—with the effects of behavior on the environment. In this chapter, we address another component of behavior change: the environmental conditions or events that *precede* the behavior. Our concern, then, will be with what happens before the student responds —with the effects of the environment on behavior.

Operant and respondent conditioning are contrasted on pages 30–31.

When describing events that affect operant behavior, it is important to remember the distinction between operant and respondent behavior described in Chapter 1. Respondent conditioning involves stimuli that elicit reflex behavior—for example, the puff of air (unconditioned stimulus) that results automatically in an eyeblink (response). This automaticity is absent in operant behavior; the relationship between antecedent events and behavior is learned rather than reflexive. Although antecedent events do not elicit operant behavior, they do exert considerable influence over such behavior.

Principles of Discrimination

Students learn to discriminate as a result of differential reinforcement.

The basis for antecedent control of operant behavior is discrimination learning (Terrace, 1966). **Discrimination** is simply the ability to tell the difference between environmental events or stimuli. Discrimination develops as a result of differential reinforcement. A certain response is reinforced in the presence of a given stimulus or group of stimuli that are said to be *discriminative stimuli* (S^D) for the response. The same response is not reinforced in the presence of a second stimulus or group of stimuli, referred to as **S-deltas** (S^Δ). After a period of time, the response will occur reliably in the presence of S^D and infrequently, if at all, in the presence of S^Δ. The S^D is then said to occasion the response (Holland & Skinner, 1961). This relationship between the S^D and the response is different from that between the unconditioned stimulus and response in respondent conditioning. The S^D does not elicit the response; it just sets the occasion for it. The response that occurs in the presence of S^D but not in its absence is said to be under stimulus control. A behavior under stimulus control will continue to occur in the presence of the S^D even when reinforcement is very infrequent. Michael (1982) suggested that care be taken to avoid saying that an event is an S^D for reinforcement—it is the behavior that is occasioned, not its reinforcement.

The development of discrimination is an important factor in much human learning. The infant learns that saying "mama" is reinforced in the presence of the adult with the glasses and the curly hair, but frequently results in the disappearance of the adult with the beard. Glasses and curly hair are an S^D for the response *mama*; a beard is an S^Δ. The first-grader learns that saying "went" in the presence of a flashcard with the letters *w-e-n-t* (S^D) results in praise, but that the same response to a flashcard with the letters *c-a-m-e* (S^Δ) does not. A group of junior-high students learns that obscene language and disruptive behavior get their math teacher's (S^D) attention, but that the social studies teacher (S^Δ) attends only to raised hands and completed assignments. Much of the everyday behavior of adults is a result of discrimination learning. We answer telephones when they ring, not when they are silent; we drive through intersections when lights are green, not when they are red.

Discriminations formed on the basis of relatively informal or imprecise patterns of reinforcement develop slowly and are often imperfect. For example, the infant may call all adults with curly hair and glasses "mama." The first-grader may say "went" when he sees a flashcard with the letters *w-a-n-t* or *w-e-t*. The junior-high students may occasionally raise their hands in math class or utter obscenities in social studies class. Adults sometimes pick up the phone when the doorbell rings. The imperfect stimulus control exerted by traffic signals provides employment for numerous police officers, tow-trucks, and ambulance attendants.

Discrimination Training

Teaching students to respond appropriately to specific stimuli is the teacher's basic job. As teachers, we want our students to obey rules, follow instructions, and perform specified academic or preacademic tasks at the appropriate time, in the appropriate place, and in response to specified instructions. A major part of the task of teaching is establishing specific times, places, instructions, and other antecedent events as discriminative stimuli for various student behaviors. In order to accomplish this, the teacher must

1. identify the target behavior.
2. identify the stimulus to be established as the S^D.
3. plan a reinforcement strategy.

This simplest form of discrimination training involves taking a behavior already in the student's repertoire but not emitted under specified conditions and establishing the desired conditions as discriminative for performance of the behavior. For example, a student may at times say "went" but not when shown a flashcard with the letters *w-e-n-t* and asked, "What is this word?" After determining that the student responds appropriately when other flashcards are presented (another instance of stimulus control), the teacher is ready to bring saying "went" under stimulus control of the letters *w-e-n-t*. The accompanying table illustrates the teaching process.

Stimulus	Response	Result
went (S^D)	"went"	"right"
go (S$^\Delta$)	"went"	no praise (extinction)

The teacher tells the student what the word is before asking for the discrimination to be made. Such prompting will be discussed later in the chapter.

In this example, the teacher establishes the letters *w-e-n-t* as the S^D for the response "went" through the process of differential reinforcement. The response is reinforced in the presence of *w-e-n-t* (S^D) and not in the presence of *g-o* (S$^\Delta$). With sufficient repetition, the student should reliably respond correctly and could then be said to have formed a discrimination. Notice that *g-o,* which functions as S$^\Delta$ in this example, will be the S^D for responding with "go." The definition, as always, depends on the function.

In order to state with any degree of confidence that the response "went" is under stimulus control, the teacher will have to establish that no other combination of letters occasions the response, including combinations such as *w-a-n-t* and *w-e-t,* whose shape and spelling (topography) closely resemble that of the S^D. The teacher will also want to determine that *w-e-n-t* is a reliable S^D for "went" when it is written in places other than on the original flashcard. One first-grade teacher thought she had finally taught one of her students to read the word *come* only to learn that the real S^D was a smudge on the flashcard.

This example illustrates that it is important to establish that the student is responding to the salient features of the stimulus. Many beginning readers identify words by their first letters alone. This strategy works well as long as *went* is the only word in the reader that starts with *w*. However, when the word *want* is introduced, the student can no longer discriminate reliably. This **stimulus overselectivity** (Lovaas, Schreibman, Koegel, & Rhen, 1971) is often a characteristic of handicapped students including those labeled *retarded, autistic,* and *learning disabled*. A student may identify his or her classroom by the scratch in the door without attending to other factors—that it is the third door down the hall or that it is opposite the water fountain. If, over the weekend, the door is painted, it is no longer an S^D for entering the room. The student gets lost, and a teacher gains support for her view that such students never retain anything you teach them.

The simple examples in this section illustrate how a discrimination is established by differential reinforcement so that an antecedent stimulus reliably occasions the appropriate response. The response is under stimulus control. The following sections offer more complex variations of discrimination training.

Discriminations Involving Multiple Stimuli

The preceding section described the process of bringing a *single* response under the control of an antecedent stimulus using differential reinforcement: the response in the presence of S^D was positively reinforced; the same response in the presence of S^Δ received no reinforcement or was extinguished. In many instances, a teacher may want to establish two or more antecedent stimuli as S^Ds for different responses. For example, teachers ask students to identify several letters of the alphabet, numerals, colors, words, and so on. In such cases, a number of different antecedent stimuli are established as S^Ds for an equal number of different responses and as S^Δs for all the others. The accompanying table shows an example using letters of the alphabet as a stimulus.

Stimulus	Response	Result
A	"A"	"Good"
B	"A"	EXT
B	"B"	"Good"
A	"B"	EXT

The vocal response of naming the letters of the alphabet must be brought under stimulus control, again by a process of differential reinforcement. Much of the activity taking place in classrooms involves just this type of discrimination training. Both the teacher of the severely handicapped, who wants students to respond differently to the instructions "Look at me" and "Touch your nose," and the regular classroom teacher, who asks students to classify plays as comedies or tragedies, are bringing responses under stimulus control.

Concept Formation

Students who are asked to make discriminations based on many different S^Ds, all of which should occasion the same response, must form an *abstraction* (Ferster, Culbertson, & Boren, 1975) or a *concept* (Becker, Engelmann, & Thomas, 1975a). To classify words as parts of speech, for example, requires such an abstraction. (See the accompanying table.)

Stimulus	Response	Result
The player made a home <u>run</u>.	"Noun"	"Right"
Put it on the <u>table</u>.	"Noun"	"Right"
<u>Run</u> over here.	"Noun"	EXT

In the grammar example shown in the table, the student must discriminate based on specific characteristics common to a large number of stimuli. Such training may be accomplished by providing many samples of positive and negative instances of the concept or abstraction involved and reinforcing correct responses. By this kind of procedure, Herrnstein and Loveland (1964), for example, were able to teach pigeons to respond differently to pictures that included people and to pictures that did not. They simply rewarded responding only when the pictures contained people. *Basic* concepts, according to Engelmann and Carnine (1982), must be taught in almost the same way to people. The authors define a basic concept as "one that cannot be fully described with other words (other than synonyms)" (p. 10).

Try to think of a way to teach a 3-year-old child the concept "red" by describing it. Obviously, it's impossible. What most parents do is provide many examples of red things, label them, and provide strong reinforcement when the child points and says "red" appropriately or when the child responds correctly to an instruction like "Give me the red block." Most children learn thousands of basic concepts in this informal manner before ever entering school. Children who don't must be taught them in a systematic fashion. For these children, we don't wait for casual opportunities to introduce "red" into the conversation. We get some red objects and some not-red objects and proceed to label, instruct, and ask for responses until the student demonstrates mastery of the concept.

While the teaching of such an abstraction as *humanness* can also be accomplished successfully in this way, such abstractions or concepts may often be taught more quickly by using some additional antecedent stimuli. It may be more efficient to provide a set of rules for identifying instances. Concepts do not need to be taught using only differential reinforcement. Most students, unlike pigeons, have some verbal skills that enable a teacher to use sets of rules as shortcuts in teaching them concepts or abstractions. Grammar rules that would supplement our earlier example are prompts; these teaching/learning aids are discussed in detail in the following section.

Prompts

A *prompt* is an additional stimulus that increases the probability that the S^D will occasion the desired response. Prompts are offered after an S^D has been presented and has failed to occasion the response. Most people are familiar with the use of prompts in the theater. An actor who fails to respond to his cues (the actress's preceding lines, for example) is prompted from the wings. The use of the word *prompt* in applied behavior analysis has a similar meaning. Students who fail to respond to S^Ds are prompted. Prompts may be presented verbally, visually, or physically. The desired response may also be demonstrated or modeled. The reading teacher who holds up a flashcard with the letters *w-e-n-t* and says, "Not came, but ..." is providing a verbal prompt. The kindergarten teacher who puts photographs as well as name tags on her students' lockers is providing a visual prompt. The mother who says to her child, "Wave bye-bye to Granny," while vigorously flapping the infant's hand, is providing a physical prompt. Each hopes that control will eventually be attached to the S^D: *went,* the student's name, or "Wave bye-bye." The prompt is a crutch to be dropped as soon as the need for it no longer exists. Prompts increase teaching efficiency. Rather than waiting for the student to emit the desired behavior, the teacher uses extra cues to increase the number of correct responses to be reinforced. It should be emphasized that when prompting is used, a reinforcer is usually delivered just as if the student had not needed prompting.

Rules as Verbal Prompts.   The English teacher who wants students to identify nouns and verbs correctly will probably not simply provide students with numerous chances to respond with "noun" or "verb" when they read sentences with the S^D underlined. Because most people have the ability to use verbal rules or definitions to form concepts, the English teacher might define *noun,* then present students with sentences and ask, "Is the underlined word a noun?" (S^D). "Is it the name of a person, place, or thing? Then it's a noun" (prompt). "Right, John, it's a noun" (S^{R+}).

Prompting using rules or definitions is not confined to academic tasks. By defining honesty, politeness, kindness, or any other concept related to social behavior, a teacher can prompt students until they are able to identify instances of each behavior. This, of course, does not ensure that students will engage in the behaviors, merely that they can label them. Engelmann (1969) provides an extended discussion of concept teaching applicable to both academic and social behavior.

Instructions as Verbal Prompts.   Instructions are often a means of prompting behavior. If, when the teacher says, "Get ready for reading," the children do not move, the teacher will probably add, "Put your materials away and go to the reading circle." If the S^D, $642 \div 24$, does not occasion correct responding, the teacher may provide step-by-step instructions. The teacher who uses instructions as prompts is making two assumptions. The first is that the instructions offered are accurate. It isn't easy to give clear, verbal instructions for a complex task. If people who ask you for directions to get from one place to another by car usually become hopelessly lost, do not be surprised if students fail to follow your instructions. The second assumption is that the student's behavior is under stimulus control of the general S^D, "Follow instructions." Many students do not follow instructions, as any experienced teacher will attest. Before depending on instructions as prompts, the wise teacher will determine that students do indeed follow them. It may be necessary to bring this response under stimulus control first.

It's not easy to give instructions.

Becker et al. (1975a) suggest that teachers reinforce following instructions to the most specific detail. Practice in this skill can be provided by specifying details arbitrarily: for example, telling students to line up with their toes on a crack in the floor or to perform activities in a specified order even though no specific order is required. This kind of practice can become a sort of game. The traditional game Simon Says provides this kind of practice in following instructions.

Hints as Verbal Prompts.   Many verbal prompts are less elaborate and more informal than rules or instructions. An example is the reading teacher who prompts the correct response to the S^D, *dog*, by saying, "This is an animal that says 'bow-wow'." When a teacher tells the class to line up, then adds, "Quietly," this too is a prompt. Such reminders, or hints, increase the probability that the correct response will be emitted, thus providing an opportunity for reinforcement.

Visual Prompts.   Many teaching strategies involve some form of visual prompting. The illustrations in most beginning readers are designed to aid students in identifying the printed word. Teachers may provide examples of correctly completed math problems to prompt students. Picture prompts have been used to help teach a wide variety of behaviors to handicapped learners, particularly complex daily living and vocational tasks (Wacker & Berg, 1983). Martin, Rusch, James, Decker, and Trtol (1974) taught moderately and severely retarded learners to prepare meals independently by using sequenced picture prompts. Wacker and Berg (1983) taught moderately and severely retarded high school students complex vocational tasks. They provided the students with books of pictures illustrating the successive steps in performing a complex assembly task and trained them to use the books, that is, to turn a page after each step and to match the correct object to its illustration. The authors found that the books greatly improved students' performance. The students were able to learn new tasks using picture prompts more quickly than they learned the initial task. Students were able to perform tasks without the books after training was completed. One advantage of this kind of procedure is the enormous saving of teacher time. Bulletin boards in classrooms can easily be used to provide picture prompts. Line drawings or photographs of the correct procedure for accomplishing some task, or a picture of the way desks should look before leaving the room, or a photograph of the class in a nice straight line on the way to the cafeteria posted just inside the doorway could all be used to prompt correct responding.

Other visual prompts are provided in written form. Classroom schedules and rules are often posted to serve as reminders or cues. Many students are easily prompted to do complex new tasks through the use of written instructions. Think of all the tasks you do which depend upon written instructions and think of how important it is that the instructions be clear and accurate. Anyone who has tried to put together children's toys late at night before a holiday can attest to the importance of clarity and accuracy. Of course, written instructions are technically verbal prompts, since the written word is a form of verbal (not vocal) communication. Since they are processed visually, however, it seems logical to consider them here.

Teachers often provide such prompts for themselves as well as for their students. A road map may help the itinerant resource teacher who is provided the S^D "Go to Oakhaven School." Some teachers post reminders to themselves to help them remember to reinforce certain student behaviors.

Modeling

"Watch me, I'll show you," says the teacher. This teacher is using yet another form of prompt. When verbal instructions or visual cues are insufficient, many teachers demonstrate, or *model*, the desired behavior. In many cases, demonstration may be the procedure of choice from the start. A home economics teacher attempting to *tell* the class how to thread a sewing machine provides a convincing demonstration of the superiority of demonstration.

The majority of students, including those who are mildly handicapped, respond to imitate the behavior of a model. Many explanations of this phenomenon have been offered (see Bandura, 1969, for a complete discussion), but the simplest is that most students' history of reinforcement includes considerable reinforcement of imitative behavior, resulting in a generalized imitation response. In other words, "Do it like this" has become an S^D for imitation in virtually any setting.

Origin of the generalized imitation response.

Several studies have established that subjects reinforced for imitating various responses will eventually also imitate unreinforced responses (Baer, Peterson, & Sherman, 1967; Brigham & Sherman, 1968; Lovaas, Berberich, Perloff, & Schaeffer, 1966). Anyone observing a parent interacting with an infant has seen examples of positive reinforcement of imitation. Such reinforcement occurs throughout the preschool years so that most children come to school already used to responding to the S^D, "Do it like this" or "Do it like Mary." Of course, children are as likely to imitate inappropriate behavior as appropriate behavior. Many a parent has at some point been startled at how quickly children imitate less admirable parental habits.

Students also imitate the behavior of their classmates. Kindergarten students spontaneously play Follow the Leader. Nonacademic behaviors, such as speech patterns, are frequently imitated. It is amazing how quickly a new student with a distinctive regional accent begins to sound exactly like his or her peers. The tendency to imitate peers perhaps reaches its zenith in the secondary school. Adolescents tend to dress alike, talk alike, and engage in the same activities. Providing handicapped students with appropriate models is one of the primary goals of the current trend toward placement of such students in settings with as many nonhandicapped students as possible. It is hypothesized that the handicapped students will imitate and thus learn from their nonhandicapped peers.

In using demonstration techniques to prompt behavior, a teacher may choose to model the behavior personally, to allow another student to provide a model, or to bring in someone from outside the group. The choice of a model is important, as certain characteristics increase models' effectiveness. Students are most likely to imitate models who

1. Have high status, particularly celebrities (Hovland, Janis, & Kelley, 1953; Lefkowitz, Blake, & Mouton, 1955).

2. Have demonstrated competence (Gelfand, 1962; Rosenbaum & Tucker, 1962).

3. Are similar to themselves (Bandura, 1969), particularly of the same sex (Goldstein, Heller, & Schrest, 1966).

4. Are reinforced for the modeled behavior (Bandura, 1969; Mayer, Rohen, & Whitley, 1969).

Modeling may be used to prompt very simple or more complex behaviors. A teacher may use modeling to prompt speech in a severely handicapped student. A math teacher may ask a competent student to demonstrate a problem on the board before other students are expected to work similar problems on their own. A physical education teacher may demonstrate a complex gymnastic performance.

Modeling can be a very effective prompting procedure, but it does have limitations. Some behaviors are difficult to imitate. Some students, particularly those who are severely handicapped, have not acquired a generalized imitation response. Although it is possible to teach students to respond to modeled prompts by reinforcing imitation (Baer et al., 1967; Lovaas et al., 1966), another form of prompting may be required in addition.

Many teaching procedures involve combining various kinds of prompts. The following anecdote illustrates a procedure combining modeling with verbal instructions.

The Students Learn to Polka

A group of children with poor gross-motor skills were enrolled in an after-school program designed to improve motor coordination. Their teacher decided that, because the students had mastered skipping and hopping, they would enjoy learning to polka. "Let's polka!" said the teacher. "Watch what I do. Step-together-step-hop! Now you do it. Step-together-step-hop." Soon when the teacher said, "Let's polka!" the students produced the desired movement. Many of them, incidentally, continued to whisper "step-together-step-hop" as they danced.

Such self-instruction is discussed in Chapter 10.

Physical Guidance

When students fail to respond to less stringent forms of prompting, they may be physically prompted. Such a procedure, often called *putting-through*, is useful for teaching many motor behaviors and some vocal behaviors that may be manually guided (Karen, 1974). Physical prompting may be a first step in the development of a generalized imitation response.

Baer et al. (1967) prompted imitation with physical guidance and ultimately brought imitation under stimulus control. Streifel and Weatherby (1973) taught a severely retarded student to follow instructions, such as "Raise your hand," by using a physical guidance procedure. The teacher provided first the verbal instruction, then guided the student so that the instruction was followed. Eventually, the student's behavior was under the control of the instructions. The implications of bringing the behavior of severely handicapped students under stimulus control of imitation and instructions are enormous. Such students become much more accessible to teaching as a result.

Physical guidance procedures are by no means limited to use with severely handicapped students. Most teachers routinely use such procedures when teaching beginning handwriting skills for example. Music teachers may guide their students' fingering. Many athletic skills are most easily taught using physical guidance. It is difficult to imagine any other way of teaching anyone to change gears in a car with a four-speed manual transmission.

When using physical guidance, teachers need to be sure that the student is cooperative. Such a procedure will be aversive (possibly to both parties) if it is performed with a resisting student. Even cooperative students may have a tendency to tighten up when physically prompted. Sulzer-Azaroff and Mayer (1977) suggest relaxation training for students who resist physical guidance because of tension.

The teacher puts his or her hand on the student's hand and pulls it up.

Fading

A prompted response is not under stimulus control. Prompts must be withdrawn, and the response must be occasioned by the S^D alone. Too abrupt removal of prompts, however, may result in termination of the desired behavior. Gradual removal of prompts is referred to as **fading**.

Any form of prompt may be gradually faded so that the response occurs and is reinforced when the SD alone is presented. Considerable skill is involved in determining the optimum rate of fading: too fast, and the behavior will not occur; too slow, and students may become permanently dependent upon the prompt. Prompts can be faded in a number of different ways. Billingsley and Romer (1983) reviewed systems for fading prompts and suggest four major categories: decreasing assistance, graduated guidance, time delay, and increasing assistance.

Decreasing Assistance

In decreasing assistance, the teacher begins with whatever level of prompting is necessary for the student to perform the desired behavior. The amount of assistance is then systematically reduced as the student becomes more competent. This procedure can be used to fade a wide variety of prompts. The English teacher using rules to help her students discriminate nouns might start with the S^D, "Is this a noun?" and the prompt, "If it's the name of a person, place, or thing, it's a noun." When her students responded reliably she might say

Is this a noun?—(S^D)

Is it the name of a person, place, or thing?—(prompt)

then

Is this a noun?—(S^D)

Remember, person, place, or thing.

then

Is this a noun?—(S^D)

Remember your rule.

and finally

Is this a noun?

The math teacher using the decreasing prompts procedure to fade a visual prompt might allow students a complete multiplication matrix to help them work problems and systematically remove easier combinations as they are mastered. The teacher who fades a modeling prompt may demonstrate less of the behavior, perhaps fading from his full demonstration to demonstrating only part of the response (Wilcox & Bellamy, 1982), perhaps finally providing only a gesture. The following anecdote provides an illustration of such a fading procedure.

The Children Learn to Hula Hoop

Coach Townsend was an elementary school physical education teacher. He decided to teach the first-graders to hula hoop, so that they could perform at open house. He began by demonstrating. "Watch me!" he said and proceeded to show the students how to do it. When the giggles died down, he handed each first-grader a hula hoop. "Ready," he said, "Hoop!" Twenty-six hoops hit the floor clattering and banging. The children looked discouraged. The coach took a hoop from one student.

"Okay," he said, "start it like this." He pushed his hoop with his hands. "Now do this." He rotated his hips. When the giggles died down, he said, "Again, just like this!"

Coach Townsend gave the hoop back to the student and just demonstrated the movements. He eventually faded the hand movement to a flick of the wrist and the hip movements to a mere twitch. Before long, the cry, "Ready, hoop!", was enough.

The great day finally arrived. Twenty-six first-graders equipped with hula hoops stood in the gym. Their parents sat in the bleachers. Coach Townsend started the music and said, "Ready, hoop!" He stood to one side of the group. The children did him great credit, but he heard suppressed giggling from the audience. The coach finally realized that, in his enthusiasm, he had gone back to modeling the necessary hip movements.

The system of decreasing prompts when using a physical guidance procedure was illustrated by Csapo (1981) in teaching severely handicapped students a discrimination task. The verbal instruction was initially followed by a full physical prompt, then by a partial prompt, and then by a gesture.

The system of decreasing prompts may also be used when combinations of prompts are used. If students learning a new math skill are initially provided with a demonstration, a sample problem, and step-by-step teacher instructions, these prompts may be removed one by one until the students are responding simply to the incomplete example.

Graduated Guidance

Graduated guidance is used in fading physical prompts (Foxx & Azrin, 1972). The teacher begins with as much physical assistance as necessary and gradually reduces pressure. The focus of the guidance may be moved from the part of the body concerned (spatial fading), or a shadowing procedure may be substituted in which the teacher's hand does not touch the student but follows his movement throughout the performance of the behavior (Foxx & Azrin, 1973). Juan learns by a graduated guidance procedure in the following vignette.

Juan Learns to Eat with a Spoon

Juan was a student in Ms. Baker's class for the severely retarded. He could feed himself with his fingers but had not learned to manage with a spoon. Ms. Baker decided that the time had come. She equipped herself with a bowl of vanilla pudding and a spoon, hoping that the pudding, which Juan loved, would be a positive reinforcer for eating using the spoon and that vanilla would make a less visible mess than chocolate on Juan, herself, the table, and the floor. (Teachers have to think of everything.)

Ms. Baker sat next to Juan, put her hand over his right hand, and guided it to the spoon. She helped him scoop up some pudding and guided his hand toward his mouth. When the spoon reached his mouth, he eagerly ate the pudding. Ms. Baker then removed Juan's left hand from the pudding dish and wiped it off with the damp cloth that she had thoughtfully provided. She repeated this procedure a number of times, praising and patting Juan whenever the spoon reached its goal. As she felt Juan making more of the necessary movements on his own, she gradually reduced the pressure of her hand until it was merely resting on Juan's. She then moved her hand away from his—to his wrist, to his elbow, and finally to his shoulder. Eventually she removed her hand entirely. Juan was using the spoon on his own.

Time Delay

Time delay differs from other fading formats in that the form of the prompt itself is not changed, just its timing. Rather than presenting the prompt immediately, the teacher waits, thus allowing the student to respond before prompting. Delays are usually only a few seconds. Time delays can be constant (the delay stays at the same length) or progressive (the interval before the prompt becomes longer as the student gains competence) (Kleinert & Gast, 1982). Time delay procedures can also be used with a variety of prompting formats. Many teachers use them instinctively.

Let's go back to our English teacher. She asks, "Is this a noun?", waits a few seconds, finds no response forthcoming, and prompts "If it's the name of a person, place, or thing, it's a noun."

Time delay can also be used to fade visual prompts. A teacher working on sight vocabulary might cover up the picture on the flashcard and wait a few seconds to allow students to identify the word without seeing the picture. If they fail to respond within the latency the teacher has identified, the picture is uncovered.

Touchette and Howard (1984) used a progressive time delay to fade a modeled prompt. Retarded students were asked to point to a specific letter or word printed on one of four cards. Initially the teacher pointed to the correct card immediately upon presentation of the verbal S^D. The teacher's pointing was gradually delayed to provide an opportunity for the student to respond without prompting. Touchette and Howard's results also indicated that responding to the S^D alone was learned somewhat more efficiently when responses made before the prompt were reinforced more heavily than responses occurring after the prompt.

Time delay is easily implemented when using a physical prompt. A teacher working with a student on dressing skills might say "Pull up your pants" and then wait for the student to perform the behavior without assistance before providing it.

Increasing Assistance

Billingsley and Romer (1983) describe the increasing assistance procedure as "similar to the reverse application of the decreasing assistance approach" (p. 6). Increasing assistance is also referred to as the system of least prompts. In using this procedure the teacher starts with the S^D, moves to the least intrusive prompt in her repertoire, and provides the students an opportunity to respond. Bear with us for one more visit to the English teacher. She says, "Is it a noun?" and gets no response. She prompts, "Remember your rule," and gets no response. She says, "Remember person, place, or thing," and gets no response. She prompts, "Is it the name of a person, place, or thing?" and gets no response. She whimpers, "If it's the name of a person, place, or thing, then it's a noun." It is sometimes difficult to implement an increasing assistance procedure without sounding either strident or frustrated.

An increasing prompts procedure can be used to fade a visual prompt. A beginning reading teacher using a set of flashcards might first provide a card with just the word *boy* and then a card with a stick figure, finally moving to a card with a representational drawing of a boy. To use the procedure with modeling, one could first provide a gesture and move toward a full demonstration. To use increasing prompts with physical guidance, one would also start with a gesture and then move toward a full "putting through" procedure.

Increasing prompts may also be implemented with combinations of prompt modalities. Van Etten, Arkell, and Van Etten (1980) describe a procedure for teaching a severely retarded student to drink out of a cup. The teacher starts with a verbal prompt, moves to a partial physical prompt, and finally uses a full physical prompt.

Effectiveness of Methods for Fading Prompts

The majority of literature on fading prompts has examined the efficacy of their use with severely and profoundly handicapped students (cf., Billingsley & Romer, 1983). Caution should be observed in generalizing the results to other populations. It appears that all methods may be used effectively but that "decreasing assistance training may be more effective than increasing in assistance training at least for severely handicapped subjects in the acquisition phase of learning, and that progressive time delay proce dures are no less effective, and may in some cases be more effective, than increasing assistance methods" (p. 7). It is interesting in the light of these tentative conclusions that increasing assistance procedures, according to our observations, are almost routinely used by teachers of normal and mildly handicapped students. It would be interesting to see the results of systematic study of increasing assistance and other prompting procedures with these populations.

Procedures for establishing and maintaining stimulus control are powerful tools. Once all prompts have been withdrawn and behaviors are under stimulus control, they will continue to occur, sometimes for years, without any reinforcement except that naturally available in the environment, and even when the person knows no reinforcement will be forthcoming. Have you ever sat at a red light at 3 A.M. at a deserted intersection and waited for the light to change? If you have, you have some idea of the power of stimulus control. Professor Grundy illustrates this power in the following anecdote.

The Yellow Tablet Episode

Professor Grundy walked cheerfully to his office. He had no classes, no appointments, no committee meetings. He planned to spend the entire day working on an article he had agreed to write for a professional journal. All his data were gathered, his calculations completed, his graphs and charts made. All that remained to do was to write the manuscript portion of the article. He opened his office door, gathered his pens, pencils, and note cards, sat down, and began to write. After a few minutes, he noticed that he had only a few pages of yellow legal tablet left. Deciding that, in any event, it was time for a break, Professor Grundy walked down to the departmental supply room to acquire a new tablet. There were none in the usual place. Grundy approached his secretary.

"Are we out of legal tablets, Ms. Cadwallader?" he asked.

"Why, no, Professor, of course not. A new shipment just came in this morning," answered Ms. Cadwallader, looking up from her typewriter. "I just haven't had a chance to put them on the shelf. Let me get one out of the box for you." Ms. Cadwallader unpacked a tablet, handed it to the professor, and turned back to her typewriter. A few minutes later, she looked up again to see Professor Grundy still standing by her desk holding the tablet.

"Why, Professor," she said, "you look ill. Is anything the matter?"

Professor Grundy answered shakily, "This tablet isn't yellow."

"Why, no, Professor, it's white. It seems that this brand is cheaper, so we ordered . . ."

"But it has margins on it, and it isn't yellow. How can I write on this thing?"

"Now, Professor," said Ms. Cadwallader soothingly, "how can it possibly make a difference whether the tablet is yellow or white? It's just paper."

"Just paper!" shouted the professor. "I'm going to talk to someone about this! I've always written on yellow tablets with no margins. There's no way I can get anything done with these!"

As Professor Grundy stalked away to find the departmental chairperson, Ms. Cadwallader murmured to herself, "Poor man. Stimulus control's got him. Yellow paper's an S^D for writing, white's an S^Δ—just like the pigeons."

Errorless Learning.   Any procedure using prompts to facilitate the establishment of stimulus control is designed to help students learn as rapidly as possible without making an excessive number of mistakes. Incorrect responses, once made, tend to be repeated (Bereiter & Engelmann, 1966). In some cases, students may practice errors to the extent that such wrong responses become well established. It is possible, however, to develop programs in which prompts are so effective and faded so slowly that virtually **errorless learning** occurs. It is possible to use any of the systems described in this section to program for errorless learning, but that term has been used most often to describe a slightly different procedure. In the systems described above, the S^D itself remained constant and prompts were added to occasion the response. Such prompts are often called **response prompts**. Many procedures termed errorless learning use alterations within the stimuli (S^Ds or S^Δs) to prompt correct responses. These prompts are often called **stimulus prompts** and the procedure itself sometimes called *stimulus shaping*. To enable the student to make a discrimination more easily some features of the S^D, the S^Δ, or both, are changed. Whaley and Malott (1971) taught a retarded student, Betty, to discriminate her name from another by pasting the name *Betty* in white letters on a black card and the name *Susan* in white letters on a white background. Reinforcing the choice of *Betty* established the correct card as the S^D. The background color on the *Susan* card was gradually darkened until both names were printed in white on black. The student made a minimal number of incorrect choices and eventually made the discrimination based on the relevant stimulus properties.

Stimulus shaping is not to be confused with the shaping procedures discussed later in this chapter.

Haupt, Van Kirk, and Terraciano (1975) used two fading procedures to teach math facts. In one procedure, the student was asked to provide the answer to subtraction facts on flashcards. At first, the answers were visible, but they were gradually covered with colored cellophane. Eventually, there were 32 pieces of cellophane over the answer. The student learned and remembered the facts. The second procedure required written answers to multiplication facts. Successive thicknesses of tracing paper covered the initially visible answers. This procedure, too, was successful.

Another example was described by Ayllon (1977). A teacher was trying to teach a group of young children to discriminate their right from their left hand. On the first day of training, each child's right hand was labeled with an x made by a felt-tipped marker. This enabled the students to discriminate the S^D (right hand) from the S^Δ (left hand) when asked to raise their hands. On the second day, the plan was again to mark right hands, but because of the permanence of the marker (or the personal habits of the children), each child still had a visible mark. As the training proceeded during the course of a week, each child's mark gradually faded, and by week's end, each child consistently raised his or her right hand when asked.

Mosk and Bucher (1984) used a combination of prompting and stimulus alteration to teach moderately and severely retarded students to hang toothbrushes and washcloths on the correct peg. At the beginning of instruction, only one peg was on the board. Distractor pegs were introduced, one at a time. At first, each distractor peg was too short to hang the item on; longer pegs were gradually introduced. This procedure was more effective than a procedure using prompting alone.

Evidence indicates (Schreibman & Charlop, 1981; Stella & Etzel, 1978) that errorless learning is most efficient if only the features of the S^D are altered, and the S^Δ is held constant. Some researchers suggest that time delay may be a more efficient and economical alternative (Bradley-Johnson, Johnson, & Sunderman, 1983 Touchette & Howard, 1984); others (Etzel & LeBlanc, 1979) disagree. This lack of agreement need not overly concern the classroom teacher. "Each has been shown to produce errorless learning under some circumstances and each should be considered as an alternative when the other fails" (Touchette & Howard, 1984, p. 187).

Fading for errorless learning provides for development of stimulus control without practicing incorrect responses. However, such a procedure has its disadvantages. Providing a completely error-free learning environment is not necessarily desirable. Spooner and Spooner (1984) suggest that optimal learning may occur when initially high error rates decrease quickly and correct responding accelerates rapidly. Terrace (1966) points out that a lack of frustration tolerance may result from errorless training. In the real world some errors are inevitable, and students must learn to handle mistakes. Krumboltz and Krumboltz (1972) suggest gradually programming students to persist after errors. Rodewald (1979) suggests that intermittent reinforcement during training may mitigate possible negative effects of errorless learning. Such intermittent reinforcement may help to develop a tolerance for nonreinforcement of responses.

Effective Prompting

The effective use of prompts to facilitate development of stimulus control requires the teacher's attention to several influences:

Guidelines for using prompts.

1. Prompts should focus students' attention on the S^D, not distract from it. Prompts that are spatially or otherwise distant from the stimulus may be ineffective (Schreibman, 1975). Cheney and Stein (1974) point out that using prompts unrelated to the stimulus may be less effective than using no prompts or trial-and-error learning. The well-meaning teacher who encourages beginning readers to use the illustrations in the preprimer as clues to the words on the page may find that overemphasis on such prompts may result in some children's developing an overdependence on the illustrations at the expense of the written word. For some students, such dependence may be so well developed as to require the use of reading materials without illustrations in order to focus attention on the relevant S^D.

2. Prompts should be as weak as possible. The use of strong prompts when weak ones will do is inefficient and may delay the development of stimulus control. In general, the best prompt is the weakest one that will result in the emission of the desired behavior. Strong prompts are often intrusive. They intrude on the environmental antecedent, the S^D, and drastically change the circumstances or conditions under which the response is to be performed. Every effort should be made to use the least intrusive prompt possible. There is very little empirical validation on which to base a hierarchy of prompts (Wilcox & Bellamy, 1982). Figure 8–1 presents a tentative hierarchy of prompts, arranged from least to most intrusive. Visual and verbal prompts are, on the whole, less intrusive then modeling, and all are less intrusive than physical guidance. This may not always be the case. A gentle push on the hand to help a preschooler slide in a recalcitrant puzzle piece is probably less intrusive than yelling "Push it the *other* way." Inefficiency is not the only undesirable effect of prompts that are stronger than necessary. Many students find strong prompting aversive (Krumboltz & Krumboltz, 1972). When students say, "Don't give me a hint. I'll figure it out myself!" the wise teacher listens.

FIGURE 8–1
Hierarchy of prompts

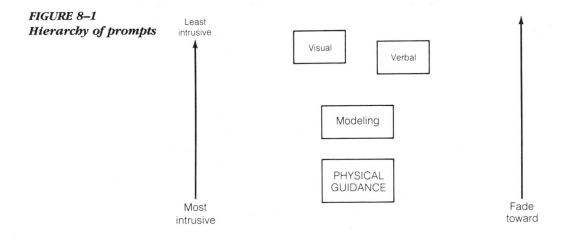

3. Prompts should be faded as rapidly as possible. Continuing to prompt longer than necessary may result in failure of the S^D to acquire control. The efficient teacher uses prompts only as long as necessary and fades them quickly, thus avoiding students' becoming dependent on prompts rather than S^Ds. Students who are allowed to use a matrix of multiplication facts for extended periods may never learn their multiplication tables.

4. Unplanned prompts should be avoided. Anyone who has observed a large number of teachers has seen students watch the teacher carefully for clues to the correct answer. A teacher may be completely unaware that students are being prompted by a facial expression or vocal inflection. Neither is an inappropriate prompt when used intentionally. But the teacher who asks, while shaking her head, "Did Johnny really want to go to the park in the story?" in such a tone that all the children answer "no" is fooling herself if she thinks that the students necessarily comprehended the story.

Chaining

So far we have discussed bringing behavior under stimulus control as if all behaviors consist of simple, discrete actions that may be occasioned by a discriminative stimulus and reinforced. Much of what we want students to learn involves many such discrete behaviors, to be performed in sequence upon presentation of the S^D. The acquisition of such sequences of behavior, or *behavioral chains*, may require a more complex teaching procedure than differential reinforcement and prompting.

Consider, for example, a classroom teacher who gives the instruction, "Get ready for math practice." The result is vague shuffling and furtive looking around among the students. Some students locate their math workbooks; others imitate them. "Come on," prompts (nags?) the teacher, "Hurry up now." One or two students locate pencils; some students still appear completely confused. The process of getting ready for math practice is actually a series of behaviors performed in a sequence.

1. Clear desk of other materials.
2. Locate math workbook.
3. Locate pencil.
4. Wait quietly for instructions.

The students in this class may be able to perform each of these behaviors, but the behaviors are not under the control of the S^D presented ("Get ready for . . ."). To set up this control, the teacher must establish a chain of behaviors that will occur when the instruction is given. The teacher might proceed by giving each instruction separately and reinforcing compliance, always starting with "Get ready for math practice." Soon two steps could be combined and reinforcement would follow only completion of the two-part chain. Finally, the teacher would have to provide only the S^D. The students would have acquired a behavioral chain. A behavioral chain is a sequence of behaviors all of which must be performed in order to earn a reinforcer.

Many complex human behaviors consist of such chains—often with dozens or even hundreds of component steps. Usually, reinforcement occurs only when the final component is performed. The instructional procedure of reinforcing individual responses for occurring in sequence to form a complex behavior is called **chaining**.

In order to understand the process involved in the development of behavioral chains, first recall that any stimulus must be defined in terms of its function and that identical stimuli may have different functions. Similarly, behaviors included in chains may simultaneously serve multiple functions. Consider the behaviors in the get-ready-for-math-practice chain. When the chain is fully established, reinforcement occurs only after the last link. The last link in the chain, however, is paired with the reinforcer and thus becomes a conditioned reinforcer, increasing the probability of occurrence of the preceding link. Each link is subsequently paired with its preceding one: each link serves as a conditioned reinforcer for the link immediately preceding it.

We can also look at the behavior chain from another perspective: each link also serves as an S^D for the link immediately following it. Consider again the chain of getting ready for math practice:

A behavior chain.

1. Clear desk of other materials (S^D for 2).
2. Locate math workbook (S^{R+} for 1; S^D for 3).
3. Locate pencil (S^{R+} for 2; S^D for 4).
4. Wait quietly for instructions (S^{R+} for 3).

Each link increases the probability of the one it follows and specifies, or cues, the one it precedes (Ferster, Culbertson, & Boren, 1975; Staats & Staats, 1963).

On another level, each of the component links of a chain, as in our example, can in turn be described as a chain; that is, clearing one's desk includes picking up books and papers, opening the desk, and putting books in the desk. Picking up books is also a chain of behaviors—raising the arm, extending it, opening the hand, grasping the books, and raising the arm. Grasping books, in fact, is still another chain, including placing the thumb . . . but wait. We can go on this way in both directions—toward increasing specificity in behaviors and toward increasing complexity. Later in the school year, the teacher in our example may say, "Boys and girls, for math practice I want you to do the first 10 problems on page 142," whereupon the students promptly

1. get ready for math practice,
2. open their books,

3. pick up their pencils, and

4. complete the assignment.

The original chain has now become merely a link in a more complex chain. The process of chaining simple behaviors into longer and more complicated sequences results in production of the most elaborate and so-phisticated forms of human behavior. There are several types of chaining procedures. Those most commonly used are *backward chaining, forward chaining,* and *total task presentation.*

For many students, behavioral chains can be acquired by having each step in the chain verbally prompted or demonstrated. The prompts can then be faded and links combined (Becker, Engelmann, & Thomas, 1975b). For other students, some or all of the separate steps of the chain may have to be taught using more elaborate prompting procedures as the chain is developed.

Backward Chaining

When backward chaining is used, the components of the chain are ac-quired in reverse order. The last component is taught first, and other com-ponents are added one at a time. Delbert and Harmon (1972) describe an elegant backward chaining procedure for teaching a retarded child to un-dress himself. The child is given the instruction, "Timmy, take your shirt off," and his shirt is pulled over his head until the arms are free and the neckband is caught just above his eyes. If the child does not automatically pull the shirt off, he is physically guided to do so. Primary and social rein-forcers are then given. During the next training session, the neckband is left at his neck, in subsequent sessions, one arm, then both arms are left in the sleeves. The S^D, "Timmy, take your shirt off," is always presented and reinforcers given only when the task is completed. The removal of each garment is taught in this manner, then the component steps are combined until the instruction "Take your clothes off" has acquired stimulus control.

Many self-care skills, such as toileting, grooming and eating, may be taught using backward chaining. Many academic and preacademic skills may also be efficiently taught using this procedure. Espich and Williams (1967) provided a model of a math program using backward chaining. At first, steps until the final one are provided, and the last step is prompted. Each frame of the program prompts one less step until students are on their own. We suggest that you work through the program, reproduced in Figure 8–2, to see how effectively backward chaining can be applied to ac-ademic subjects.

Forward Chaining

When forward chaining is used, the teacher starts with the first link in the chain, trains it to criterion, and then goes on to the next. The student may be required to perform all the steps previously mastered each time (Banks & Baer, 1966), or each step may be separately trained to criterion and then the links made (Patterson, Panyon, Wyatt, & Morales, 1974). To use forward chaining to teach undressing skills, the teacher would start with the stu-

Figure 8–2
Backward chaining
academic program

Source: From Developing
programmed instructional
materials by J.E. Espich & B.
Williams. Copyright © 1967
by Fearon Publishers.
Reprinted by permission of
Pitman Learning, Inc.,
Belmont, CA 94002.

This is not a test. It is a program.
It is designed to *teach you something* by making you apply the
knowledge you already have plus new information that this program will
supply.

Do whatever the program asks. If you come to a blank (_____),
put the correct word, number, symbol, or whatever, in the blank.

When a response is required, you will find the correct response on the
lower part of the page. Check your answer, then go on to the next bit of
information.

Program page 1

---

1. To find the square of a two-digit number ending in 5:
   A. Multiply the first digit by the next higher consecutive number.
   B. Write 25 to the right of the result.

   NO RESPONSE REQUIRED.

Program page 2

---

2. To find the square of 35:
   A. Multiply the first digit (3) by the next higher consecutive
      number (4).   12
   B. Write 25 to the right of the result (12).   12 25
   C. The square of 35 is 1225.

   NO RESPONSE REQUIRED.

Program page 3

---

3. To find the square of 25:
   A. Multiply the first digit (2) by the next higher consecutive
      number (3).   6
   B. Write 25 to the right of the result.   6 25
   C. The square of 25 is _____ .

   $2 \times 3 = 6$
   C. The square of 25 is 625.

Program page 4

---

4. To find the square of 75:
   A. Multiply the first digit (7) by the next higher consecutive number.
   B. Write _____ to the _____ of the result.
   C. The square of 75 is _____ .

   $7 \times 8 = 56$
   B. Write 25 to the right of the result.
   C. The square of 75 is 5625.

Program page 5

---

5. To find the square of 15:
   A. Multiply _____ by _____ .
   B. Write _____ to the _____ of _____ .
   C. The square of 15 is _____ .

   A. Multiply 1 by 2.
   B. Write 25 to the right of 2.
   C. The square of 15 is 225.

Program page 6

---

6. What is the square of:
   A. 65? _____
   B. 95? _____

   A. 4225
   B. 9025
                              The End

Program page 7

dent fully dressed, deliver the instruction, "Timmy, take your shirt off," and then provide whatever prompting was required to get Timmy to cross his arms and grab the bottom of his tee-shirt. When Timmy reliably performed this behavior, she would add the next step until Timmy's shirt was off.

Many academic applications of forward chaining come to mind. The first grade teacher who wants her students to print the letters of the alphabet in sequence may start with *A* and add a letter a day until the children can print all 26 capital letters in sequence. The chemistry teacher who wants his students to know the elements on the periodic table in order starts with a few at a time and adds a few each day. A teacher who wants students to recite a poem may have them recite the first line until it is mastered and then add a line at a time until the students can recite the entire poem. A conscientious teacher who requires students to write a report (an extremely complex chain) teaches them to find references, then take notes, then outline, then prepare a rough draft, then turn in a final copy.

Total Task Presentation

When using total task presentation, the teacher requires that the student perform all of the steps in sequence until the entire chain is mastered. Total task presentation may be particularly appropriate when the student has already mastered some or all of the components of a task but has not performed them in sequence. However, it is also possible to teach completely novel chains in this manner (Spooner, 1981; Spooner & Spooner, 1983; Walls, Zane, & Ellis, 1981). The following anecdote illustrates teaching a skill using total task presentation. Notice Mr. Grist's use of a variety of prompting strategies. Can you label each one?

Carole Learns to Go Through the Lunch Line

It was Carole's first day in a regular school. She had previously attended a center for moderately retarded students but was now placed in a self-contained class in a regular elementary school. Her teacher, Mr. Grist, knew that meals at the center were served family-style and that the self-service line in the school cafeteria might be confusing to Carole. He put her in line behind Linda and walked beside the girls, saying, "Look, Carole, Linda took a fork. Now you take a fork." If necessary, he added a gesture or a physical prompt. He repeated this strategy for each of the steps required. When Carole reached the table safely, he praised her. He repeated this strategy daily, waiting to see whether she needed prompting before he offered it. By the end of the week Carole was managing nicely.

Combining verbal instructions, modeling, and chaining.

Many academic chains are forged using a total task presentation. The arithmetic teacher working on long division usually requires her students to solve an entire problem, with whatever coaching is required, until they have mastered the process. Students finding the longitude and latitude of a given location in geography practice the entire process until it is mastered as do biology students learning to operate a microscope.

As with prompting techniques, the jury is still out as to which chaining technique is most effective, although there is some indication (Spooner & Spooner, 1984) that total task presentation may be most effective in teaching complex assembly tasks to retarded students. Classroom teachers are again advised to try what seems in their professional judgment to be the best procedure, and if that doesn't work, to try something else. This may be a good place to reiterate our statement in Chapter 5 that a teacher must regularly read the professional literature. New conclusions about effective management and instructional techniques are published constantly. In the following anecdote Ms. Cadwallader uses chaining (what kind is it?) to help Professor Grundy learn a new skill.

The Computerized Professor

Having failed to convince his department chairperson to reorder yellow tablets (she assured him of her confidence in his ability to form new discriminations), Professor Grundy reluctantly concluded that he might as well learn to use the word processor. Several microcomputers were available in the department, and he had seen several of his colleagues typing away as their words appeared in green letters on the screen. He waited until Friday afternoon when the department was normally deserted except for secretaries and graduate assistants and crept into one of the computer rooms. His only previous experience with computers had been in statistics courses during his graduate work when his carefully punched stacks of cards disappeared into the computer center and returned the next day with only a single sheet of green-and-white striped paper bearing the words, "Invalid logon, program dump." He stood for a moment moodily staring at the computer, then sat down and pushed a red toggle switch at the left of the machine. He jumped as the printer clicked and began to hum, a fan began to purr, and a little box with a glowing red light began to rattle and whirr. Grundy sat bemused for a moment and then began to type "The quick brown fox . . ." Nothing happened. The printer hummed, the fan purred, the light glowed, and the box whirred. He waited a few minutes and headed for the secretarial office and said, "Ms. Cadwallader, could I see you in the computer room for a moment?" Shrugging a who knows? response to the other secretaries' incredulous looks, she followed the professor.

"No, no, professor, that's not the way to boot the word processing program."

Ms. Cadwallader was delighted at the thought of no longer having to decipher Grundy's dreadful handwriting and happily agreed to help him learn to use the word processor.

"Okay, Professor," she chirped, "first you get the word processor program out of this little case and put it in disk drive 1."

"Disk drive 1?" asked the professor.

"Watch, I'll show you," said Ms. Cadwallader, "just like this. Now you do it. See, it's simple. Then you put a formatted disk in disk drive 2."

"Formatted disk?" muttered the Professor.

"Just one of these in this case, Professor. I fix them all up for you professors so you won't have to bother. Just write your name on it with a felt pen. Don't ever use a ball point, you'll bomb your text files."

"Bomb?" growled the professor. "Text files?"

"Don't worry about that, just use a felt pen." said Ms. Cadwallader hastily, "Now you're ready to boot it."

"Boot it," snorted Grundy, "I'll boot it!"

"No, no, Professor," said Ms. Cadwallader, "that just means to turn the computer on so it will load the program into its memory."

Mrs. Cadwallader patiently took the professor through all the steps necessary to operate the word processor. When she left, he was typing briskly and smiling with glee at the green words appearing on the screen.

Two hours later the professor emerged, triumphantly brandishing his disk.

"I did it, Ms. Cadwallader!" he crowed. "I even figured out how to turn the machine off by myself. I know you told me to call you, but I didn't need to. I'll be able to finish this report in an hour on Monday. I can't believe I waited so long to learn to use that machine."

"Professor," asked Ms. Cadwallader fearfully, "did you save what you wrote on your disk before you turned off the computer?"

"Save?" whispered the professor.

Task Analysis

The most exacting task facing the teacher who wants students to acquire complex behavioral chains is determining exactly what steps, links, or components must be included and their sequence. The process of breaking complex behavior into its component parts is called **task analysis**. Task analysis requires considerable practice but can be applied to behaviors ranging from eating with a spoon to writing a term paper. It is perhaps easier, in general, to analyze motor tasks than those related to academic and social behavior, but the analysis is equally important for teaching all complex behaviors. A complete task analysis of the process of task analysis is beyond the scope of this book. Moyer and Dardig (1978) provide excellent basic procedures for teachers.

In order to acquire a general idea of what is involved in task analysis, take a simple task, such as putting on a jacket, and list its component parts in the correct sequence. Then read your steps, in order, to a tolerant friend while she or he does exactly what you have written. Don't worry, you'll do better next time. One of the authors assigns this task as part of a

To task analyze:
pinpoint the terminal
behavior; list necessary
prerequisite skills; list
component skills in
sequence.

midterm exam. Fs are given only to task analyses that result in feet being placed through armholes.

Task analysis is the basis for programs teaching complex vocational skills to people who have severe and profound handicaps. It is theoretically possible, by breaking a task into sufficiently small components, to teach anybody anything. Time limitations, however, make it impractical to teach some students some things. The technology nevertheless exists. Teachers can even teach students to perform behaviors that those teachers cannot perform, so long as the teachers can recognize and reinforce the terminal behavior and its components (Karen, 1974).

DIFFERENTIAL REINFORCEMENT FOR SHAPING

The behavioral procedures so far described in this chapter assume that students are able to perform with some degree of prompting the components of the target behavior. The emphasis has been on differential reinforcement to bring the desired behavior under the control of a specified stimulus. Many behaviors that teachers want students to perform are not a part of the students' behavioral repertoire. For such behaviors, a different approach is required.

Shaping

Shaping is defined as differential reinforcement of successive approximations to a specified target behavior. Becker et al. (1975b) list two essential elements of shaping: differential reinforcement and a shifting criterion for reinforcement. Differential reinforcement, in this case, requires that responses that meet a certain criterion are reinforced, while those that do not meet the criterion are not. The criterion for reinforcement shifts ever closer to the target behavior.

Although the term *differential reinforcement* is used in both stimulus control and shaping, the usage is somewhat different. In developing stimulus control, a response in the presence of the S^D is reinforced; the same response in the presence of S^Δ is not reinforced. The differentiation of reinforcement depends on the antecedent stimulus. In shaping procedures, differential reinforcement is applied to *responses* that successively approximate (or become increasingly closer to) the target behavior. It is easy to confuse shaping with fading, because both involve differential reinforcement and gradual change. The following guidelines should clarify the differences:

1. Fading is used to bring an already learned behavior under the control of a different stimulus; shaping is used to teach a new behavior.

2. The behavior itself does not change when fading is used; only the antecedent stimulus varies. In shaping, the behavior itself is changed.

3. In fading, the teacher manipulates antecedents; in shaping, consequences are manipulated.

Shaping is not a stimulus control procedure; it is included in this chapter because it is an integral part of many teaching strategies that combine elements of stimulus control, prompting, fading, and chaining.

Suppose his teacher wants Harold to remain in his seat for an entire 20-minute work period. She has observed that so far Harold has never remained in his seat longer than 5 minutes, with an average of 2 minutes. A program in which Harold earns a reinforcer for remaining in his seat for 20 minutes is doomed—Harold will never come into contact with the reinforcer. Instead of this approach, the teacher defines her target behavior as Harold's remaining in his seat for the full 20 minutes but sets up a graduated sequence of criteria:

1. Harold remains in his seat for 3 minutes.
2. Harold remains in his seat for 5 minutes.
3. Harold remains in his seat for 10 minutes.
4. Harold remains in his seat for 15 minutes.
5. Harold remains in his seat for 20 minutes.

Each step in the sequence will be reinforced until established; then the criterion for reinforcement will be shifted to the next step. Shaping procedures may be used to establish new behaviors of many kinds, from verbal behavior in severely handicapped students to study behaviors in college students.

Shaping appears deceptively simple. Its efficient use requires great skill on the part of the teacher. Particularly important is the precise definition of the target behavior. A teacher who does not know exactly where he or she is going will have great difficulty in defining carefully graded steps toward getting there. The second skill required in planning a shaping program is the ability to choose a place to start. The teacher must choose a behavior that the student already performs and that can be shaped. A further consideration is the size of the steps toward the goal. If the steps are too small, the procedure is needlessly time-consuming and inefficient. If the steps are too large, the student's responses will not be reinforced and the behavior will be extinguished (Reese, 1966). Finally, the teacher must consider how long to remain at each plateau—just long enough to establish the behavior solidly, but not so long that the student becomes stuck at that level.

It is not always possible to make all of these decisions before beginning a program. For example, Harold's teacher might find that even after Harold has consistently remained in his seat for 5 minutes for a full week, he fails to meet the criterion of sitting for 10 minutes. The teacher would then have to drop back to 5 (or even 4) minutes and gradually work back up to 10, using smaller increments this time. The ability to evaluate and adjust ongoing programs is vital to the success of shaping procedures.

In shaping Harold's behavior, the teacher was able to define rather simply a criterion for reinforcement at each level. Such simplicity is not always

*"I wonder how he'll
react when he finds
out about our class
behavior-shaping
project.*

possible. When shaping such behaviors as the production of vocalizations, the teacher must carefully judge whether a criterion has been met. A shaping procedure for teaching vocal imitation to a severely handicapped child involves presenting the S^D—"ah" for example—and reinforcing successive approximations to correct imitation. The teacher might reinforce any vocalization at first, then only vowellike sounds, then only close approximations to "ah." In the heat of a language-training session with an autistic child, it is frequently difficult to determine whether a given vocal utterance is closer to the target behavior than the preceding one. Only extended practice, supervised by an instructor who shapes the teacher's behavior, will result in the development of such a skill.

Shaping is an extremely useful teaching tool. It provides a means of developing new behaviors in students ranging from the profoundly handicapped to the gifted. It is by no means always the procedure of choice, however. Shaping should be used only when no combination of reinforcement, prompting, or chaining will result in the desired behavior. One of the authors, as a college supervisor, once observed a student teacher in a regular second-grade classroom who was, she said, shaping a child's behavior toward the terminal objective of sitting in a chair. She was currently reinforcing the behavior of standing within 5 feet of it. ETA of his bottom on the chair was the end of her student teaching assignment, 6 weeks later. Having observed the child sitting for 30 minutes in the cafeteria, the college supervisor offered expert consultation. The student teacher, with great trepidation, gently led the student to his chair and issued the instruction, "Sit down, Mike." Mike sat. A program much like Harold's kept him sitting. Save shaping for when it's necessary. Common sense and applied behavior analysis are not mutually incompatible.

Sam Develops Work Habits

Sam was a 9-year-old third-grader who spent an hour a day in a resource class for children with learning disabilities. He was able to keep up academically with his classmates due to the resource teacher's help, but his regular class teacher, Ms. Walker, complained that he had poor work habits. Mr. Finch, the resource teacher, determined that Sam's poor work habits included failing to bring pencils and paper, failing to complete class and homework assignments, and keeping a messy desk. The two teachers defined the target behavior as developing good work habits—bringing materials, completing assignments, and keeping a tidy desk. The successive approximations to the goal were defined as follows:

1. Sam brings at least one pencil to school.

2. Sam still has a pencil at noon.

3. Sam still has a pencil at 3:00.

4. Sam meets criterion #3 and also has a notebook containing at least 10 sheets of clean paper in the morning.

5. Sam meets criterion #4 and still has paper at noon.

6. Sam meets criterion #5 and still has paper at 3:00.

7. Sam meets criterion #6 and completes at least ½ of his class assignments.

8. Sam meets criterion #6 and completes at least ¾ of his class assignments.

9. Sam meets criterion #6 and completes all class assignments.

10. Sam meets criterion #9 and has no more than three extraneous items (trash, toys, etc.) in his desk on a daily spot check.

11. Sam meets criterion #9 and has no more than two extraneous items in his desk.

12. Sam meets criterion #9 and has no more than one extraneous item in his desk. (The teachers agreed to stop here—everyone is entitled to one extraneous item.)

13. Sam meets criterion #12 and has books, notebook, and folders neatly stacked in his desk.

14. Sam meets criterion #13 and shows Ms. Walker his list of homework assignments at 3:00.

15. Sam meets criterion #14 and turns in at least ½ of his homework.

16. Sam meets criterion #14 and turns in at least ¾ of his homework.

17. Sam meets criterion #14 and turns in all of his homework.

Mr. Finch had some difficulty persuading Ms. Walker that it was necessary to reward Sam (as she saw it) for doing less than other students did without rewards. She also stated that good work habits had to be learned at home and that students who had not developed them by third grade were unlikely ever to do so. She agreed to try, however.

Mr. Finch provided a checklist and Ms. Walker recorded Sam's behavior. Sam was to earn the privilege of helping Mr. Finch clean up the resource room from 3:00 to 3:15 each day that he met criterion.

In spite of Ms. Walker's skepticism, the program was successful. It took Sam only a few days to begin bringing (and keeping) pencils and paper. He earned his privilege for 5 days just by having materials but, to Ms. Walker's surprise, completed all of his class assignments on two of those days. (It is much easier to complete assignments when you have pencil and paper.) The teachers agreed to drop step 7 and require Sam to complete ¾ of his class assignments. Ms. Walker wanted to require all classwork, but Mr. Finch argued that this was expecting too much and that Sam might be unable to each criterion. "Let's not blow it now," he suggested, "he's doing so well." This criterion was maintained for a week. Then reinforcement was made contingent upon completing all classwork.

At this point, the teachers agreed that homework had a higher priority than a neat desk and went straight to step 14. Anyway, reported Ms. Walker, since Sam was completing his assignments, the number of paper airplanes, doodles, and other junk in his desk had diminished drastically, to within normal limits for third graders.

The greatest surprise in the entire program was that when Sam was required to write down his assignments in order to earn his privilege, he immediately began bringing all of his homework. No further shaping was required; Sam had good work habits. When last observed, Ms. Walker, with a strange gleam in her eye, was muttering, "I'll bet I could shape up his sloppy handwriting too, and Melanie's tardiness, and Jack's loud voice, and . . ." Mr. Finch, meanwhile, praised Ms. Walker's cooperative attitude to the principal at a faculty meeting and decided that the next step in her shaping program was for her to design and reproduce her own checklist.

As we stated earlier, shaping and fading are often used in combination. The following example illustrates a combined procedure.

Ms. Wallace's Class Learns to Print the Letter A

Ms. Wallace was trying to teach her students to print the letter *A*. At first she just told them, "Make a capital *A*." They made no response to this S^D. "Look at the one on the chart," she said. "Make it just like this one." To this visual prompt, some of the students responded by producing a creditable *A*. But,

"Does this look like the one on the chart, Harold?" asked Ms. Wallace. She then tried some verbal instructions: "Make two slanted lines that look like a tee-pee, then make another line across the middle." This verbal prompt resulted in success for some students, but,

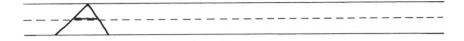

"Ralph, your teepee is a little flat," sighed Ms. Wallace. In desperation, Ms. Wallace walked around the room guiding her students' hands through the correct movements. Physical prompting resulted in success for many students, but,

"Melissa, relax your hand, for heaven's sake. I'm only trying to HELP you," wailed Ms. Wallace.

In the teacher's lounge that afternoon, Ms. Wallace sobbed, "I can't go through this 25 more times." An unkind colleague pointed out that she had forgotten the lowercase letters, making 51 more to go. Before Ms. Wallace became completely hysterical, an experienced first-grade teacher showed her a worksheet like this:

--A--A--A--A--A-------------

"You see," said Ms. Weatherby, "you just reinforce successive approximations to the terminal behavior of writing an *A* independently, tracing with fewer and fewer cues until the prompt just fades away."

SUMMARY

This chapter discussed the process of differential reinforcement for bringing students' behavior under stimulus control. The process for bringing both simple and complex behaviors under the control of antecedent stimuli has been described.

Verbal, visual, modeling and physical prompts have been discussed, as well as systematic procedures for fading these prompts. Backward chaining, forward chaining, and total task presentation have been described. Finally, procedures for shaping behaviors that students cannot initially perform were suggested.

REFERENCES

AYLLON, T. 1977. Personal communication.

BAER, D.M., PETERSON, R.F., & SHERMAN, J.A. 1967. The development of imitation by reinforcing behavioral similarity to a model. *Journal of the Experimental Analysis of Behavior, 10*, 405–416.

BANDURA, A. 1969. *Principles of behavior modification*. New York: Holt, Rinehart, & Winston.

BANKS, M.E., & BAER, D.M. 1966. Chaining in human learning. Working paper #138. Unpublished manuscript. Parsons Research Center.

BECKER, W.C., ENGELMANN, S., & THOMAS, D.R. 1975a. *Teaching 1: Classroom management*. Chicago: Science Research Associates.

BECKER, W.C., ENGELMANN, S., & THOMAS, D.R. 1975b. *Teaching 2: Cognitive learning and instruction*. Chicago: Science Research Associates.

BEREITER, C., & ENGELMANN, S. 1966. *Teaching disadvantaged children in the preschool*. Englewood Cliffs, N.J.: Prentice-Hall.

BILLINGSLEY, F.F., & ROMER, L.T. 1983. Response prompting and the transfer of stimulus control: Methods, research, and a conceptual framework. *Journal of the Association for Persons with Severe Handicaps, 8*, 3–12.

BRADLEY-JOHNSON, S., JOHNSON, C., & SUNDERMAN, P. 1983. Comparison of delayed prompting and fading for teaching preschoolers easily confused letters & numbers. *Journal of School Psychology, 21*, 327–335.

BRIGHAM, T.A., & SHERMAN, J.A. 1968. An experimental analysis of verbal imitation in preschool children. *Journal of Applied Behavior Analysis, 1*, 151–158.

CHENEY, T., & STEIN, N. 1974. Fading procedures and oddity learning in kindergarten children. *Journal of Experimental Child Psychology, 17*, 313–321.

CSAPO, M. 1981. Comparison of two prompting procedures to increase response fluency among severely handicapped learners. *Journal of the Association for the Severely Handicapped, 6*, 39–47.

DELBERT, A.N., & HARMON, A.S. 1972. *New tools for changing behavior*. Champaign, Ill.: Research Press.

ENGELMANN, S. 1969. *Conceptual learning*. Sioux Falls, S.D.: Adapt Press.

ENGELMANN, S., & CARNINE, D. 1982. *Theory of instruction: Principles and applications*. New York: Irvington Publishers, Inc.

ESPICH, J.E., & WILLIAMS, B. 1967. *Developing programmed instructional materials*. Palo Alto, Calif.: Fearon.

ETZEL, B.C., & LEBLANC, J.M. 1979. The simplest treatment alternative: The law of parsimony applied to choosing appropriate instructional control and errorless-learning procedures for the difficult-to-teach child. *Journal of Autism and Developmental Disorders, 9*, 361–382.

FERSTER, C.B., CULBERTSON, S., & BOREN, M.C.P. 1975. *Behavior principles,* 2nd ed. Englewood Cliffs, N.J.: Prentice-Hall.

FOXX, R.M., & AZRIN, N.H. 1972. Restitution: A method of eliminating aggressive-disruptive behavior of retarded and brain damaged patients. *Behavior Research and Therapy, 10*, 15–27.

FOXX, R.M., & AZRIN, N.H. 1973. *Toilet training the retarded*. Champaign, Ill.: Research Press.

GELFAND, D.M. 1962. The influence of self-esteem on rate of verbal conditioning and social matching behavior. *Journal of Abnormal and Social Psychology, 65*, 259–265.

GOLDSTEIN, A.P., HELLER, K., & SCHREST, L.G. 1966. *Psychotherapy and the psychology of behavior change*. New York: Wiley.

HAUPT, E.J., VAN KIRK, M.J., & TERRACIANO, T. 1975. An inexpensive fading procedure to decrease errors and increase retention of number facts. In E. Ramp & G. Semb (Eds.), *Behavior analysis: Areas of research and application* Englewood Cliffs, N.J.: Prentice-Hall.

HERRNSTEIN, B.J., & LOVELAND, D.H. 1964. Complex visual concept in the pigeon. *Science, 146*, 549–550.

HOLLAND, J.G., & SKINNER, B.F. 1961. *The analysis of behavior*. New York: McGraw-Hill.

HOVLAND, C.I., JANIS, I.C., & KELLEY, H.H. 1953. *Communication and persuasion*. New Haven, Conn.: Yale University Press.

KAREN, R.L. 1974. *An introduction to behavior theory and its applications*. New York: Harper & Row.

KLEINERT, H.L., & GAST, D.L. 1982. Teaching a multihandicapped adult manual signs using a constant time delay procedure. *Journal of the Association of the Severely Handicapped, 6*(4), 25–32.

KRUMBOLTZ, J.D., & KRUMBOLTZ, H.D. 1972. *Changing children's behavior*. Englewood Cliffs, N.J.: Prentice-Hall.

LEFKOWITZ, M.M., BLAKE, R.R., & MOUTON, J.S. 1955. Status factors in pedestrian violation of traffic signals. *Journal of Abnormal and Social Psychology, 51*, 704–706.

LOVAAS, O.I., BERBERICH, J.P., PERLOFF, B.F., & SCHAEFFER, B. 1966. Acquisition of imitative speech by schizophrenic children. *Science, 151*, 705–707.

LOVAAS, O.I., SCHREIBMAN, L., KOEGEL, R.L., & RHEN, R. 1971. Selective responding by autistic children to multiple sensory input. *Journal of Abnormal Psychology, 77*, 211–222.

MARTIN, J., RUSCH, F., JAMES, V., DECKER, P., & TRTOL, K. 1974. The use of picture cues to establish self control in the preparation of complex meals by mentally retarded adults. *Applied Research in Mental Retardation, 3*, 105–119.

MAYER, G.R., ROHEN, T.H., & WHITLEY, A.D. 1969. Group counseling with children: A cognitive behavioral approach. *Journal of Counseling Psychology, 16*, 142–149.

MICHAEL, J. 1982. Distinguishing between discriminative and motivational functions of stimulus. *Journal of the Experimental Analysis of Behavior, 37*, 149–155.

MOSK, M.D., & BUCHER, B. 1984. Prompting and stimulus shaping procedures for teaching visual-motor skills to retarded children. *Journal of Applied Behavior Analysis, 17*, 23–34.

MOYER, J.R., & DARDIG, J.C. 1978. Practical task analysis for educators. *Teaching Exceptional Children, 11*, 16–18.

PATTERSON, E.T., PANYON, M.C., WYATT, S., & MORALES, E. September 1974. Forward vs. backward chaining in the teaching of vocational skills to the mentally retarded: An empirical analysis. Paper presented at the 82nd Annual Meeting of the American Psychological Association, New Orleans.

REESE, E.P. 1966. *The analysis of human operant behavior*. Dubuque, Iowa: William C. Brown.

RODEWALD, H.K. 1979. *Stimulus control of behavior*. Baltimore: University Park Press.

ROSENBAUM, M.E., & TUCKER, I.F. 1962. The competence of the model and the learning of imitation and nonimitation. *Journal of Experimental Psychology, 63*, 183–198.

SCHREIBMAN, L. 1975. Effects of within-stimulus and extra-stimulus prompting on discrimination learning in autistic children. *Journal of Applied Behavior Analysis, 8*, 91–112.

SCHREIBMAN, L., & CHARLOP, M.H. 1981. S^+ versus S^- fading in prompting procedures with autistic children. *Journal of Experimental Child Psychology, 34*, 508–520.

SPOONER, F. 1981. An operant analysis of the effects of backward chaining and total task presentation. *Dissertation Abstracts International, 41*, 3992A (University Microfilms No. 8105615).

SPOONER, F., & SPOONER, D. 1983. Variability: An aide in the assessment of training procedures. *Journal of Precision Teaching, 4*, no. 1, 5–13.

SPOONER, F., & SPOONER, D. 1984. A review of chaining techniques: Implications for future research and practice. *Education and Training of the Mentally Retarded, 19*, 114–124.

STAATS, A.W., & STAATS, C.K. 1963. *Complex human behavior*. New York: Holt, Rinehart, & Winston.

STELLA, M.E., & ETZEL, B.C. 1978. Procedural variables in errorless discrimination learning: Order of S^+ and S^- manipulation. *American Psychological Association*. Toronto, Canada.

STRIEFEL, S., & WETHERBY, B. 1973. Instruction-following behavior of a retarded child and its controlling stimuli. *Journal of Applied Behavior Analysis, 6*, 663–670.

SULZER-AZAROFF, B., & MAYER, G.R. 1977. *Applying behavior-analysis procedures with children and youth*. New York: Holt, Rinehart, and Winston.

TERRACE, H.S. 1966. Stimulus control. In W.K. Honig (Ed.) *Operant behavior: Areas of research and application*. New York: Appleton-Century-Crofts.

TOUCHETTE, P.E., & HOWARD, J.S. 1984. Errorless learning: Reinforcement contingencies and stimulus control transfer in delayed prompting. *Journal of Applied Behavior Analysis, 17*, 175–188.

VAN ETTEN, G., ARKELL, C., & VAN ETTEN, C. 1980. *The severely and profoundly handicapped: Programs, methods, and materials*. St. Louis. C. V. Mosby.

WACKER, D.P., & BERG, W.K. 1983. Effects of picture prompts on the acquisition of complex vocational tasks by mentally retarded adolescents. *Journal of Applied Behavior Analysis, 16*, 417–433.

WALLS, R.T., ZANE, T., & ELLIS, W.D. 1981. Forward chaining, backward chaining, and whole task methods for training assembly tasks. *Behavior Modification, 5*, 61–74.

WHALEY, D.C., & MALOTT, R.W. 1971. *Elementary principles of behavior.* New York: Appleton-Century-Crofts.

WILCOX, B., & BELLAMY, G.T. 1982. *Design of high school programs for severely handicapped students.* Baltimore: Paul H. Brookes.

Section Four

Maintaining Behavior Change

9 Providing for Generalization of Behavior Change

- Identifying and defining a behavioral objective to guide a student's behavior change can take several hours of observation, consultation, and contemplation.
- Planning a behavioral program that will work for that student can take considerable concentration as well.
- Executing the program you've planned takes patience, perseverance, faith, and trust.

Imagine then the frustration when, after months of careful programming, thinning, fading, and so on, you see a student go right back to performing the same old inappropriate behavior as soon as she sits down in a new classroom. The potential for this kind of disappointment emphasizes the importance of programming for generalization and the importance of this chapter.

Preceding chapters described principles and procedures related to strengthening appropriate behaviors, reducing or eliminating maladaptive behaviors, and teaching new behaviors. The technology of behavior change presented in these earlier chapters has been thoroughly investigated and its efficacy demonstrated beyond doubt. Applied behavior analysts have not, however, been as thorough in demonstrating that behaviors changed using their technology are changed permanently or, indeed, that the changed behaviors are displayed in any setting except that in which training programs are executed. It is meaningless to change behavior unless the change can be made to last and unless behavior will occur in settings other than the original training site and in the absence of the original trainer.

Many of the criticisms we discussed in Chapter 2 have resulted from the short-lived results of many behavior change programs. Fortunately, these criticisms have resulted in a radical increase in concern for documenting generalization of behavior change.

In order to program meaningful behavior change, teachers must utilize the behavioral principle of generality. Baer, Wolf, and Risley (1968) stated that "a behavioral change may be said to have generality if it proves durable over time, if it appears in a wide variety of possible environments, or if it spreads to a wide variety of related behaviors" (p. 96). Baer et al. describe three ways a behavior may show generality: over time, across settings, and across behaviors. Each of these types of generalization will be discussed in more detail later in this chapter. The following examples illustrate behavior changes that do not have generality:

The Quartz County school system has a special class for students with behavior problems. An elaborate token economy is used, and all students behave well and complete their assignments. The special education supervisor is worried, however, because as soon as the students are returned to the regular classroom, they revert to the patterns of disruptive behavior and poor academic performance that resulted in their original referral to the special program.

Ms. Kitchens is a resource teacher for children with learning disabilities. Her students spend approximately 2 hours a day in her classroom and the remainder of their time in regular classes. Ms. Kitchens has noticed that, while her students do very well in the resource room, they do not perform in the regular classroom those academic tasks she knows they can do.

Mr. Foote's first graders have learned to recognize a large number of words using flashcards and a look-say approach. However, when confronted with an unfamiliar word, the students merely guess. They have not learned to decode new words based on relationships between symbols and sounds.

The preceding examples describe situations in which some academic or social behavior has been successfully changed. The change, however, did not last once the contingencies leading to the original behavior change were removed. There is no question that applied behavior analysis procedures often cause situation-specific behavior changes that do not last. Indeed, several of the research designs described in Chapter 5 depend upon this phenomenon (Hartmann & Atkinson, 1973). The classic ABAB (reversal) design demonstrates functional relationships between behavior and consequences by successfully applying and withdrawing such consequences and demonstrating that the dependent variable (the behavior) changes according to the condition. If the behavior failed to return to its baseline rate under one of these designs, the experimenter would have failed to demonstrate a functional relationship.

The multiple baseline design enables an experimenter to demonstrate functional relationships by successfully applying contingencies to several different behaviors, to behavior in several different settings, or to the same behavior displayed by different students. Again, a functional relationship is shown only when the behaviors are not changed until contingencies are implemented. Baer et al. (1968) stated that "generalization should be programmed rather than expected or lamented" (p 97). The experimenter attempting to establish functional relationships between procedures and behavior may, indeed, lament the occurrence of generalization. The classroom teacher is more likely to expect it and to lament its absence. With few exceptions (Kifer, Lewis, Green, & Phillips, 1974; Patterson, 1965), it is the professional expecting generalization who is disappointed. Because generalization does not automatically result when behavior is changed (Axelrod, 1977; Kazdin, 1977; Stokes & Baer, 1977), does this mean that applied behavior analysis procedures are useless? If you have stuck with us this far, you know that we don't think so. To most behavior analysts, the lack of automatic generalization simply indicates the need for developing a technology of generalization as efficient as the technology of behavior change. Such a technology need not interfere with the necessity for demonstrating functional relationships; it may be applied after these relationships have been established (Kallman, Hersen, & O'Toole, 1975).

As Baer et al. (1968) suggested, generalization must be programmed. This chapter presents the principles of generalization to indicate the basis of this programming and suggests specific ways that teachers can increase the odds that the behaviors their students learn will be maintained even when all the charts and graphs have been discarded.

GENERALIZATION

Before suggesting guidelines for facilitating generalization, we need to differentiate several types. The first of these occurs when a response that has been trained in a specific setting with a specific instructor occurs in a different setting or with a different instructor. This phenomenon, called simply *generalization* (Koegel & Rincover, 1977), *transfer of training* (Kazdin, 1977), or *stimulus generalization* (Barton & Ascione, 1979), should be distinguished from *response maintenance* (Kazdin, 1977), which refers to the tendency of a learned behavior to occur after programmed contingencies have been withdrawn. Response maintenance may also be labeled *maintenance* (Koegel & Rincover, 1977), *resistance to extinction, durability*, or *behavioral persistence* (Atthowe, 1973). Finally, the term *response generalization* is used when referring to unprogrammed changes in similar behaviors when a target behavior is modified (Twardosz & Sajwaj, 1972). This phenomenon may also be termed *concomitant* or *current behavior change* (Kazdin, 1973). It is easy to see why some confusion may result from varying terminology or from researchers' failure to differentiate among the varieties of generalization.

In this chapter, the term *generalization* will refer to any of the three types and the terms *stimulus generalization, maintenance*, and *response generalization* will be used when distinctions are made among them. Table 9–1 illustrates the relationships among the terms used to describe generality of behavior change.

Stimulus Generalization

Chapter 8 discussed the development of discrimination. You will recall that discrimination training involves bringing a response under stimulus control so that, for example, a student reads "went" only when the stimulus *w-e-n-t* is presented and not when the teacher shows a flashcard with *w-a-n-t* or *w-e-t*. The student whose reading behavior is under faulty stimulus control is demonstrating stimulus generalization. Such faulty stimulus control is also demonstrated when young children call all four-legged animals "doggie" or when the "big man on campus" calls the president of the women's liberation organization "baby."

Stimulus generalization occurs when responses that have been reinforced in the presence of a specific stimulus (S^D) occur in the presence of different but similar stimuli (Guttman & Kalish, 1956). A group of stimuli that should occasion the same response may be considered members of a *stimulus class*. In general, the more similar the stimuli, the more likely it is that stimulus generalization will occur. As the stimuli become increasingly different from S^D, the likelihood decreases that the response will be emitted in their presence. The pattern of decreased responding as stimuli become less similar to the S^D has been called a *generalization gradient*. Figure 9–1 illustrates a hypothetical generalization gradient for a student who is learning to identify the letter *b*. Responding with "b" to letters similar to *b* is more frequent than responding to letters whose configuration is different. In this example, stimulus generalization is working against the teacher, so efforts must be directed at tightening stimulus control.

On the other hand, we could illustrate a very similar generalization gradient using the letter *b* as originally presented and the letter *b* in different sizes, typefaces, colors, and contexts. In this instance, the teacher wants to

Table 9–1   Terminology used to describe types of generalization

	Generalization		
	Stimulus generalization	Maintenance	Response generalization
Also called:	Generaiization Transfer of training	Response maintenance Resistance to extinction Durability Behavioral persistence	Concomitant behavior change Current behavior change

Figure 9–1
Generalization
gradient

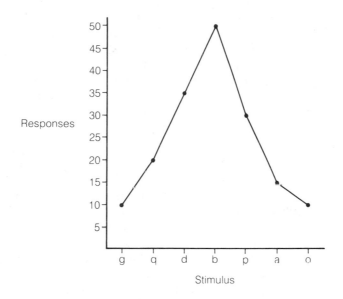

promote stimulus generalization, not prevent it. Unfortunately, the same teaching process that facilitates appropriate discrimination may result in too fine a discrimination for practical purposes. The students described at the beginning of this chapter who performed differently in the resource room and the regular classroom illustrate this problem. Those students formed a discrimination that resulted in responses that were too situation specific. The resource room had become an S^D for appropriate behavior; the regular classroom, an S^Δ. It is paradoxical that teachers must expend so much effort in bringing students' behavior under stimulus control but must also work towards avoiding too great a degree of the same control.

Maintenance

If you were unable to remember this, perhaps your behavior, when no longer reinforced by passing tests, has undergone extinction.

Most behaviors that teachers want students to perform should occur even after systematic applied behavior analysis procedures have been withdrawn. Teachers want their students to read accurately in class and also to continue reading accurately after they are no longer in school. Math problems in school are merely a means to an end—we want students eventually to balance checkbooks, fill out income tax forms, or multiply recipes. Appropriate social behaviors, too, while adaptive in the classroom, are also necessary when a specific program for their systematic reinforcement no longer exists. Chapter 7 detailed what happens when positive reinforcement is abruptly withdrawn from a behavior that previously has been reinforced on a continuous schedule: the behavior decelerates and is eventually extinguished. Extinction may be a very useful phenomenon when a teacher withdraws attention from a maladaptive behavior. On the other hand, it may be frustrating when the teacher has systematically developed some appropriate behavior only to see that it has disappeared when the student is observed a year later in another class.

Assuring that behavior will be maintained is an important part of teaching. It is impossible for the teacher to follow students around forever, reinforcing them with cereal or a smile and praise. Behaviors that extinguish rapidly when the artificial contingencies used to develop them are withdrawn can hardly be considered learned in any meaningful sense. Most experimental evidence indicates that extinction occurs unless specific measures are taken to prevent it. (Kazdin, 1977 Kazdin & Bootzin, 1972; Rincover & Koegel, 1975; Stokes, Baer, & Jackson, 1974).

Response Generalization

Sometimes changing one behavior will result in changes in other similar behaviors. Such similar behaviors are often referred to as a *response class,* and changes in untrained members of the response class, as **response generalization**. For example, if students receive reinforcers for completing multiplication problems and subsequently increase their rates of completing both multiplication and division problems, response generalization has occurred to untrained members of the response class: completion of arithmetic problems.

Unfortunately, this kind of generalization does not happen often. Usually only the specific behavior being reinforced will change. "Behavior, unlike the flower, does not naturally bloom" (Baer & Wolf, 1970, p. 320). We cannot count on response generalization.

TRAINING GENERALIZATION

Assuring the generality of behavior change is particularly important to the teacher of handicapped students. Since recent legislation requires that all such students be educated in the least restrictive environment, large numbers of mildly handicapped students are in special classes only temporarily or for only part of the school day. The special educator cannot count on being able to apply systematic applied behavior analysis procedures for extended periods of time or even for the entire school day. Even the teacher of more severely handicapped students must be aware that these students, too, will be living in an environment that is least restrictive—that is, one that as much as possible resembles that of their nonhandicapped peers. Special educators must prepare their students to perform in situations where systematic contingency management programs may not be available.

The regular educator also must be aware of techniques for promoting generalization. Regular teachers will teach large numbers of mildly handicapped students who have been taught appropriate academic and social behaviors using applied behavior analysis procedures. In order to help these students perform optimally in the regular classroom, regular teachers must be aware not only of the techniques used to teach these students but also of techniques that will encourage generalization to less-structured settings.

The procedures for promoting generalization described in the following sections include some that would not meet more technical or stringent definitions of generalization of behavior change (see Johnston, 1979; Keller & Schoenfeld, 1950; Skinner, 1953). Traditionally, generalization has been noted only when behavior occurs spontaneously in circumstances where no contingencies are in effect. For practical purposes, we shall also consider changes in behavior that can be facilitated by relatively minor changes in the setting in which generalization is desired. If such changes can be made fairly effortlessly and gains made in the training setting can be maintained, for all practical purposes the behavior has generalized.

Stokes and Baer (1977) reviewed the literature on generalization assessment and training in applied behavior analysis and categorized the techniques for assessing or programming generalization as follows:

1. Train and Hope
2. Sequential Modification
3. Introduce to Natural Maintaining Contingencies
4. Train Sufficient Exemplars
5. Train Loosely
6. Use Indiscriminable Contingencies
7. Program Common Stimuli
8. Mediate Generalization
9. Train "To Generalize" (p. 350)

The following sections review relevant research on each of the types of generalization training described by Stokes and Baer and then provide examples illustrating possible classroom uses.

Train and Hope

Unplanned generalization does sometimes happen. It may be likely to happen in cases where the skill trained is particularly useful to the student or where the skill becomes reinforcing in itself. For example, Patterson (1965) used a DRO procedure to eliminate aggressive/disruptive classroom behavior in a hyperactive boy. The inappropriate behavior was reduced on the playground and at home (where no formal contingencies were in operation) as well as in the classroom. Other researchers have reported that students who are given token reinforcers for appropriate behavior during some part of the school day may show changes in that behavior during other parts of the day or even in other classrooms (Walker, Mattsen, & Buckley, 1971).

Appropriate behaviors may also last after programs are withdrawn. Hewett, Taylor, and Artuso (1969) used a structured classroom design including token reinforcement for academic and social behavior. They reported maintenance of high rates of academic behavior and low rates of disruption even after the tokens were withdrawn. Many behaviors taught to normal and mildly handicapped children do generalize. Students who learn to read in school delight (and sometimes drive to distraction) their parents by reading street signs. This spontaneous generalization however, is much less likely to occur with more severely handicapped students (Horner, McDonnell, & Bellamy, undated).

In spite of reported evidence that some behaviors are generalized, it is important to remember that most are not. When they are, we do not usually know why (Kazdin, 1977). It has been suggested that some aspect of the generalization setting may have acquired conditioned reinforcing characteristics (Baer & Wolf, 1970; Bijou, Peterson, Harris, Allen, & Johnston, 1969; Chadwick & Day, 1971; Medland & Stachnik, 1972). Another possibility is that the behavior of teachers or parents has been permanently altered by the implementation of the applied behavior analysis procedure and that reinforcing consequences, though no longer formally programmed, may still occur more frequently than before intervention (Greenwood, Sloane, & Baskin, 1974; Kazdin, 1977).

Although there is hope that behaviors acquired or strengthened through formal contingency management programs may generalize, there is no certainty. Hoping will not make generalization happen. The teacher who expects it should be prepared to monitor students' behavior closely and to instigate more effective procedures immediately upon learning that early hopes have been dashed. The following example provides an illustration of a behavior that generalized for no discernible reason.

Ms. Andrews Works a Miracle

Ms. Andrews, who did private tutoring to supplement her income, was asked to work with Brandon, who had completed the first semester of the seventh grade with 2 Cs, 2 Ds, and an F. Because Brandon had previously been an excellent student, his parents were frantic. Ms. Andrews tested Brandon, found no learning problem, and decided that Brandon simply had a particularly bad case of seventh grade slump. He was not doing homework or classwork, and he didn't study for tests. He seemed to have efficient study skills—he just wasn't using them. Ms. Andrews suggested that once-a-week tutoring was not the answer and urged the parents to implement some contingency management. She explained the low probability that study behavior would generalize beyond her living room. They insisted on tutoring and implemented no program.

For two weeks, Ms. Andrews worked with Brandon on grade-level materials —not the books he was using at school because he naturally forgot to bring them. She provided very high rates of verbal praise for reading, studying, and completing math problems, as well as access to a gamelike vocabulary development program if Brandon completed other tasks.

During the third week, Brandon's mother called. Three of his teachers had written notes indicating that Brandon's work in their classes had improved drastically. Tests and papers from all classes were As and Bs. Ms. Andrews, said Brandon's mother, had worked a miracle. Agreeing modestly before hanging up, Ms. Andrews spent the next few minutes staring into space and wondering, "Did he just coincidentally decide to kick into gear, or was it really something I did?"

Sequential Modification

Review Chapter 5 for an extended discussion of the multiple baseline design.

A procedure that allows stimulus generalization or transfer of training across settings is *sequential modification*. In this procedure, generalization (in the practical sense) is promoted by applying the same techniques that successfully changed behavior in one setting to all settings where the target behavior is desirable. Exactly the same process is undertaken when using a multiple-baseline–across-settings design to demonstrate a functional relationship between independent and dependent variables. A similar procedure to ensure maintenance would necessitate training those responsible for the student's education and care after the student is dismissed from a special training program. Teachers would be trained to carry out the same applied behavior analysis procedures employed in the training situation. In some cases, it may be unrealistic to program exactly the same contingencies in the generalization setting. For example, a classroom teacher may not be able to monitor disruptive behavior as closely, or deliver reinforcers as often, as a special-class teacher. Similarly, parents may be unable or unwilling to structure programs as closely as may be done in institutional settings. In such cases, modified versions of programs may still provide enough environmental control to maintain the target behaviors at rates close to those established during training.

Examination of any published study using a multiple-baseline–across-settings experimental design provides an example of sequential modification. Hall, Cristler, Cranston, and Tucker (1970) changed the behavior of a group of fifth-grade students who were consistently late coming back from three different recesses during the school day. A contingency applied to being in their seats within 4 minutes of the end of the noon recess resulted in decreased tardiness in coming back from that recess but no change in the tardiness in returning from the other two. The contingency was then applied to return from the morning recess as well, whereupon tardiness diminished for that situation but not for the afternoon recess. Not until reinforcement depended upon rapid return from all three recesses did the rates of tardiness decline for all.

Long and Williams (1973) reinforced appropriate behavior in a junior-high class using contingent free time. Behavior was monitored during both math and geography classes. No change in behavior occurred in the second class when the program was implemented in the first, but when contingent free time was used in both classes, appropriate behavior occurred at high rates in both.

Walker and Buckley (1968) changed the behavior of a 9-year-old child in a special class for children labeled behaviorally disturbed. The student received points for attending to academic tasks while in the special class and displayed high rates of attention in that class. When the student was returned to the regular classroom, a similar point system was employed and attending behavior remained at a high level for both settings.

A study by Lovaas, Koegel, Simmons, and Long (1973) demonstrated the effectiveness of applying sequential modification in assuring both transfer and maintenance. Parents whose autistic children had acquired basic communication, self-care, and social skills in a special program were trained to reinforce these skills with their children. On a 1- to 4-year follow-up, children who had remained at home and whose parents had continued to reinforce appropriate behavior had maintained the gains made in the special program and in some cases had made additional progress. Those children who had been institutionalized where personnel were not trained in applied behavior analysis procedures had lost all the gains made in the special program.

It is not always necessary to employ identical procedures in order to obtain stimulus generalization or maintenance. Walker and Buckley (1972) studied the effects of several ways of promoting maintenance in students whose behavior was changed in a special class setting using a token economy. After 2 months in the special class, the students were divided into four groups and sent back to regular classes. For one group, the authors used a procedure they referred to as *peer reprogramming*. Each experimental subject had the opportunity to earn points for appropriate behavior during two 30-minute periods a week. The points could be exchanged for prizes such as field trips, which were then awarded to the entire class. An interesting double contingency was in effect, because the target student had to earn the right to the 30-minute reinforcement period by appropriate behavior during the rest of the week. A rather elaborate electronic signaling system involving lights and buzzers was used during the reinforcement period to indicate to the subject and classmates that points were or were not accumulating. The authors theorized that, because the student's peers were being rewarded for his or her appropriate behavior, they would reinforce such behavior rather than disruption or inattentiveness.

The procedure described by Walker and Buckley for the second group was more closely related to exact sequential modification. This procedure used a duplication of many conditions in the special class. The same curriculum materials, systematic social reinforcement, and token economy were implemented in the regular class. The only difference in procedure was the absence of time-out and response-cost contingencies, which the authors feared might be used punitively by inexperienced teachers. An adapted rating system was used when the students were in classes other than their homeroom, and points based on ratings were awarded by the homeroom teacher.

For the third group, the classroom teachers in whose classes the target students were to be placed received systematic training in applied behavior analysis. The teachers' implementation of the procedures in their classrooms was monitored, and course credit and free tuition was contingent on correct implementation (another double contingency!). A final group of students was simply returned to the regular class.

The authors reported that peer reprogramming and similar reinforcement procedures maintained the students' appropriate behavior at over 70% of the rate shown in the special class. The teacher-training and no-treatment procedures were less effective, though still maintaining over 60% of the gains. A reanalysis of the data by Cone (1973) showed smaller percentages of maintenance and no clear difference among the three treatment procedures, which were, however, more effective than simply training and hoping—the no-treatment procedure. In addition, students who received the programmed maintenance procedures continued to show more appropriate behavior 4 months after all modifications were withdrawn.

Martin, Burkholder, Rosenthal, Tharp, and Thorne (1968) used systematic sequential modification in changing the behavior of severely disruptive adolescents placed in an educational program at a community mental health center. A five-phase system was used, beginning with full-time special-class placement, token reinforcement, and home-based reinforcement and eventually moving to full-time attendance in regular classes with very little tangible reinforcement based on grades and attendance. The students maintained high rates of academic performance and low rates of disruptive behavior. Drabman (1973), working with students in a state hospital classroom, used feedback in the form of ratings given by either a teacher or a peer to maintain low levels of disruptive behavior after reinforcement was withdrawn.

Anderson Inman, Walker, and Purcell (1984) used a procedure they call *transenvironmental programming* to increase generalization from resource rooms to regular classrooms. This procedure involves assessment of the target environment, providing students in the resource room with skills identified as critical in the regular classroom, using techniques to promote transfer of the skills acquired, and evaluating student performance in the regular classroom. Specific techniques for facilitating transfer included reinforcing the newly acquired skills in the regular classroom.

The transfer and maintenance of behaviors by sequential modification may not technically qualify as generalization, as we discussed earlier. However, the provision of identical or similar applied behavior analysis procedures in alternate settings is a practical and frequently successful procedure. Even if the alterations are necessary for extended periods of time, their effectiveness may make their implementation well worth the trouble. The following vignettes illustrate the use of sequential modification.

Connie Learns to Do Her Work

Modification 1

Connie was a second-grade student in Ms. Gray's resource room for children with learning disabilities. Connie performed very well in the resource room, where she earned points exchangeable for free time when she completed academic tasks, but did no work in the regular classroom, where she spent most of the school day. Instead, she spent her time wandering around the classroom bothering other students.

Modification 2

After conferring with the regular classroom teacher, Ms. Wallace provided her with slips of paper preprinted with Connie's name, a place for the date, and several options to check regarding Connie's academic work and class-

Figure 9–2
Connie's chart

CONNIE

Assignments complete:
Yes _____
No _____
Partly _____

DATE: _____

Behavior:
Good _____
Fair _____
Poor _____

_____ Initials

room behavior. (See Figure 9–2.) Ms. Gray then awarded Connie bonus points for her work in the regular classroom. Although Connie continued to be less productive in the regular classroom than in the resource room, her behavior was acceptable, and the amount of academic work completed was comparable to that of most of the students in the class. The regular classroom teacher was so impressed with the procedure that she made several rating slips for several of her problem students and awarded them special privileges when they completed work and behaved properly.

Whose behavior has generalized here?

Introduce to Natural Maintaining Contingencies

An ideal applied behavior analysis program seeks to change behaviors that receive reinforcement in the student's natural environment. In other words, as a result of the program the student would behave appropriately for the same reasons that motivate students who had never been referred because of inappropriate behavior. The student would work hard at academic tasks in order to earn good grades, behave well in the classroom in order to receive the approval of the teacher, or perform a job for money. For example, although fairly complicated procedures—including shaping, chaining, and graduated guidance—may be necessary in order to teach severely handicapped students to feed themselves, self-feeding may generalize to other settings and be durable after training is withdrawn (Browning, 1980). Such a skill has a built-in positive reinforcer in that children who feed themselves efficiently can control their own intake of food. Toilet training of both retarded persons (Azrin, Sneed, & Foxx, 1973) and nonhandicapped young children (Foxx & Azrin, 1973) may be maintained after contingencies are withdrawn because discomfort is avoided. Similarly, students who are taught such skills as reading or math may maintain these skills without programmed generalization because the skills are useful. Some social behaviors may also generalize.

An increasing emphasis in the education of the severely handicapped has been on training functional skills; that is, skills useful to the student in his environment—school or workplace and community (Brown, Nietupski, & Hamre-Nietupski, 1976). Rather than teach these students meaningless schooltype skills like sorting blocks by color, we teach them skills they need for maximum independence—riding the bus, using the laundromat, cooking a meal. These skills by their nature are more prone to be maintained by the natural environment. Choosing behaviors to change that will be maintained by the natural environment applies the Relevance of Behavior Rule. This rule was first described by Ayllon and Azrin (1968).

Baer and Wolf (1970) conceptualize the Relevance of Behavior Rule as a form of *trapping*. Baer and Wolf assert that if applied behavior analysts can generate behaviors that are reinforced by the natural environment, a situation analogous to catching a mouse in a trap will be created. The mechanism of trapping works this way:

> Consider, for example, that very familiar model, the mouse trap. A mouse trap is an environment designed to accomplish massive behavior modification in a mouse. Note that this modification has thorough generality: The change in behavior accomplished by the trap will be uniform across all environments, it will extend to all of the mouse's behaviors, and it will last indefinitely into the future. Furthermore, a mouse trap allows a great amount of behavioral change to be accomplished by a relatively slight amount of behavioral control. A householder without a trap can, of course, still kill a mouse: He can wait patiently outside the mouse's hole, grab the mouse faster than the mouse can avoid him, and then apply various forms of force to the unfortunate animal to accomplish the behavioral change desired. But this performance requires a great deal of competence: vast patience, supercoordination, extreme manual dexterity, and a well-suppressed squeamishness. By contrast, a householder with a trap needs very few accomplishments: If he can merely apply the cheese and then leave the loaded trap where the mouse is likely to smell that cheese, in effect he has guaranteed general change in the mouse's future behavior.
>
> The essence of a trap, in behavioral terms, is that only a relatively simple response is necessary to enter the trap, yet once entered, the trap cannot be resisted in creating general behavioral change. For the mouse, the entry response is merely to smell the cheese. Everything proceeds from there almost automatically. The householder need have no more control over the mouse's behavior than to get him to smell the cheese, yet he accomplishes thorough changes in behavior. (Baer & Wolf, 1970, p. 321)

Some behaviors do lend themselves to trapping. If behaviors can be generated that result in increased peer reinforcement, they are particularly likely to be maintained in the natural environment. Social and communication skills, grooming skills, and even assertiveness may need only to be generated in order to be maintained. The network of reinforcement available for such behaviors may form an irresistible environmental trap that, like the mousetrap, once entered is inescapable.

Unfortunately, it is often difficult to pinpoint behaviors that will be reinforced by the natural environment (Kazdin, 1977). Most natural environments seem to ignore appropriate behavior and concentrate attention on inappropriate behavior. Few drivers are stopped by policemen for compliments; workers are seldom praised for getting to work on time and attending regularly. Even in the classroom, teachers tend to pay little or no attention to students who are doing well but instead correct students whose behavior is disruptive or inattentive. It is unwise for the applied behavior analyst to assume that any behavior will be maintained by the student's natural environment. However, we can say that the maintenance or transfer of behaviors to reinforcement contingencies in the natural environment may be facilitated by the following:

1. Observing the student's environment. What parents, teachers, or other adults describe as desirable behavior for the student may or may not be what is reinforced.

2. Choosing behaviors that are subject to trapping as determined by observation. For example, if teachers in a given school heavily reinforce pretty handwriting, students may be taught pretty handwriting by a resource teacher, even if it would not otherwise be a priority.

3. Teaching students to recruit reinforcers from the environment (Seymour & Stokes, 1976). Students can be taught to call adults' attention to appropriate behavior and thus to receive praise or other reinforcers.

4. Teaching students to recognize reinforcement when it is delivered. It has been the authors' experience that many students who have difficulty in the regular classroom may not recognize more subtle forms of social reinforcement. This may be a function of what Bryan and Bryan (1978) call a lack of social perception. Some students cannot pick up subtleties of nonverbal communication that may be the only reinforcers available. Teaching students to recognize such subtleties may increase the reinforcing potential of the natural environment.

The teacher who wants the natural environment to take over the reinforcement should be aware that this is by no means an automatic process. Careful monitoring should take place in order to assess the natural environment and to determine how well the behavior change is being maintained. The first vignette illustrates a behavior that is maintained because it receives naturally occurring reinforcers. The second anecdote illustrates a failure of the environment to provide sufficient reinforcers.

Alvin Learns to Read

Alvin was an adjudicated juvenile delinquent in Mr. Daniel's class at the detention center. He was a virtual nonreader when he came into the class, but with a systematic phonic reading method using token reinforcement for correct responses, Mr. Daniel taught Alvin to read phonetically regular one- and two-syllable words. Mr. Daniel wondered whether Alvin would ever read after being discharged from the center, for the boy certainly showed little enthusiasm for any of the high-interest, low-vocabulary books available in the classroom for recreational reading. Alvin appeared to read only when the token condition was in effect. However, about a year after Alvin's release, Mr. Daniel happened to meet him emerging from an adult book store. He had several paperback books under his arm, one open in his hands, and a look of intense concentration on his face.

Marvin Fails the Sixth Grade

Mr. Cohen, a resource teacher, had worked with Marvin for 2 years, while he was attending regular fourth- and fifth-grade classes. Marvin had done well in both his regular classes because his teachers had used high rates of verbal praise and free time contingent upon completion of work. Marvin was dismissed from the resource room at the end of fifth grade. Mr. Cohen had not even considered that Marvin might have trouble in the sixth grade, but the sixth-grade teacher, Ms. Roach, was, to put it bluntly, a hard-nose. She did not believe in praising students for good behavior or in providing any consequence for academic work except grades. As she put it, "That's why they come to school. I don't believe in coddling them." Marvin went back to the behaviors that had resulted in his original referral: he disrupted the classroom, did not complete assignments, and ultimately failed the sixth grade. Mr. Cohen felt he had learned two things at Marvin's expense: never assume that the same conditions exist in all classrooms and always follow up students whose behavior appears to have been permanently changed.

Train Sufficient Exemplars

General Case Programming. *General case programming* promotes generalization by training sufficient exemplars. It was developed to teach language, academic, and social skills to mildly handicapped or disadvantaged young children (Becker & Engelmann, 1978). General case programming emphasizes using sufficient members of a class of stimuli to ensure that students will be able to perform the task on any member of the class of stimuli. If we want a child to identify red objects, we do not have to expose him to every red object in the world to ensure that he can perform this task. We just have to expose him to enough red objects having enough variety in their redness, and he will reliably identify any red object as red. For many children it is not necessary to be very systematic in picking our red objects—we just label whatever objects come our way. With handicapped learners, however, careful attention must be given to selecting objects that will facilitate their acquisition of this skill.

Engelmann and Carnine (1982) state that examples used to train the general case must teach *sameness*—the characteristics of a stimulus that are the same for all members of a class—and *difference*—the range of variability within the members of a class. In other words, what do all red things have in common and how different can things be and still be red? The selection of the training stimuli is the critical factor in general case programming. If all the stimuli used in training sessions are red plastic objects of the same shade, the student asked to fetch the red book from the teacher's desk (it's her copy of Engelmann and Carnine's *Theory of Instruction*) may very well fail to do so because the book is an orangish-red, not the pinkish-red of the training stimuli. General case programming has been very successful in teaching academic behaviors to normal and mildly handicapped learners. It has also been used to teach appropriate social behaviors (Engelmann & Colvin, 1983).

General case programming (though not usually labelled as such) has also been used to teach prelanguage and language skills to severely handicapped learners. The study discussed in Chapter 8 (Garcia, Baer, & Firestone, 1971) in which students were taught a generalized imitation response is an example. You will remember that the students, systematically taught to imitate movements made by the trainer, also imitated movements whose imitation was never reinforced. Similar training has resulted in generalized correct use of many verbal responses including plurals (Guess, Sailor, Rutherford, & Baer, 1968), present and past tense (Schumaker & Sherman, 1970), and appropriate syntax (Garcia, Guess, & Byrnes, 1973).

Recent applications of general case programming emphasize teaching functional skills to severely handicapped learners (Wilcox & Bellamy, 1982). If, for example, a teacher wanted to train a student to be able to use any vending machine in a section of a city, the teacher would have to determine what variations in vending machines existed (where coins are put, how the machine is activated, etc.) and provide training on machines having all these variations (Sprague & Horner, 1984). The procedure has been equally successful in teaching telephone skills (Horner, Williams, & Stevely, 1984), street crossing (Horner, Jones, & Williams, 1984), and many other skills.

Training in Multiple Settings.   Another approach to using sufficient exemplars is to train the behavior in a number of settings or with several different trainers. Such a practice often results in generalization of the behavior change into settings where no training has taken place. Considerable evidence exists to indicate that training novel responses under a wide variety of stimulus conditions increases the probability of generalization under conditions where no previous training has occurred. This procedure differs from sequential modification in that change is targeted and assessed in settings (or behaviors) in which no intervention has taken place.

For example, Stokes et al. (1974) determined that a greeting response in retarded children occurred only in the presence of the trainer who had systematically reinforced the response. The introduction of a second trainer resulted in the generalization of the response to 14 staff members.

Emshoff, Redd, and Davidson (1976) trained delinquent adolescents to use positive interpersonal comments. Those delinquents who were trained under multiple conditions (activities, trainers, locations) showed greater generalization in a nontreatment setting than did those trained under constant conditions.

A large number of studies have demonstrated that the effects of suppressing deviant behaviors in severely handicapped individuals are situation- and experimenter-specific unless trained across several conditions (for example, Birnbrauer, 1968; Corte, Wolf, & Locke, 1971; Garcia, 1974; Lovaas & Simmons, 1969; Risley, 1968). The effects of punishment contingencies are apparently even less likely to generalize than behavior changes resulting from positive reinforcement.

The training of sufficient exemplars is a productive area for teachers concerned with increasing generalization. It is not necessary to teach students to perform appropriate behaviors in every setting nor in the presence of every potential teacher or other adult. The teacher need train only enough to ensure that a generalized response has been learned. Nor is it necessary to teach every example of a response class that we want students to perform. Imagine the difficulty of teaching students to read if this were so. We expect students to generalize letter or syllable sounds and thus to decode words they have never read before. Most teaching of academic skills is based on the assumption that students will be able to use these skills to solve novel problems or to perform a variety of tasks. The following examples show how teachers may use training a number of exemplars to program response generalization as well as generalization across trainers and across settings.

Carol Learns to Use Plurals

Carol was a language-delayed 5-year-old in Ms. Sims' preschool special-education class. Carol had learned the names of many common objects but did not differentiate the singular from the plural forms. Ms. Sims equipped herself with several sets of pictures showing both single and multiple examples of objects whose names Carol knew. Using a discrimination training procedure, she brought Carol's verbal responses under stimulus control of the correct form of a number of singular and plural pairs:

cow	cows
shoe	shoes
dog	dogs
bird	birds
plane	planes

She then tested for generalization. She showed Carol a new set of pictures:

chair	chairs

Carol responded correctly. Her response using singular and plural forms had generalized to an untrained example. Carol used singular and plural forms correctly in a variety of untrained examples. Her response generalization was so broad that when the principal and superintendent visited the class, Ms. Sims heard Carol say to another student, "Mans come." Ms. Sims restrained herself from tearing her hair out by remembering that such overgeneralization is not uncommon among young children and that she would just have to teach irregular plurals as a separate response class.

Ms. Almond Has a Student Teacher

After much effort, Ms. Almond had taught her junior-high remedial English class to behave appropriately and to complete the assignments she gave them. She had used a point system but now depended primarily on social reinforcement and free time on Friday, awarded if the class had had what she rather loosely called "a good week." They almost always earned their free time, because their behavior was generally very good.

At the beginning of the winter quarter, Ms. Almond was assigned a student teacher. To her horror, although the students behaved appropriately as long as

Ms. Almond remained in the room, they went wild when only the student was there. She concluded that her students had formed a discrimination—that she was an S^D for appropriate behavior, but that teachers in general were not.

Ms. Almond reinstituted the token system that she had used at the beginning of the year and made certain that her student teacher was as involved in giving points and awarding free time as she was. The students maintained their appropriate behavior for either teacher, and the point system was subsequently phased out with little loss of behavior change.

Soon after the student teacher departed, Ms. Almond became ill and a substitute was hired. Ms. Almond returned to school after a week, fully expecting the usual horror stories from her principal and colleagues about the misbehavior of her students during her absence. To her surprise, the students had behaved beautifully. Apparently generalization had taken place. Teaching the students to behave appropriately in English class in the presence of two teachers resulted in their behaving in English class in the presence of any teacher.

Ms. Almond's Class Become Model Students

Ms. Almond was so intrigued with her students' generalized appropriate behavior that she asked Mr. Rich, the math teacher about their behavior in his class. She was disappointed to learn that in his class they were as disruptive and inattentive as ever. Mr. Rich, however, was so impressed by Ms. Almond's story that he decided to start a point system of his own and soon found that the students were well behaved and completed assignments. Soon after, Ms. Almond met the eighth-grade social studies teacher in the lounge and was pleased to hear her say that although she had not done anything differently, the students were behaving much better in her class. The science and health teachers told the same story. Training the desired behavior in two settings had resulted in generalization into settings where training had never occurred.

Train Loosely

Teaching techniques based on principles of behavior have usually emphasized tight control of many teaching factors (see Becker, Engelmann, & Thomas, 1975; Stephens, 1976; White & Haring, 1980). Teaching procedures for handicapped students are often rigidly standardized, adhering to the same format, presenting items in a predetermined sequence, and requiring mastery of one skill before training on another has begun. Though this may be an efficient means of instruction, varying the conditions of training could lead to greater response generalization. Schroeder and Baer (1972) found that when teaching vocal imitation skills to retarded students, these skills were better generalized to untaught members of the class when the training stimuli were less tightly structured. Rather than tightly restricting the vocal skills being taught (serial training), the researchers allowed a number of different imitations to be taught within a single session (concurrent training).

In another example of varying training conditions, Panyan and Hall (1978) investigated the effects of concurrent as opposed to serial training in the acquisition, maintenance, and generalization of two different response classes: tracing and vocal imitation. The serial training procedure required mastery of one response class (tracing) before instruction began on a second response class (vocal imitation). In the concurrent training procedure, training on the two different response classes was alternated within a single training session before either task was mastered. The procedure used did not affect the time required for acquisition or maintenance of the response trained, but generalization to untrained responses was greater under the concurrent training condition. These studies have definite implications for the teacher. Apparently, there is no added efficiency to training students to mastery on one skill before beginning instruction on another, unless of course the first skill is prerequisite to the second. Alternating teaching within sessions does not interfere with learning, rather it leads to greater generalization. Thus, statements like "I can't worry about Harold's math until I get him reading on grade level" may not be justified.

Although research in this area is not extensive, it does indicate a possible way to improve teaching. Certainly, teachers should be aware of the probable effects of the sequencing of training on potential response generalization. These studies demonstrate once again that teaching procedures should never be based on assumptions but should instead be validated by analysis of their effects.

Use Indiscriminable Contingencies

As we described in Chapter 6, resistance to extinction or maintenance of behavior is greatly prolonged by intermittent schedules of reinforcement. Intermittent reinforcement may be used to maintain behaviors at a high rate, or it may be a step towards eliminating reinforcement entirely. It is possible to thin a schedule of reinforcement to such a degree that very few reinforcers are used. If reinforcement is then withdrawn altogether, the behavior will continue. This resistance to extinction is not permanent, because behavior, if unreinforced, will eventually be extinguished. However, it may be possible that eventually will be so far in the future as to make no practical difference. The behavior will be maintained as long as necessary.

Considerable evidence indicates that intermittent reinforcement schedules lead to increased maintenance of behavior change (Kale, Kaye, Whelan, & Hopkins, 1968; Kazdin & Polster, 1973; Phillips, Phillips, Fixsen, & Wolf, 1971). Teachers should consider this evidence when planning and implementing behavior change strategies. Even if intermittent reinforcement must be continued indefinitely, if schedules are very lean, this may be a fairly efficient and economical means of providing for maintenance.

Other procedures besides intermittent reinforcement make it difficult for students to discriminate which responses will be reinforced. One strategy that may lead to generalization across settings is delaying the delivery of reinforcers. Schwarz and Hawkins (1970) videotaped the behavior of a student during math and spelling classes. After the end of the school day, the student was shown the videotape, and appropriate behaviors in math class were reinforced. In the next days, a behavior change was evident in spelling class as well as math class. The authors hypothesized that the generalization across settings was due to the delayed reinforcement, which made it more difficult for the student to determine when contingencies were in effect.

Fowler and Baer (1981) used a delayed reinforcement procedure to modify a variety of behaviors of preschool children. The children received tokens exchangeable for toys either immediately after the period during which they earned them or at the end of the day after other periods, during which no contingency was in effect, had intervened. The children generalized the appropriate behavior; that is, they behaved better all day when the reinforcer was not given until the end of the day.

Another way of delivering reinforcers that resulted in decreased discriminability of contingencies was demonstrated by Koegel and Rincover (1977). The authors taught autistic children to perform simple nonverbal imitations or to follow simple instructions. After the behavior was learned (using continuous reinforcement), schedules were thinned. Once training was concluded, the children were observed in order for the researchers to assess maintenance of behavior change. The behaviors were ultimately extinguished. (The thinner the schedule during training, the more responses occurred before extinction.) However, the use of noncontingent rewards after extinction resulted in recovery of the behavior. At random intervals, the children were given candy identical to that earned in the original training setting—whether the child's response was correct or not. The noncontingent rewards delayed extinction considerably. Apparently, the reinforcer had acquired the properties of a discriminative stimulus. It served as a cue to the students that, in this setting, reinforcement was available. The students were unable to discriminate which responses would be reinforced, so they produced larger numbers of the correct (previously reinforced) responses before extinction.

Very thin intermittent schedules of reinforcement are the most frequently used means of making reinforcement contingencies indiscriminable. However, evidence indicates that any procedure that makes it difficult for students to determine when contingencies are in effect is likely to result in greater durability of behavior change, either in the original training setting or in other settings. The following examples illustrate procedures that make it difficult for students to make such determinations.

Suzie Learns to Comply with An Instruction

Suzie was a 4-year-old student in Mr. Wethered's class for developmentally delayed preschool children. Suzie was very active, and Mr. Wethered was concerned because she responded neither to her name nor to "Come here." He was afraid that she would run into the street while waiting for the bus or playing on the playground.

In order to teach Suzie to come when called, Mr. Wethered started by issuing the instruction, "Suzie, come, here," while Suzie was only a few steps from him. He physically prompted Suzie by pulling gently on her hand. He gradually faded the prompt and then began increasing the distance that Suzie had to move. Throughout the initial training procedure, Suzie received a small bite of cookie whenever she responded correctly.

Once the behavior was well established, Mr. Wethered thinned the schedule of reinforcement: Suzie got a bite of cookie on a VR5, then VR10, then VR20, and so on. As time passed, Mr. Wethered was able to keep the behavior occurring at a high rate by very occasionally giving Suzie a bite of cookie when she complied with his instruction. He felt that he could depend on her to come when called and that occasional reinforcement of the response could be kept up indefinitely without disrupting the classroom. After Suzie's mother began calling Suzie occasionally to come get a cookie, she found that Suzie consistently came when called at home as well as at school.

Ms. Bell's Class Learns to Complete Assignments

Ms. Bell's morning group in her intermediate-level resource room for mildly retarded students consistently failed to complete their assignments. Each student was expected independently to complete a reading comprehension, a math, and a spelling activity, while Ms. Bell worked with small groups on other academic skills. Ms. Bell began giving her students tokens worth 5 minutes free time to be exchanged at the end of the morning for each assignment completed. This resulted in almost 100% completion of assignments. Ms. Bell then announced that tokens would not be given until the end of the morning and that only two assignments could earn tokens. She put the words *reading, spelling*, and *math*, on slips of paper and allowed a student to draw two slips. The students did not know until the end of the morning which two assignments would earn free time but continued to complete their assignments. Ms. Bell, who wanted to move to a very lean reinforcement schedule, then announced that there would be the possibility of two drawings daily: one for a "yes"/"no" card, which would determine whether reinforcement would be available, and a second if the "yes" card was drawn. The second drawing would determine which assignment would earn free time. At first, one "yes" and one "no" card was available. She then gradually added "no" cards to the pool until there was only a 20% probability that free time would be available. The students continued to complete their assignments in all three subject areas and seemed to enjoy the suspense of never knowing when free time would be available or for which specific behavior.

Program Common Stimuli

Walker and Buckley (1972) assert that "intra-subject behavioral similarity across different settings is probably a function, in part, of the amount of stimulus similarity which exists between such settings" (p. 209). Thus a possible method of achieving either maintenance or stimulus generalization is deliberate programming of similar stimuli in the training setting and in the setting in which generalization is desired. This may be accomplished by either increasing the similarity of the training situation to the natural environment or by introducing elements of the training situation into the natural environment.

Several studies have investigated the effects of introducing elements of the natural environment into the training situation in order to increase the probability of generalization. For example, Ayllon and Kelly (1974) restored speech in an electively mute retarded girl. After speech occurred frequently in the training situation (a counselor's office), elements similar to those present in the classroom were introduced. Other children, a blackboard, and desks were installed in the room, and the trainer began to function more as a traditional teacher by standing in the front of the room, lecturing, and asking questions. Training was also continued in the classroom. Increases in speech occurred in the classroom, and a follow-up 1 year later indicated that speech was maintained in several novel settings. Although the specific effects of the increase in similarity between training and natural settings is difficult to assess because of the package of treatment employed, generalization to the classroom did occur and was maintained.

Jackson and Wallace (1974) used tokens to increase vocal loudness in a retarded adolescent girl. After satisfactory levels of loudness were achieved within the training setting, procedures similar to those employed by Ayllon and Kelly were introduced. Generalization to the classroom setting occurred.

Koegel and Rincover (1974) trained autistic children to respond to instructions in a one-to-one situation. Generalization to a classroom setting was programmed by gradually introducing more children into the training situation, so that it resembled the classroom.

The effectiveness of procedures like these has recently led to a wide use of simulations with handicapped students. For example, van den Pol et al. (1981) successfully used pictures of fast-food items to teach moderately retarded students to use community fast-food establishments. Simulated instruction for more severely handicapped students is generally less successful (Foxx, McMorrow, & Mennemeier, 1984; Marchetti, McCartney, Drain, Hooper, & Dix, 1983). Horner et al. (no date) suggest that simulations employing real rather than representational items from the environment may be more successful. When real telephones are used, for example, students learn a generalized telephone response (Horner, Williams, & Stevely, 1984). Simulations using real stimuli have been used to teach a variety of skills including video game use (Sedlack, Doyle, & Schloss, 1982) and menstrual care (Richman, Reiss, Bauman, & Bailey, 1984).

A limited number of studies have attempted to increase generalization by introducing elements of the training situation into the setting where generalization was desired. Rincover and Koegel (1975) trained autistic children to imitate nonverbal behaviors modeled by a therapist. When the children responded correctly on 20 consecutive trials without a prompt, a transfer test was made. For children who emitted no correct responses on the transfer test, an assessment of stimulus control was made. Stimuli from the training environment were introduced into the extratherapy setting, one at a time. If the child did not respond correctly in the presence of the first stimulus, that stimulus was removed and another one introduced. This process was continued until the stimulus controlling the behavior was identified, and the responses occurred in the extratherapy setting. It was found that each of the children was responding selectively to some incidental stimulus in the treatment room. When this stimulus was provided in the extratherapy setting, each child responded correctly. The amount of responding in the extratherapy situation was, however, consistently less than in the training situation.

A similar attempt to provide a discriminative stimulus present in the treatment situation in an extratherapy situation was reported by Bachrach, Erwin, and Mohr (1965). A hospitalized anorexic girl (anorexia is a disorder characterized by a refusal to eat) earned reinforcers for eating and weight gain. All meals were eaten on a table covered with a purple tablecloth. The patient was discharged after gaining 33 pounds and was given the purple tablecloth and told to eat all meals at a table covered with it. It was postulated that the table cloth had become an S^D for eating, because eating had been consistently reinforced in its presence. The patient continued to gain weight after discharge and maintained the gains after 2 years.

Troutman (1977) used an analogous procedure to program generalization from resource rooms to regular classrooms. The students showed high rates of academic task completion and low rates of disruptive behavior in the resource room but did not perform appropriately in the regular classroom. Each student was allowed to bring from home some object,

"If it worked on the bananas, it'll work on the keys. As soon as he leaves, it's back to the jungle for us."

such as a figurine, to decorate his or her desk in the resource room, where they earned points for completing academic tasks and appropriate behavior. After a few weeks, the students were instructed to take their objects to the regular classroom. Both academic task completion and appropriate behavior then generalized to the new setting.

Several researchers have investigated the use of students' peers as stimuli common to the training and generalization settings. Stokes and Baer (1976) taught word-recognition to students with learning disabilities through a peer-tutoring procedure. The students did not display the skills in other settings until the peer tutor and pupil were brought together. Then both students showed increased generalization.

Johnston and Johnston (1972) used a similar procedure to maintain and generalize correct articulation in two students. Each was trained to monitor and correct the other's speech, and monitoring was reinforced with tokens. The students consistently showed more correct speech when the monitor was present, even when monitoring no longer occurred.

The potential effectiveness of stimulus control for stimulus generalization and maintenance is a factor that should be considered by all teachers. Relatively simple and economical measures can help ensure reliable generalization in many settings and help maintain gains long after training has been terminated. The following vignettes illustrate generalization achieved by increasing stimulus similarity.

Ms. Statler's Class Gets Ready for the Sheltered Workshop

Ms. Statler was a teacher of moderately-to-severely retarded secondary students. Most of her students were placed in a sheltered workshop when they graduated from her program. Ms. Statler was sure that her students had adequate prevocational skills and that they knew how to behave appropriately, but she was concerned that they would not display these behaviors in a new and very different setting.

After observing the sheltered workshop, Ms. Statler quietly borrowed two long tables from the cafeteria and replaced the desks in her classroom. She also began playing tape-recorded music so that the noise level in the classroom approximated that of the sheltered workshop. She continued to reinforce task completion and appropriate social interaction on a variable-interval schedule and began using the same kind of punchcard system used at the sheltered workshop. When she checked on her students after their transfer, the supervisor of the sheltered workshop stated that this was by far the best group that had ever come from the secondary program.

Sammy Learns to Behave in the Second Grade

Sammy was a student in Mr. Reddy's class for severely emotionally disturbed students. His academic work was superior, but he displayed bizarre behavior, such as shouting gibberish and making strange movements with his hands and arms. This behavior was controlled in the special class by positive reinforcement on a DRO schedule, but Mr. Reddy had noticed that outside the special class Sammy continued to shout and gesture. Because the goal of the special program was to return students to the regular classroom, Mr. Reddy was concerned about Sammy's behavior outside the special class. Mr. Reddy decided to borrow a second grader from the regular class. After getting parental permission, he invited Brad, a gifted student, to visit his class for half an hour three times a week. He taught Brad basic learning principles to help with a project Brad was doing for his enrichment class. He also allowed Brad to give reinforcers to Sammy during his observation periods. When Sammy began spending short periods of time in the regular class, Brad came to get him and brought him back. Sammy consistently behaved appropriately in the second-grade class, even though no reinforcers were ever given there. Brad's presence had become an S^D for appropriate behavior for Sammy.

Mediate Generalization and Train to Generalize

We shall consider together these last two procedures for facilitating generalization: mediating generalization and training to generalize. It is possible to increase the probability of generalization by reinforcing generalization as a response class (Stokes & Baer, 1977). In other words, if students receive reinforcers specifically for displaying behavior in settings other than the training setting, performing learned behaviors in a novel setting may become a generalized response class. The students have thus been trained to generalize. It is particularly appropriate to explain the contingency to the student; that is, students may be told that if they perform a particular behavior in a new setting, reinforcement will be available.

In *mediating generalization,* students are taught to monitor and report on their own generalization of appropriate behavior. Such a program involves self-control or self-management, possibly the most promising of all techniques for ensuring the generalization and maintenance of behavior change. Procedures for teaching students such skills are described in Chapter 10. For many students, the ultimate objective of the applied behavior analyst is to bring behavior under the control of self-monitoring, self-administered contingencies, and even self-selected goals and procedures.

Many procedures for promoting generalization have been described in this chapter. Decide for yourself which one results in a domestic problem for Professor Grundy in the following vignette.

Professor Grundy Takes up Jogging

Professor Grundy looked at the chart above his bathroom scales. "Big bones or not," he muttered, "something's got to be done. Maybe I need to jog. Argh!"

He began to listen to the large number of his colleagues who ran daily. (He learned that the correct term, for those in the know, was *running* not *jogging*.) Professor Grundy read all about the benefits of running. He began to be a believer.

One Saturday morning, the professor visited his local sporting goods store. He purchased expensive shoes, a tee-shirt and a pair of extremely brief running shorts. Sunday morning he donned his outfit and presented himself to Mrs. Grundy.

"Well, Minerva, my dear," he asked, "how do I look?"

Ten minutes later, as Mrs. Grundy was still gasping for breath and wiping

her eyes, he left the house and proceeded to run (albeit slowly and stiffly) around the block. He made it for an entire mile, before deciding that perhaps tennis. . . . Then he remembered the expense and decided to stick with it for a month. Three weeks later, he was hooked. He could hardly wait to hit the street. When he walked downstairs one morning about this time, Mrs. Grundy looked at him appraisingly. "Why, Oliver," she said thoughtfully, "you have very nice legs."

"Thank you, my dear," said Professor Grundy. That very morning several young ladies leaned out of a passing car and whistled at the professor. His dignity was somewhat offended, but still. . . .

The next Saturday he wore his running shorts all day. He noticed admiring glances at the shopping center and at the public library (where he had gone to return several overdue books on running). He bought more shorts and wore them during all his leisure time. Mrs. Grundy balked, however, when he prepared to wear them on a visit to her mother. "Oliver," she said, "I think this is carrying generalization *just* a bit far!"

SUMMARY

As we have seen, many techniques for promoting stimulus generalization, response generalization, and maintenance of behavior have been investigated. In the first edition of this text, we stated that the technology of generalization was in its infancy. It has now reached the toddler stage. Its status is still well described by Buchard and Harig (1976):

Formulating questions about generalization is somewhat like a game that might go something like this: At the start, there are two things to worry about.

Will the behavioral change be maintained in the natural environment or won't it? If it won't, then there are two more things to worry about. Does the lack of generalization reflect the type or level of the reinforcement schedule that produced the behavioral change in the first place, or did the behavior fail to generalize because of a lack of supporting contingencies in the community? If the problem pertains to the schedule you have little to worry about. You go back to the treatment setting and strengthen the desirable behavior, preferably through a positive reinforcement schedule. However, if the problem is with the supporting contingencies in the natural environment, then you have two more things to worry about. Should you reprogram the natural environment to provide supporting contingencies through intermittent reinforcement, fading, or overlearning? And so on. . . .

Obviously the game is one in which these questions can be formulated ad infinitum. With generalization, the unfortunate part is that there will probably always be something to worry about; this is the nature of the beast! It's hard enough to try to determine whether or not behavior has changed, and if so why, let alone trying to determine whether or not the change also occurred in a completely different setting.
(pp. 428–429)

Applied behavior analysts who are also teachers may never stop worrying about generalization. However, given the techniques described in this chapter, there appears to be less and less excuse for mere worrying. If applied behavior analysis procedures are to become an accepted part of the repertoire of every teacher, it is time, as Baer et al. suggested in 1968, to stop lamenting and start programming.

REFERENCES

ANDERSON-INMAN, L., WALKER, H.M., & PURCELL, J. 1984. Promoting the transfer of skills across settings: Transenvironmental programming for handicapped students in the mainstream. In W. Heward, T.E. Heron, D.S. Hill, & J. Trap-Porter (Eds.), *Focus on behavior analysis in education.* Columbus, Ohio: Charles E. Merrill.

ATTHOWE, J.M. 1973. Token economies come of age. *Behavior Therapy, 4,* 646–654.

AXELROD, S. 1977. *Behavior modification for the classroom teacher.* New York: McGraw-Hill.

AYLLON, T., & AZRIN, N. 1968. *The token economy: A motivational system for therapy and rehabilitation.* New York: Appleton-Century-Crofts.

AYLLON, T., & KELLY, K. 1974. Reinstating verbal behavior in a functionally mute retardate. *Professional Psychology, 5,* 385–393.

AZRIN, N.H., SNEED, T.J., & FOXX, R.M. 1973. Drybed: A rapid method of eliminating bedwetting (enuresis) of the retarded. *Behavior Research and Therapy, 11,* 427–434.

BACHRACH, A.J., ERWIN, W.J., & MOHR, J.P. 1965. The control of eating behavior in an anorexic by operant conditioning techniques. In L.P. Ullman & L. Krasner (Eds.), *Case studies in behavior modification* (pp. 153–163). New York: Holt, Rinehart & Winston.

BAER, D.M., & WOLF, M.M. 1970. The entry into natural communities of reinforcement. In R. Ulrich, T. Stachnik, & J. Mabry (Eds.), *Control of human behavior,* vol. 2. Glenview, Ill.: Scott, Foresman & Co.

BAER, D.M., WOLF, M.M., & RISLEY, T.R. 1968. Some current dimensions of applied behavior analysis. *Journal of Applied Behavior Analysis, 1,* 91–97.

BARTON, E.J., & ASCIONE, F.R. 1979. Sharing in preschool children: Facilitation, stimulus generalization, response generalization, and maintenance. *Journal of Applied Behavior Analysis, 12,* 417–430.

BECKER, W.C., & ENGELMANN, S.E. 1978. Systems for basic instruction: Theory and applications. In A. Catania & T. Brigham (Eds.), *Handbook of applied behavior analysis: Social and instructional processes,* pp. 57–92. Chicago: Science Research Associates.

BECKER, W.C., ENGELMANN, S., & THOMAS, D.R. 1975. *Teaching 2: Cognitive learning and instruction.* Chicago: Science Research Associates. ·

BIJOU, S.W., PETERSON, R.F., HARRIS, F.R., ALLEN, K.E., & JOHNSTON, M.S. 1969. Methodology for experimental studies of young children in natural settings. *Psychological Record, 19,* 177–210.

BIRNBRAUER, J.S. 1968. Generalization of punishment effects—A case study. *Journal of Applied Behavior Analysis, 1,* 201–211.

BROWN, L., NIETUPSKI, J., & HAMRE-NIETUPSKI, S. 1976. The criterion of ultimate functioning. In M.A. Thomas (Ed.), *Hey, don't forget about me!* pp. 2–15. Reston, Va.: CEC Information Center.

BROWNING, R.M. 1980. *Teaching the severely handicapped child: Basic skills for the developmentally disabled.* Boston: Allyn & Bacon.

BRYAN, T., & BRYAN, J. 1978. *Understanding learning disabilities.* Sherman Oaks, Calif.: Alfred.

BUCHARD, J.D., & HARIG, P.T. 1976. Behavior modification and juvenile delinquency. In H. Leitenberg (Ed.), *Handbook of behavior modification and behavior therapy* Englewood Cliffs, N.J.: Prentice-Hall.

CHADWICK, B.A., & DAY, R.C. 1971. Systematic reinforcement: Academic performance of underachieving students. *Journal of Applied Behavior Analysis, 4,* 311–319.

CONE, J.D. 1973. Assessing the effectiveness of programmed generalization. *Journal of Applied Behavior Analysis, 6,* 713–718.

CORTE, H.E., WOLF, M.M., & LOCKE, B.J. 1971. A comparison of procedures for eliminating self-injurious behavior of retarded adolescents. *Journal of Applied Behavior Analysis, 4,* 201–213.

DRABMAN, R.S. 1973. Child versus teacher administered token programs in a psychiatric hospital school. *Journal of Abnormal Child Psychology, 1,* 68–87.

EMSHOFF, J.G., REDD, W.H., & DAVIDSON, W.S. 1976. Generalization training and the transfer of treatment effects with delinquent adolescents. *Journal of Behavior Therapy and Experimental Psychiatry, 7,* 141–144.

ENGELMANN, S., & CARNINE, D. 1982. *Theory of instruction: Principles and applications.* New York: Irvington Publishers, Inc.

ENGELMANN, S., & COLVIN, G. 1983. *Generalized compliance training: A direct-instruction program for managing severe behavior problems.* Austin, Texas: Pro-Ed.

FOWLER, S.A., & BAER, D.M. 1981. "Do I have to be good all day?" The timing of delayed reinforcement as a factor in generalization. *Journal of Applied Behavior Analysis, 14,* 13–24.

FOXX, R.M., & AZRIN, N.H. 1973. *Toilet training the retarded: A rapid program for day and nighttime independent toileting.* Champaign, Ill.: Research Press.

FOXX, R.M., MCMORROW, M.J., & MENNEMEIER, M. 1984. Teaching social/vocational skills to retarded adults with a modified table game: An analysis of generalization. *Journal of Applied Behavior Analysis, 17,* 343–352.

GARCIA, E. 1974. The training and generalization of a conversational speech form in nonverbal retardates. *Journal of Applied Behavior Analysis, 7,* 137–149.

GARCIA, E., BAER, D.M., & FIRESTONE, I. 1971. The development of generalized imitation within topographically determined boundaries. *Journal of Applied Behavior Analysis, 4,* 101–112.

GARCIA, E., GUESS, D., & BYRNES, J. 1973. Development of syntax in a retarded girl using procedures of imitation, reinforcement, and modelling. *Journal of Applied Behavior Analysis, 6,* 299–310.

GREENWOOD, C.R., SLOANE, N.H. JR., & BASKIN, A. 1974. Training elementary aged peer-behavior managers to control small group programmed mathematics. *Journal of Applied Behavior Analysis, 7,* 103–114.

GUESS, D., SAILOR, W., RUTHERFORD, G., & BAER, D.M. 1968. An experimental analysis of linguistic development: The productive use of the plural morpheme. *Journal of Applied Behavior Analysis, 1,* 297–306.

GUTTMAN, N., & KALISH, H.I. 1956. Discriminability and stimulus generalization. *Journal of Experimental Psychology, 51,* 79–88.

HALL, R.V., CRISTLER, C., CRANSTON, S.S., & TUCKER, B. 1970. Teachers and parents as researchers using multiple baseline designs. *Journal of Applied Behavior Analysis, 3,* 247–255.

HARTMANN, D.P., & ATKINSON, D. 1973. Having your cake and eating it too: A note on some apparent contradictions between therapeutic achievements and design requirements in N = 1 studies. *Behavior Therapy, 4*, 589–591.

HEWETT, F.M., TAYLOR, F.D., & ARTUSO, A.A. 1969. The Santa Monica Project: Evaluation of an engineered classroom design with emotionally disturbed children. *Exceptional Children, 35*, 523–529.

HORNER, R.H., JONES, D., & WILLIAMS, J.A. 1984. *Teaching generalized street crossing to individuals with moderate and severe mental retardation.* Manuscript submitted for publication.

HORNER, R.H., MCDONNELL, J.J., & BELLAMY, G.T. (undated). *Teaching generalized skills: General case instruction in simulation and community settings* (Contract No. 300–82–0362). Unpublished manuscript, University of Oregon.

HORNER, R.H., WILLIAMS, J.A., & STEVELY, J.D. 1984. *Acquisition of generalized telephone use by students with severe mental retardation.* Manuscript submitted for publication.

JACKSON, D.A., & WALLACE, R.F. 1974. The modification and generalization of voice loudness in a fifteen-year-old retarded girl. *Journal of Applied Behavior Analysis, 7*, 461–471.

JOHNSTON, J.M. 1979. On the relation between generalization and generality. *The Behavior Analyst, 2*, 1–6.

JOHNSTON, J.M., & JOHNSTON, G.T. 1972. Modification of consonant speech sound articulation in young children. *Journal of Applied Behavior Analysis, 5*, 233–246.

KALE, R.J., KAYE, J.H., WHELAN, P.A., & HOPKINS, B.L. 1968. The effects of reinforcement on the modification, maintenance, and generalization of social responses of mental patients. *Journal of Applied Behavior Analysis, 1*, 307–314.

KALLMAN, W.H., HERSEN, M., & O'TOOLE, D.H. 1975. The use of social reinforcement in a case of conversion reaction. *Behavior Therapy, 6*, 411–413.

KAZDIN, A.E. 1973. Methodological and assessment considerations in evaluating reinforcement programs in applied settings. *Journal of Applied Behavior Analysis, 6*, 517–531.

KAZDIN, A.E. 1977. *The token economy: A review and evaluation.* New York: Plenum Press.

KAZDIN, A.E., & BOOTZIN, R.R. 1972. The token economy: An evaluative review. *Journal of Applied Behavior Analysis, 5*, 343–372.

KAZDIN, A.E., & POLSTER, R. 1973. Intermittent token reinforcement and response maintenance in extinction. *Behavior Therapy, 4*, 386–391.

KELLER, F.S., & SCHOENFELD, W.N. 1950. *The principles of psychology.* New York: Appleton-Century-Crofts.

KIFER, R.E., LEWIS, M.A., GREEN, D.R., & PHILLIPS, E.L. 1974. Training predelinquent youths and their parents to negotiate conflict situations. *Journal of Applied Behavior Analysis, 7*, 357–364.

KOEGEL, R.L., & RINCOVER, A. 1974. Treatment of psychotic children in a classroom environment: I. Learning in a large group. *Journal of Applied Behavior Analysis, 7*, 45–59.

KOEGEL, R.L, & RINCOVER, A. 1977. Research on the difference between generalization and maintenance in extra-therapy responding. *Journal of Applied Behavior Analysis, 10*, 1–12.

LONG, J.D., & WILLIAMS, R.L. 1973. The comparative effectiveness of group and individually contingent free time with inner-city junior high school students. *Journal of Applied Behavior Analysis, 6*, 465–474.

LOVAAS, O.I., KOEGEL, R., SIMMONS, J.Q., & LONG, J.S. 1973. Some generalization and follow-up measures on autistic children in behavior therapy. *Journal of Applied Behavior Analysis, 6*, 131–166.

LOVAAS, O.I., & SIMMONS, J.Q. 1969. Manipulation of self-destruction in three retarded children. *Journal of Applied Behavior Analysis, 2*, 143–157.

MARCHETTI, A.G., MCCARTNEY, J.R., DRAIN, S., HOOPER, M., & DIX, J. 1983. Pedestrian skills training for mentally retarded adults: Comparison of training in two settings. *Mental Retardation, 21*, 107–110.

MARTIN, M., BURKHOLDER, R., ROSENTHAL, R., THARP, R.G., & THORNE, G.L. 1968. Programming behavior change into school milieux of extreme adolescent deviates. *Behavior Research and Therapy, 6*, 371–383.

MEDLAND, M.B., & STACHNIK, T.J. 1972. Good-behavior game: A replication and systematic analysis. *Journal of Applied Behavior Analysis, 5*, 45–51.

PANYAN, M.C., & HALL, R.V. 1978. Effects of serial *versus* concurrent task sequencing on acquisition, maintenance, and generalization. *Journal of Applied Behavior Analysis, 11*, 67–74.

PATTERSON, G.R. 1965. An application of conditioning techniques to the control of a hyperactive child. In L.P. Ullmann & L. Krasner (Eds.), *Case studies in behavior modification* New York: Holt, Rinehart and Winston.

PHILLIPS, E.L., PHILLIPS, E.A., FIXSEN, D.L., & WOLF, M.M. 1971. Achievement place: Modification of the behaviors of pre-delinquent boys within a token economy. *Journal of Applied Behavior Analysis, 4*, 45–59.

RICHMAN, G.S., REISS, M.L., BAUMAN, K.E., & BAILEY, J.S. 1984. Teaching menstrual care to mentally retarded women: Acquisition, generalization, and maintenance. *Journal of Applied Behavior Analysis, 17*, 441–451.

RINCOVER, A., & KOEGEL, R.L. 1975. Setting generality and stimulus control in autistic children. *Journal of Applied Behavior Analysis, 8*, 235–246.

RISLEY, T.R. 1968. The effects and side effects of punishing the autistic behaviors of a deviant child. *Journal of Applied Behavior Analysis, 1*, 21–34.

SCHROEDER, G.L., & BAER, D.M. 1972. Effects of concurrent and serial training on generalized vocal imitation in retarded children. *Developmental Psychology, 6*, 293–301.

SCHUMAKER, J., & SHERMAN, J.A. 1970. Training generative verb usage by imitation and reinforcement procedures. *Journal of Applied Behavior Analysis, 3*, 273–287.

SCHWARZ, M.L., & HAWKINS, R.P. 1970. Application of delayed reinforcement procedures to the behavior of an elementary school child. *Journal of Applied Behavior Analysis, 3*, 85–96.

SEDLAK, R.A., DOYLE, M., & SCHLOSS, P. 1982. Video games: A training and generalization demonstration with severely retarded adolescents. *Education Training for the Mentally Retarded, 17*, 332–336.

SEYMOUR, F.W., & STOKES, T.F. 1976. Self-recording in training girls to increase work and evoke staff praise in an institution for offenders. *Journal of Applied Behavior Analysis, 9*, 41–54.

SKINNER, B.F. 1953. *Science and human behavior.* New York: Macmillan.

SPRAGUE, J.R., & HORNER, R.H. 1984. The effects of single instance, multiple instance, and general case training on generalized vending machine use by

moderately and severely handicapped students. *Journal of Applied Behavior Analysis, 17*, 273–278.

STEPHENS, T.M. 1976. *Directive teaching of children with learning and behavioral handicaps.* Columbus, Ohio: Charles E. Merrill.

STOKES, T.F., & BAER, D.M. 1976. Preschool peers as mutual generalization-facilitating agents. *Behavior Therapy, 7*, 549–556.

STOKES, T.F., & BAER, D.M. 1977. An implicit technology of generalization. *Journal of Applied Behavior Analysis, 10*, 349–367.

STOKES, T.F., BAER, D.M., & JACKSON, R.L. 1974. Programming the generalization of a greeting response in four retarded children. *Journal of Applied Behavior Analysis, 7*, 599–610.

TROUTMAN, A.C. 1977. *Stimulus control: A procedure to facilitate generalization from resource rooms to regular classrooms.* Unpublished doctoral dissertation, Georgia State University.

TWARDOSZ, S., & SAJWAJ, T. 1972. Multiple effects of a procedure to increase sitting in a hyperactive, retarded boy. *Journal of Applied Behavior Analysis, 5*, 73–78.

VAN DEN POL, R.A., IWATA, B.A., IVANCIC, M.T., PAGE, T.J., NEEF, N.A., & WHITLEY, F.P. 1981. Teaching the handicapped to eat in public places: Acquisition, generalization and maintenance of restaurant skills. *Journal of Applied Behavior Analysis, 14*, 61–69.

WALKER, H.M., & BUCKLEY, N.K. 1968. The use of positive reinforcement in conditioning attending behavior. *Journal of Applied Behavior Analysis, 1*, 245–250.

WALKER, H.M., & BUCKLEY, N.K. 1972. Programming generalization and maintenance of treatment effects across time and across settings. *Journal of Applied Behavior Analysis, 5*, 209–224.

WALKER, H.M., MATTSEN, R.H., & BUCKLEY, N.K. 1971. The functional analysis of behavior within an experimental class setting. In W.C. Becker (Ed.), *An empirical basis for change in education.* Chicago: Science Research Associates.

WHITE, O.R., & HARING, N.G. 1980. *Exceptional teaching* 2nd ed. Columbus, Ohio: Charles E. Merrill.

WILCOX, B., & BELLAMY, G.T. 1982. *Design of high school programs for severely handicapped students.* Baltimore: Paul H. Brookes.

10

Teaching Students to Manage Their Own Behavior

- Who is the only person who will be with a student through his or her entire educational experience, in school and out?
- Who is most tuned in to a student's day-to-day behavior and learning?
- Who knows best what kind of reinforcement a student desires at any one time?
- So who should learn to bear the primary responsibility for maintenance, monitoring, and reinforcement of behavioral programs?

That's right, the answer to all of the questions above is *the student*. Teachers must build into the programs they initiate some provision for students to learn to monitor and reinforce personal growth toward and beyond behavioral objectives and long-term goals.

Throughout this book, we have described procedures for teachers to use in changing the behavior of their students. Chapter 9 discussed some ways of increasing the generalizability of behavior-change procedures and thus minimizing the necessity for continued teacher support. This chapter further examines techniques that enable teachers to make students less dependent upon teachers' environmental manipulation. The procedures discussed in this chapter place the responsibility for change upon the student. The focus in all of the approaches is on self-management, on teaching students to become effective modifiers of their own behavior. John Dewey (1939) suggested that "the ideal aim of education is the creation of self control" (p. 75). Students who have self-control are able to learn and to behave appropriately even when adult supervision is not available. The better able students are to manage themselves, the more time teachers can spend designing effective learning environments.

Lovitt (1973) noted "that self-management behaviors are not systematically programmed (in the schools) which appears to be an educational paradox, for one of the expressed objectives of the educational system is to create individuals who are self-reliant and independent" (p. 139). If we, as teachers, agree that our goals include independence for our students, then we must make provisions for making students independent. One of our colleagues tells prospective special education teachers that they should always be trying to do themselves out of their jobs—to teach their students to manage without special education programs. Although total independence is not possible for all handicapped students, most can be taught to be more self-reliant. Kazdin (1975) offered several additional reasons for preferring self-management to that controlled by external change agents:

1. The use of external change agents sacrifices consistency, since teachers or others may "miss" certain instances of behavior.

2. Problems associated with communication between agents in different settings (such as teachers and parents) can also undermine the success of a program.

3. The change agents themselves can become an environmental cue for the performance or lack of performance of a behavior.

4. An individual's contribution to the development of a personal behavior-change program may increase performance.

5. External agents are not always available in the environment where the target behavior is occurring or should occur. (pp. 189–190)

Both typical and handicapped learners can be taught to monitor and alter their own behavior. In order for the procedures to be used effectively, students must be trained in their use (as, for example, in O'Leary & Dubey, 1979; Rosenbaum & Drabman, 1979; Rueda, Rutherford, & Howell, 1980). The students are trained in procedures of self-management much as they are in other behaviors are—according to the principles of learning.

Any technique a teacher can use to change behavior can also be used by students to change their own behavior. Students can be taught self-management of behavior by instructing them in certain procedures: monitoring their behavior through self-data recording, evaluating their behavior, and providing their own consequences through self-reinforcement and self-punishment. Students can also learn to manipulate behavioral antecedents by using self-instruction.

Self-management usually comes in packages.

Although each of these self-management techniques is reviewed separately in this chapter, in practice they have most often been employed in packages, where combinations of the procedures were put into effect: for example, self-recording with self-reinforcement or self-instruction with self-reinforcement. Examples of self-management packages are provided later in this chapter.

A COMMON EXPERIENCE

Self-management procedures are as much a part of the natural environment as are all the behavioral procedures described in this book. Many people use self-recording, reinforcement, punishment, and instruction in managing their everyday behavior. Noted author Irving Wallace (1977) has described how he and other authors have practiced self-recording techniques. Of his own work habits, he relates

> I maintained work charts while writing my first four published books. These charts showed the date I started each chapter, the date I finished it, and the number of pages written in that period. With my fifth book, I started keeping a more detailed chart, which also showed how many pages I had written by the end of every working day. I am not sure why I started keeping such records. I suspect that it was because, as a free-lance writer, entirely on my own, without employer or deadline, I wanted to create disciplines for myself, ones that were guilt-making when ignored. A chart on the wall served as such a discipline, its figures scolding me or encouraging me. (p. 516)

Wallace then quotes from the autobiography of novelist Anthony Trollope, published in 1883:

> When I have commenced a new book, I have always prepared a diary, divided into weeks, and carried on for the period which I have allowed myself for the completion of the work. In this I have entered, day by day, the number of pages I have written, so that if at any time I have slipped into idleness for a day or two, the record of that idleness has been there, staring me in the face, and demanding of me increased labour, so that the deficiency might be supplied. According to the circumstances of the time—whether my other business might then be heavy or light, or whether the book I was writing was or was not wanted with speed,—I have allotted myself so many pages a week. The average number has been about 40. It has been placed as low as 20, and has risen to 112. And as a page is an ambiguous term, my page has been made to contain 250 words; and as words, if not watched, will have a tendency to straggle, I have had every word counted as I went. . . . There has ever been the record before me, and a week passed with an insufficient number of pages has been a blister to my eye and a month so disgraced would have been a sorrow to my heart.
>
> I have been told that such appliances are beneath the notice of a man of genius, I have never fancied myself to be a man of genius, but had I been so I think I might well have subjected myself to these trammels. Nothing surely is so potent as a law that may not be disobeyed. It has the force of the waterdrop that hollows the stone. A small daily task, if it be really daily, will beat the labours of a spasmodic Hercules. (p. 518)

Finally, Wallace quotes from an article written about Ernest Hemingway by George Plimpton that appeared in the *Paris Review* in 1965:

> He keeps track of his daily progress—"so as not to kid myself"—on a large chart made out of the side of a cardboard packing case and set up against the wall under the nose of a mounted gazelle head. The numbers on the chart showing the daily output of words differ from 450, 575, 462, 1250, back to 512,

the higher figures on days Hemingway puts in extra work so he won't feel guilty spending the following day fishing on the gulf stream. (p. 518)

The use of self-reinforcement is probably familiar to most people. Consider the following teacher's internal monologue as she heads home from a day at school:

What a day!

Bus duty at 7:00 A.M. . . . Jenny Lind fell, sprained her ankle, screamed like a banshee . . . her mother says she's going to sue . . .

Clifford refused to believe that 6 × 4 was 24 today just because it was yesterday . . .

Two fights over whose turn it was to use the blue paint in art class. Pat ended up with the blue; Mark ended up blue . . .

The lunch money didn't balance . . .

Why does that Velma Johnson always sit next to me at lunch and talk my ear off?

I kept my temper all day though—I deserve a stop at Baskin Robbins!

If the same teacher also practices self-punishment, she may remember the three scoops of rocky road at lunchtime tomorrow and confine herself to two pieces of lettuce and a diet drink.

Many of us also practice **self-instruction**, providing ourselves with verbal prompts. We talk to ourselves as we do complex or unfamiliar tasks. Many children also use such self-instruction naturally. For example, Kohlberg, Yaeger, and Hjertholm (1968) recorded the self-instruction processing of a 3½-year-old child during solitary play with a set of Tinker Toys.

The wheels go here, the wheels go here. Oh, we need to start it all over again. We have to close it up. See, it closes up. We're starting it all over again. Do you know why we wanted to do that? Because I needed it to go a different way. Isn't it pretty clever, don't you think? But we have to cover up the motor just like a real car. (p. 695)

Training students to manage their behavior is a mechanism for systematizing and making more powerful these naturally occurring phenomena. Some students may be effective self-managers without training, others unready to manage their own behavior to even a small extent. The wise teacher will remain alert for signs that students are ready to begin managing their own behaviors and take advantage of this readiness.

PREPARING STUDENTS TO MANAGE THEIR OWN BEHAVIOR

The teacher who uses some systematic behavior management program can use a number of techniques to increase students' potential for taking the responsibility for managing their own behavior.

1. When giving reinforcers, the teacher should explain to the student what behavior resulted in reinforcement.

   "Sammy, you did 10 math problems correctly. You get 10 points—1 for each problem."

2. The teacher may ask the student to relate part of the contingency.

   "Sammy, you get 10 points. Why do you get 10 points?" or "Sammy, you did 10 problems right. How many points have you earned?"

3. The teacher may ask the student to state the entire contingency.

   "Sammy, how many points? Why?"

4. The teacher may involve students in choosing reinforcers and in determining their cost in terms of behavior.

Students who have been exposed to such techniques will frequently volunteer statements about their behavior and its consequences. It is a small step from asking a student how many points he has earned and why, to allowing him to record the points himself with teacher supervision. Ultimately, Sammy may be allowed simply to count the number of problems he has done correctly and record his points on his card. Such self-recording is the foundation for many self-management procedures.

Self-recording of Data

Chapter 4 discusses techniques for data collection.

The purpose of data collection is to provide a record that reflects the occurrence of a behavior. If a student is taught recording procedures and asked to collect the data, that student is managing at least one part of her or his program for behavior change. Systems of data collection can be adapted so that the student simply records the occurrence or nonoccurrence of the target behavior. The student is thus asked to assess the quantity of a behavior. For example, a student may be asked to note each time she raises her hand before speaking in class or, conversely, each time she calls out without raising her hand. In addition, the student may be asked to assess the quality of her behavior. For example, she may be instructed to count and record each correctly solved arithmetic problem or each one of a series of assigned tasks.

Self-recorded data provide the student and teacher with concrete feedback regarding behavior. This information may be used to determine what reinforcers are available. In some cases, a student's collecting data on a behavior may have what is termed a *reactive effect* on the behavior. Simply as a function of the self-recording process, the behavior may change in the desired direction. In this capacity, self-recording functions as a behavior-change technique (Rosenbaum & Drabman, 1979). Consider the effect of a student's recording each instance of daydreaming while reading a chapter on self-management. In many cases, tallying each instance on a 3 × 5 card will reduce the rate of daydreaming. (If 3 × 5 cards are unavailable, just make a pencil mark in the margin.)

As a self-management and behavior-change technique, self-recording has been successfully used with a variety of behaviors. Increases have been noted in paying attention (Broden, Hall, & Mitts, 1971; Kneedler & Halla-

The reactive effect of self-recording may be only temporary.

han, 1981), academic responses (Hundert & Bucher, 1978; James, Trap, & Cooper, 1977; Lovitt, 1973; Sagotsky, Patterson, & Lepper, 1978), class attendance (McKenzie & Rushall, 1974), appropriate verbalizations (Nelson, Lipinski, & Boykin, 1978), and vocational skills (Connis, 1979). Decreases have been recorded in talking out in class (Broden et al., 1971; Lovitt, 1973), aggression (Lovitt, 1973), off-task behavior (Glynn, Thomas, & Shee 1973; Sagotsky et al., 1978), out-of-seat behavior (Sugai & Rowe, 1984), generally disruptive behavior (Stevenson & Fantuzzo, 1984), and overeating (Snoy & van Benten, 1978). Although self-recording may produce early behavior changes, these changes may tend to dissipate over time unless supported with additional self-management procedures, such as self-reinforcement (Kanfer, 1975; McLaughlin, 1976; O'Leary & Dubey, 1979).

Although self-recording has been used for initial behavior-change programs, recent findings strongly suggest that self-recording facilitates maintenance of behavior change once initial teacher programming is terminated and therefore may be used most effectively with target behaviors that have already been modified by traditional teacher-managed strategies (O'Leary & Dubey, 1979). Teaching students to use self-recording should include the following components:

1. selecting a target behavior
2. operationally defining the behavior
3. selecting an appropriate system of data collection (Data-collection systems that have been successfully used include adaptations of event recording, time-sampling, and permanent product recording; notation methods include tally sheets, wrist counters, graphs, and charts.)
4. instructing the student in the use of the data-collection system selected
5. monitoring of at least one data practice recording session
6. students' employing self-recording independently

Self-recording Using Event Recording

Broden et al. (1971) investigated the use of self-recording with two eighth-grade students. The first student, Liza, was earning a D − in history. Due to the lecture format of his class, the history teacher felt he could not participate in a suggested behavior-change program, because he could not consistently attend to Liza during the class period. Liza's counselor therefore decided to employ a self-recording procedure to increase the time she spent attending and studying. The counselor met with Liza and gave her a slip of paper that had 3 rows of 10 squares (see Figure 10–1). They discussed a definition of studying and examples of what constituted studying. Liza was instructed to take the slip of paper to class each day and, whenever she thought of it, to record a + in the square if she had been studying for the last few minutes and a − if she had not.

At a weekly conference, Liza and the counselor discussed the self-recording slips and the counselor praised Liza for studying. Before the self-

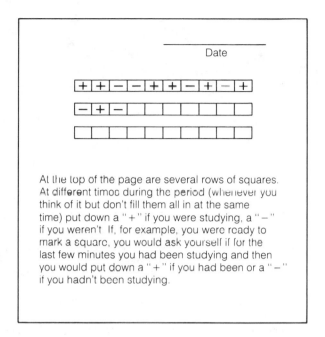

FIGURE 10–1   *Liza's self-recording data form*

Note. From "The effect of self-recording on the classroom behavior of two eighth-grade students" by M. Broden, R.V. Hall, & B. Mitts, *Journal of Applied Behavior Analysis*, 1971, *4*, 191–199. Copyright 1971 by Society for the Experimental Analysis of Behavior, Inc. Reprinted by permission.

recording strategy was used, Liza's study rate was 30% of the history class period despite two conferences with the counselor and promises from Liza that she would "really try." After self-recording was instituted, Liza's study behavior increased to 78% of the class period. After 30 days, when the teacher noted the change in Liza's study behavior, he agreed to praise Liza and give her increased attention when she studied. Liza continued to carry slips to class, sometimes filling them out and sometimes not. Her study time continued to increase and finally reached 88% of the class period.

The second eighth-grade student in the study, Stu, was referred by his math teacher who wanted to find "some means to shut Stu up" (Broden et al., 1971, p. 195). The teacher was asked to give Stu a slip of paper at the beginning of each of the 2 periods of class (25 minutes before lunch and 20 minutes after lunch). On each slip of paper was a 2 × 5 inch box, a place for his name and the date, and the statement "record a mark every time you talk out without permission" (see Figure 10–2). No further instructions or contingencies were given to Stu.

During baseline, Stu talked out on the average of 1.1 times per minute during the first period of class and 1.6 times a minute during the second period of math class. When self-recording was employed during both session, talk-outs were recorded 0.3 times per minute during the first period and 1.0 times per minute during the second period. When self-recording

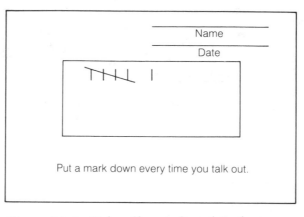

Figure 10–2   Stu's self-recording data form

Note. From "The effect of self-recording on the classroom behavior of two eighth-grade students" by M. Broden, R.V. Hall, & B. Mitts, *Journal of Applied Behavior Analysis*, 1971, *4*, 191–199. Copyright 1971 by Society for the Experimental Analysis of Behavior, Inc. Reprinted by permission.

was reinstituted following a return to baseline conditions, the first period talk-outs returned to a low level but not the talk-outs during the period after lunch.

A similar self-recording procedure was used by Sagotsky et al. (1978) with fifth- and sixth-grade students working on math modules. The students were given a sheet with a space to mark the page and problem where they stopped working each day and a grid of 12 empty boxes. The students were told that during the math period each day, they should note from time to time whether or not they were actually working on their math units. If the student was not working, he or she was to put a − in one of the boxes. It was suggested to the students that this might serve as a reminder to them to resume studying. The percentage of time spent studying increased.

In place of the use of tally sheets, such as those used by Liza and Stu, Nelson et al. (1978) chose to use a mechanical adaptation. In a study employing self-recording of instances of appropriate classroom verbalizations, retarded adolescents were given either hand-held counters or ones worn on the belt to count the number of times they talked appropriately. Results were similar to those obtained in the other studies.

Self-recording Using Time-sampling

Glynn et al. (1973) employed self-recording using time-sampling to increase on-task behavior with a class of second-graders. In one condition, an audio tape was prerecorded so that an audible tone sounded on a random schedule. The 30-minute lesson was divided into intervals ranging from 1–5 minutes. One of the class members was selected and instructed to make marks on the chalkboard beside the names of any of three reading groups in which all members were on-task within the 10-second period following every interval tone. In another condition, the taped signal was maintained, but each student in the class was given a 10 × 2 inch card with his or her name and 5 rows of 20 squares, 1 row for each day of the week. The students were instructed to place a check in one of the squares if they were on-task whenever the signal sounded. The students' rate of on-task behavior increased in both conditions.

Essentially the same procedure was used with three hyperactive 7–10-year-old-boys (Barkley, Copeland, & Sivage, 1980). Each of the students had a card for self-recording. On a wall facing the boys was a poster of the rules for individual work time that they had assisted in developing. The rules were (1) stay in your seat, (2) work quietly, (3) don't bug others, (4) don't space out, and (5) raise your hand if you need help. A tape recorder signaled an opportunity for the students to record their on-task behavior. At randomly programmed intervals, a bell would sound on the tape player. At this signal, each boy was to ask himself if he had been on-task as determined by the rules on the poster. If so, he was to place a check mark on his card. If not, no check was to be given.

For the first week, the bell rang at random intervals on an average of once every minute during a 30-minute work period. This was therefore a variable interval 1-minute schedule. For the second and third weeks, a VI3 min. schedule was used. Bonus points for accuracy (honesty) of recording were awarded to the students, in addition to the points earned and self-recorded, if a boy's tally agreed with an observer's record of his behavior. This procedure was successful in increasing study behavior.

The following vignette shows how one teacher used a time-sampling procedure combined with self-recording to help her students learn self-management.

Ms. Dietrich's Students Learn to Work Independently

Ms. Dietrich was a resource teacher for elementary students with learning problems. She arranged her schedule so that each group of students received direct instruction for the first 20 minutes of each scheduled resource hour. That group then worked independently while Ms. Dietrich taught another group. She was concerned that she could not give tokens to the groups working independently without disrupting the lesson she was teaching with another group. She decided to teach the students self-recording. She acquired a child's noisemaker, a "cricket." At first, she observed the students working independently and clicked the cricket only if all of them appeared on-task. They awarded themselves a point when they heard the click. After a while, she just clicked randomly during her direct teaching and told the students to give themselves a point if they were working hard when they heard the sound. She found her procedure very effective. It was almost as good as being in two places at once.

Self-recording Using Permanent Product Recording

James et al. (1977) employed self-recording using a form of permanent product recording. The students were to record the quality of their manuscript letter strokes. Students copied letters from a model sheet and used a transparent overlay to correct their work. Pretraining for the accuracy of their self-recording included proper alignment of the overlay, using the overlay to determine which letter strokes met the definition of the behavior, and using practice sheets and student samples to self-record correct letter strokes. After completing each worksheet, the students returned the model sheet and were given transparent overlays of the letters copied. Each letter they judged to conform to the letter on the transparent overlay was traced (dots connected) on the self-recording overlay to indicate those strokes they considered correct. They entered the total correct on the right edge of their self-recording sheets. The students received no feedback from the teacher on the accuracy of their self-recording, but their copying improved.

Connis (1979) used a form of permanent product self-recording with four mentally retarded adults to sequence steps for food service vocational skills, such as dishwashing. The training stations were part of a public restaurant facility. Photographs representing assigned tasks were taped in correct sequence on a wall in the trainees' work area. Below each picture was taped a small, blank square of paper. The self-recording procedure required the trainees to (1) walk to the photographs, (2) look at the appropriate photograph in sequence, (3) mark an X on the square of paper beneath the photograph, and (4) begin the task represented in the photograph with a 40-second time limit. The students learned the sequence more rapidly than with traditional procedures.

Similar monitoring or self-recording systems are used daily in the classroom and at home. For example, the teacher of a learning disabilities class used a checklist similar to the one in Figure 10–3 for a language arts activity.

In another example, a teacher who was also the mother of twin 7-year-old boys devised a self-recording checklist for morning tasks. The checklist recording system was initiated because either she or the boys were late getting out of the house every morning. So that she wouldn't have to remind the boys constantly of the tasks they needed to do first thing in the morning, the mother placed a checklist on the refrigerator (see Figure 10–4).

FIGURE 10–3
Self-recording checklist for a language arts activity

Name: _____   Date: _____

Check yourself off!

Have I capitalized the correct letters?   _____
Have I put a period at the end of sentences?   _____
Have I spaced between words?   _____
Have I put my name on the paper before handing it in to the teacher?   _____
Have I answered all the questions?   _____
Have I checked my answers in the paragraphs I read?   _____

FIGURE 10–4
Self-recording
checklist for home
use

Have I?

	Terry	Todd
1. Washed my face and hands?	—————	—————
2. Brushed my teeth?	—————	—————
3. Brushed my hair?	—————	—————
4. Picked up my dirty clothes?	—————	—————
5. Made my bed?	—————	—————
6. Put my lunch money in my pocket?	—————	—————
7. Gotten my homework and my books?	—————	—————

Two reasons for the success of self-recording procedures.

Several suggestions have been made as to why self-recording changes behavior (Hayes & Nelson, 1983; Kanfer, 1975; Rachlin, 1974). All agree that self-recording forces students to monitor their behavior. Kanfer emphasizes the role of self-reward and self-punishment, which, he states, inevitably accompany self-monitoring. Thus, a student who indicates on a chart that he has completed his daily tasks or reviews his tally of hand-raising may, by his act of self-recording, be saying to himself, "I am a good boy" (Cautela, 1971). Rachlin (1974) and Hayes and Nelson (1983) emphasize that self-monitoring provides environmental cues that increase the students' awareness of potential consequences. As the preceding examples show, self-recording must have some reinforcing qualities if there is a behavior change in the desired direction.

Self-recording does not always have reinforcing effects however. Lovitt (1973) describes a procedure used with a second-grade boy who "had the irksome habit of hitting other children when he first arrived at school" (p. 144). The boy hit other children 10–15 times in the 5-minute period allowed for hanging up coats and settling in. The boy was told to record each time he hit someone by marking it down on a tally sheet. At the end of each day, the teacher interviewed him regarding his record. His hitting behavior dropped to approximately 5 hits per 5-minute period. On the 13th day of the project, the boy tore up his tally sheet and announced he would no longer record his hits. For the 5 days following, during which the teacher continued to record data, the hit rate was 0. Lovitt notes that "recording in most of the other instances appeared to be reinforcing, whereas in this instance self-recording was apparently a punishing event. His performance was far better when he was not self-recording than when he was" (p. 144). The improvement in behavior, however, did not occur until after the self-recording activity had taken place.

But won't they cheat?

An issue often raised when considering a self-recording procedure is the accuracy of students' records. The teacher may worry that students may be untruthful. Israel (1978) has indicated that there is a higher correspondence between what a child says and then does, than between what a child does and then reports. If heightened awareness is accepted as the important functional factor in the success of self-recording procedures, however, accuracy may not matter. Indeed, Rosenbaum and Drabman

(1979) state that self-recording does not have to be accurate in order to produce desirable changes in target behaviors. Nevertheless, researchers have reported high reliability of self-recording. Santogrossi, O'Leary, Romanczyk, and Kaufman (1973) reported 95% reliability of recording among adolescent boys in a psychiatric hospital. Broden et al. (1971) reported that Liza's mean estimate of her study behavior was 81%, whereas the independent observer reported a mean estimate of 80% for her behavior. Bolstad and Johnson (1972) found that elementary-school children "were capable of self-observing their frequency of disruptive behavior with respectable accuracy" (p. 451). Glynn et al. (1973) provided mean percentages of on-task behavior, comparing the recording of eight of their second-graders and that of independent observers. Their data are presented in Figure 10–5. The general trend was that students were harder on themselves when judging the quality of their behavior than were the observers.

Increased student accuracy in self-recording has been successfully taught (Hundert & Bucher, 1978; Nelson et al., 1978; Wood & Flynn, 1978). Students received reinforcers contingent on matching their self-recording of behavior to records made by independent observers. Checking was then faded so that it occurred only occasionally. However, increased accuracy did not necessarily lead to better performance (O'Leary & Dubey, 1979).

Self-reinforcement or Self-punishment

In most classrooms, teachers arrange contingencies. They specify what behaviors are expected and the consequences for performing those behaviors. Contingencies are stated in the form of "if ..., then ..." statements: "If you complete your composition, then you may have 5 minutes extra free time in the school yard"; or "For each correct answer to the reading comprehension questions, you will earn one token." Students may be involved in the process of contingency management in a number of ways. They may be allowed to choose reinforcers, to assist in the process of determining the cost of the reinforcers in relation to behavior, or even to choose behaviors to be modified. The ultimate goal in allowing students to participate in contingency management is to encourage their use of the

	David	Clifford	Billy	Robbie	Dean	Deborah	Wayne	Chris
Student Record	93	100	86	92	65	62	67	69
Observer's Record	77	96	82	95	82	81	94	96

Mean percentages of on-task behavior recorded

FIGURE 10–5   On-task behavior recorded by students and independent observers

procedures they have been taught to manage their own behavior. As with self-recording, the transition from teacher-managed to student-managed programs must be gradual, and students must be explicitly taught to use self-reinforcement or self-punishment.

Contingencies that students set themselves may be more effective than those set by teachers

It has been repeatedly demonstrated that self-determined reinforcement and teacher-determined reinforcement can be equally effective in producing behavior change. In fact, self-determined reinforcers may sometimes be more effective than those externally determined. Lovitt and Curtiss (1969) compared the academic response rate of a 12-year-old student under a teacher-determined contingency for reinforcement with his rate under a self-determined contingency for reinforcement. The self-selected contingencies resulted in a 44% increase in academic response rate. The teacher-selected contingencies had been explained, written out, and attached to the student's desk. As he completed each academic assignment, he was shown how many responses had been made and was asked to calculate the corresponding points he had earned. The teacher had set the following reinforcement value for each assignment (p. 51):

Math	10 problems	1 min. free time
Reading		
(no errors)	1 page	2 mins.
(errors)	1 page	1 min.
Spelling	18 words	1 min.
Writing	20 letters	1 min.
Language arts	10 answers	1 min.
Library book	1 story	3 mins.
	3 questions	1 min.

Under the self-determined contingency for reinforcement, the student was asked to specify his own reinforcer in each area and to record his selections on a card that was then attached to his desk. The student selected the following reinforcers for each task:

Math	10 problems	2 mins. free time
Reading		
(no errors)	1 page	3 mins.
(errors)	1 page	2 mins.
Spelling	5 words	1 min.
Writing	10 letters	2 mins.
Language arts	10 answers	2 mins.
Library book	1 story	6 mins.
	3 questions	2 mins.

Rhode, Morgan, and Young (1983) used a self-reward system to generalize appropriate behavior from resource rooms to regular classrooms. Students were first trained to evaluate their own behavior in the resource room and then asked to do so in the regular classroom. O'Brien, Riner, and Budd (1983) successfully taught the parent of a disruptive child to implement a self-reward system and thus to teach her child to evaluate his own behavior.

When allowing students to select their own contingency for tasks to be completed, specific instructions should be given as to the procedure to be followed. The following set of instructions (Felixbrod & O'Leary, 1974) has served as a model for many investigations of self-reinforcement:

1. When people work on a job, they get paid for what they do. I am going to pay you points which you can use to buy these prizes (pointing to prizes and point-exchange values). YOUR job is to answer these arithmetic questions. Answer the questions in order. In order to earn the points, only correct answers will count (repeated). You will have 20 minutes to do these. But you can stop before 20 minutes are up if you want to . . .

2. I am going to let YOU decide how many points you want to get paid for each right answer. Take a look at the numbers on the next page (pointing to a separate page on which the subject is to choose a performance standard). I want YOU to decide how many points you want to get paid for each right answer. (Experimenter points to each possible choice in a list of 10 possible performance standards: "I want to get paid 1 point; 2 points . . . 10 points for each right answer." After I leave the room, draw a circle around the number of points you want to get paid for each right answer. (p. 846)

Several studies (for example, Felixbrod & O'Leary, 1974; Frederiksen & Frederiksen, 1975) conducted in classrooms have shown that students tend to select more lenient performance standards than those selected by teachers. Because stringent standards—whether self-determined or teacher-determined—produce significantly better academic performance than do lenient standards, students should be taught to set fairly stringent standards for themselves. A prompt for selecting stringent standards can be used successfully if it is followed by social reinforcement. The following instructional prompt has been used to encourage students to select stringent standards (Brownell, Coletti, Ersner-Hershfield, Hershfield, & Wilson, 1977):

If you pick 10 points per problem and you do 2 problems then you get 20 points. But, if you pick 2 points per problem and you get 10 problems correct you still get 20 points and the same good prize. So, if you pick a lower number you can get the same good prize just doing more problems. We would really like you to pick 1, 2, or 3 points per problem. Let's take a look at the prizes and I'll show you what I mean. (Starting with the lowest value prize describe to the child how much each prize is worth, and then concentrate on the highest.) Now, this prize costs 250 points. If you pick 1 point per problem then you have to do more than 12 pages of problems to win this prize. But, if you pick 2 points per problem then you only have to do 6 pages of problems to get the same prize. Do you understand? I'll leave the room and you circle one of these numbers. Remember, we would like you to pick 1, 2, or 3 points. (p. 445)

Start with teacher-controlled contingencies.

Contingency management systems are most often implemented with the teacher controlling the selection and administration of reinforcers. We recommend that a period of teacher-controlled contingency management precede any effort to teach students self-reinforcement. After the students become accustomed to the mechanics of the system, contingencies can more effectively be managed by the students themselves. Drabman, Spitalnik, and O'Leary (1973) and Turkewitz, O'Leary, and Ironsmith (1975) used the following successive steps to transfer gradually both recording and reinforcement responsibilities from the teacher to the students in a token economy. Desired changes in behavior were maintained throughout the transfer process to teach students self-reinforcement and finally were maintained by the students themselves. In addition, behavior changes generalized to control periods in which token reinforcement was not in effect, although student work received teacher praise.

1. Students recorded points awarded by the teacher;

2. Students were awarded bonus points for matching teacher ratings;

3. Matching was gradually faded in four phases, with successively smaller numbers of students being required to match the teacher;

4. Students rated their own behavior and determined their own reinforcement independently of the matching criterion. (Drabman et al., 1973, p. 11)

Drabman et al. (1973) suggested several factors that may have contributed to the maintenance of the desired changes in behavior:

1. Continuous teacher praise for appropriate student behavior may have become more reinforcing over the course of the investigation.

2. Peer reinforcement for appropriate behavior also increased.

3. Accurate self-evaluation was praised by the teacher and may have acquired conditioned reinforcing properties, thereby strengthening appropriate behavior.

4. Improved academic skills incompatible with inappropriate behavior were developed.

The following vignette illustrates a combined self-recording and self-reinforcing procedure.

DeWayne Passes Intro to Behavior Mod

DeWayne was panic stricken. At midterm, his average in his behavior mod course was 67. If he failed, his grade-point average would drop below the point where he would be retained in school. After listening to a lecture on self-reinforcement (remarkably similar . . .), he decided to try it on himself.

DeWayne acquired an inexpensive kitchen timer and a supply of 3 × 5 cards. He decided that he needed short-term and long-term reinforcers. He began setting the timer for an hour and making a check on a 3 × 5 card whenever he sat at his desk or in the library without getting up or talking to anyone for the entire hour. He allowed himself a 10-minute break for talking to his roommate, making a cup of coffee, or taking care of bodily necessities. When he had four checks, he decided to allow himself to go out with his friends for pizza and a soft drink. He saved the cards and when he had accumulated five cards with at least four checks, he called his girlfriend and asked her to go to a movie. Although he was somewhat skeptical about the effective-

"What do you mean, you've got another date? I didn't know until just now if I'd have enough cards to go out tonight!"

ness of his plan, he found that it worked. When he studied for an average of 4 hours a day, his grades began to improve. He finished the semester with a 3.0 grade point average and a conviction that applied behavior analysis was more than a gimmick for getting children to stay in their seats.

Self-punishment

Most self-management procedures have emphasized self-reinforcement. Some investigations, however, have analyzed the effectiveness of teaching students to punish rather than reinforce behaviors. In classrooms, the form of self-punishment most often investigated has been the use of response cost in conjunction with token reinforcement systems (Humphrey, Karoly, & Kirschenbaum, 1978; Kaufman & O'Leary, 1972).

Humphrey et al. (1978) worked in a second-grade reading class that was "initially a chaotic one as evidenced by the teacher having referred 13 of 22 children to a special program for children with behavioral problems" (p. 593). A self-reinforcement system and a self-punishment system were put into place in order to compare their relative effectiveness and also to assess potential side effects of the self-punishment. The students using reinforcement placed tokens from banks into cups kept at their desks, contingent upon accurate performance on reading assignments. Criteria for reinforcement were printed on the answer keys accompanying each reading paper. The teacher checked the work of six randomly selected students each day to minimize inaccuracy and cheating. Periodically, the teacher read aloud a list of classroom conduct rules that were also posted on a bulletin board. The students using response cost began each morning with their cups filled with the maximum number of tokens that could be earned on that day. They fined themselves for inaccurate work by referring to the response cost criteria printed on the answer keys. The response cost criteria were the converse of the reinforcement criteria; that is, if a correct answer was worth two tokens in reinforcement, an incorrect answer cost two tokens in response cost. Students also fined themselves for failing to complete the daily allotment of reading assignments. Under both types of self-management systems, the students greatly accelerated their rate of attempted reading assignments and maintained the accuracy of their work. In addition, disruptive behaviors, which were not under the contingencies, decreased slightly.

Self-reinforcement is more effective than self-punishment.

Although self-punishment (response cost) was seen as effective, self-reinforcement appeared to be slightly superior as a procedure. Several explanations for the relative superiority of self-reinforcement compared to self-punishment are possible.

1. Self-response cost may have been more difficult to use because it required children to subtract rather than add and to adjust the quantity of tokens they fined themselves daily according to the amount of work they had not attempted. Decreased persistence and effectiveness of performance would be expected if response cost, relative to self-reward, was in fact more difficult or perceived as more difficult.

2. A second explanation is the fear of failure hypothesis. Response cost may maximize fear of failure because errors must be minimized in order to avoid losing reinforcers. Response cost could, therefore, have caused children to work more slowly (cautiously). Furthermore, if response cost accentuated children's awareness of their failure, the children may have been prone to evaluate more negatively their behavior. Negative self-evaluation or selective attention to failures or inadequacies can produce performance decrements and ultimate disuse of behavior control strategies. (Humphrey et al., 1978, p. 599)

In the following vignette, observe the combination of self-recording, self-reinforcement, and self-punishment.

Professor Grundy Completes a Book

Professor Grundy was panic-stricken. He had just received a telephone call from the editor in charge of production on his textbook manuscript (which, as you may remember, was considerably overdue). Because he had heard the words *breach of contract* and *by the end of the month* used in the same sentence, he concluded that he had best accelerate his rate of writing. He decided that he needed to write at least 10 pages every day and that he would need some motivation for doing so. He had bought a personal computer to use for word processing but had spent hours playing Zap the Barbarians and other games and very little time processing words. He therefore provided himself with a supply of chocolate chip cookies, two styrofoam cups, and a number of marbles borrowed from his nephew Dennis. He figured out exactly how many bits of memory equalled a printed page. Before he sat down to work, he filled one styrofoam cup with marbles and set the other next to it. He also made a neat stack of 10 cookies. Mrs. Grundy agreed to cooperate by keeping his coffee cup constantly full. (She was motivated by the hope that the decor of her living room might eventually consist of something besides stacks of books, notecards, crumpled yellow paper, and yards of little paper strips with holes in them.)

Professor Grundy's preparations for writing had consumed almost an hour. He was tempted to take a break before beginning but restrained himself. He began putting words on the screen and at the end of each page ate 1 of the 10 cookies in his stack. Whenever he found his mind wandering, he transferred a marble from the full to the empty cup. He had decided that he would have to write an extra page for each marble over 10 that accumulated in the previously empty cup.

Professor Grundy was very pleased with the effectiveness of his self-management system. "It's amazing," he thought to himself. "I've been using applied behavior analysis procedures on other people for years and even teaching students to do it. Why didn't I ever think to try it on myself?" He spent considerable time in self-congratulation and then, with a guilty start, transferred a marble and got back to work.

Self-instruction

Self-instruction is a process of providing one's own verbal prompts. Prompting, as we discussed in Chapter 8, is necessary when discriminative stimuli are insufficient to set the occasion for the required response. Prompts are often supplied by others; self-instruction involves providing prompts for oneself. Many adults provide prompts for themselves when engaging in difficult or unfamiliar tasks. We talk ourselves through such activities as starting a new car or performing a complicated dance step. We provide prompts like "*i* before *e* except after *c*" when we encounter a hurdle in letter writing. Teaching students to use self-instruction tactics enables them also to provide verbal prompts for themselves, rather than remain dependent on the teacher.

Teaching students to engage in self-instruction helps them learn self-management of the process of completing a task, rather than academic content. Self-instruction enables students to identify and guide themselves through the process necessary to solve problems. Training in self-instruction occurs before students are given problems to solve, questions to answer, or tasks to perform. Students who are taught self-instruction procedures may be able to generalize these strategies to other settings: for example, from one-to-one tutoring to the classroom (Bornstein & Quevillon, 1976). They may also be able to generalize across tasks: for example, from arithmetic or printing tasks to a phonics task not specifically trained but requiring similar process management for completion (Burgio, Whitman, & Johnson, 1980).

Teaching students self-instruction tactics has been effective in

1. training hyperactive and impulsive children to increase attending and on-task behaviors (Barkley et al., 1980; Bornstein & Quevillon, 1976; Meichenbaum & Goodman, 1971; Palkes, Stewart, & Kahana, 1968; Peters & Davies, 1981).

2. increasing students' ability to demonstrate academic skills (Bem, 1967; Bryant & Budd, 1982; Burgio et al., 1980; Douglas, Parry, Marton, & Garson, 1976; Leon & Pepe, 1983; Robin, Armel, & O'Leary, 1975).

3. increasing appropriate social behaviors (Burron & Bucher, 1978; O'Leary, 1968).

Most studies investigating the use of self-instruction have used adaptations of a training sequence developed by Meichenbaum and Goodman (1971). A five-step training program for self-instruction was successfully employed to increase the self-control of hyperactive second-graders in order to increase their ability to attend to a task and to decrease errors. Students were taught individually using the following sequence (Meichenbaum, 1977; Meichenbaum & Goodman, 1971):

> 1. An adult model performed a task while talking to himself out loud (*cognitive modeling*);
>
> 2. The student performed the same task under the direction of the model's instructions (*overt, external guidance*);
>
> 3. The student performed the task while instructing himself aloud (*overt self guidance*);
>
> 4. The student whispered the instructions to himself as he went through the task (*faded, overt self-guidance*); and finally,
>
> 5. The student performed the task while guiding his performance via private speech (*covert self-instruction*). (Meichenbaum & Goodman, 1971, p. 32)

The cognitive model provided by the teacher and then rehearsed, overtly and covertly, by the student is shown in the following example for a task requiring the copying of line patterns (Meichenbaum & Goodman, 1971):

Cognitive modeling.

> Okay, what is it I have to do? You want me to copy the picture with the different lines. I have to go slowly and carefully. Okay, draw the line down, down, good; then to the right, that's it; now down some more and to the left. Good. I'm doing fine so far. Remember, go slowly. Now back up again. No, I was supposed to go down. That's okay. Just erase the line carefully.... Good. Even if I make an error I can go on slowly and carefully. I have to go down now. Finished. I did it! (p. 117)

For students to learn to imitate an effective and complete strategy, the teacher must include in the initial modeling several performance-relevant skills that guide the task process. These skills include

1. Problem definition ("What is it I have to do?");

2. Focusing attention and response guidance ("Carefully . . . draw the line down");

3. Self-reinforcement ("Good, I'm doing fine"); and

4. Self-evaluative coping skills and error-correction options ("That's okay.... Even if I make an error I can go on slowly."). (Meichenbaum, 1977, p. 23)

The same basic strategy has been used to teach kindergarten children with handwriting deficiencies to print the letters of the alphabet (Robin et al., 1975). The following sequence also fades from a model by a teacher to the students' performance while providing their own instructional guidance:

1. The teacher modeled correct letter copying and self-instructing aloud while the student observed;

2. The teacher modeled correct performance while the student self-instructed along with the teacher;

3. The student copied the letter while the teacher self-instructed along with the student;

4. The student copied the letter instructing himself aloud;

5. The student copied the letter whispering self-instruction. (Robin et al., 1975, p. 182)

Self-instruction modeled by the teacher in the first step of training included questions about the nature of the task, answers to these questions in the form of planning to overcome performance deficiencies, comments guiding the student's pen while in motion, correction of errors as they occurred, and self-reinforcing statements. While the teacher was consistent in presentation of the self-instruction paradigm, the students were allowed some flexibility in developing individualized patterns of self-instruction. After each correct self-instructional sequence, the students' behavior was reinforced with praise. The following examples of instructional statements illustrate those modeled by the teacher during training:

1. Question about the task: "What is it I have to do?"

2. Answer in form of planning: "I have to make a 'p'."

3. Appropriate directive comment; "I have to go down, down, slow, stop at the bottom, stop."

4. Correction of error: "No, that's not straight. I have to make a straight line, like a stick."

5. Self-reinforcement: "It's done and I made a good letter." (Robin et al., 1975, p. 182)

When working with educable mentally retarded students to increase on-task behavior during arithmetic and printing tasks, Burgio et al. (1980) rated a student as self-instructing when she or he performed any one of the following behaviors:

1. Asked a question (e.g., "What does the teacher want me to do?");

2. Answered the question (e.g., "She wants me to copy this word.");

3. Provided direction on how to do the task (e.g., "First, I should look at both numbers," "I take away 4.");

4. Reinforced himself for completing the task (e.g., "I did a good job," "I'm doing real good so far.");

5. Provided a cue to ignore distraction (e.g., "I hear people talking but I'm not gong to let them bother me.");

6. Specified how to cope with task-failure (e.g., "Oh, I was messy in printing that word. That's okay, I'll be even more careful on the next word.").

As in the previous examples, the task (arithmetic or printing) was initially modeled by the teacher while verbalizing the self-instructions. The teacher would sometimes purposely make an error and then verbalize the failure-coping statements. The teacher then said to the student, "Now it's your turn. First you add the numbers (print the letters) while I say the words." The student was then asked to verbalize the self-instructions while performing the tasks. During this step, the teacher whispered the self-instructions along with the student. Finally, the student self-instructed while the teacher remained silent.

To train impulsive, mentally retarded adolescent students to be more reflective and thus better able to finish a task, a teacher modeled the following strategy:

> Now let's see, what am I supposed to do? I have to find which of these (pointing to the six alternatives) goes into this space (pointing to the blank space in the rectangular figure). Good. Now I have to remember to go slowly and be sure to check each one of these carefully before I answer. Is this the one (pointing to the first alternative)? It's the same color but it looks different because the lines are thicker. Good, now I know it isn't this one. Now I have to check this next one (pointing to the second alternative). It looks different because there aren't any lines on it and this one (pointing to the standard) has lines on it. Good, now I know it isn't his one. Next I have to check his one (pointing to alternative three). It looks the same to me. The colors are the same and the lines are the same too. I think it might be this one but I have to check the other ones slowly and carefully before I choose . . . (continues to check remaining three alternatives one at a time). Good, I have checked them all and I have gone slowly and carefully. I think it is this one (points to the correct variant). (Peters & Davies, 1981, p. 379)

When the teacher performed the second problem, an incorrect answer was intentionally selected in order to illustrate how to cope with errors. The teacher would say

> Oops, that is not the right answer. I made a mistake. It's o.k., just be careful. I should have checked more carefully. I must follow the plan to check each one. Good. Now I'm going slowly and carefully. (Peters & Davies, 1981, p. 379)

In this case, the training sequence in self-instruction consisted of the following steps:

1. The teacher modeled two sample items.

2. Students were instructed to perform the next item themselves, doing it exactly as the teacher had done while talking out loud. Students were corrected

when their verbalization deviated from what had been modeled. The teacher provided social reinforcement for using the self-instructional technique. If students did not verbalize, the teacher modeled the strategy again, one sentence at a time, directing the student to repeat each.

3. The students were asked to continue as before, but now to whisper instead of talking out loud.

4. The students were instructed to perform the task while talking to themselves—to say the words "inside your head." Frequent reminders were given to go slowly and carefully and to check each alternative before answering. (p. 379)

In a study to develop self-control for hyperactive 7–10-year-old boys with chronic behavior problems during math and reading lessons, Barkley et al. (1980) added a group modeling process to self-instruction training. The teacher presented a problem and modeled its solution via self-instruction. She then divided the group of boys into smaller groups and presented a series of similar problems for each boy to solve using self-instruction techniques. Thus, modeling of the self-instruction was provided by both the teacher and the other boys in the group. The four steps of self-instruction used were

1. Listen to the directions;

2. Repeat the problem/question out loud; that is, repeat what you were asked to do;

3. Describe the problem in your own words and talk yourself through the solution out loud;

4. Check your answer to see if it solves the problem. (p. 379)

Four similar self-instruction steps were used in training 9-year-old boys who were under psychiatric care for hyperactive behavior (Palkes et al., 1968). With these students, visual cues were added to encourage and remind them to use the self-instruction statements. These cues consisted of four 5 × 7 inch cards printed with instructions and self-directed commands (see Figure 10–6) and kept on the student's desk in front of him during task performance.

Self-instruction can be a useful procedure for helping students become more independent and for maintaining and generalizing behavior change. Several factors appear to influence the effectiveness of self-instruction.

Factors influencing the effectiveness of self-instruction.

1. Actual implementation of the procedure during task performance.

2. The ability of the students to perform the response in question. Higa, Tharpe, and Calkins (1978) found that unless kindergarten and first-graders had practiced making a motor response, self-instructions actually interfered with performance. No amount of self-instruction will enable student to perform tasks not in their repertoire.

3. Reinforcement for adhering to self-instructions.

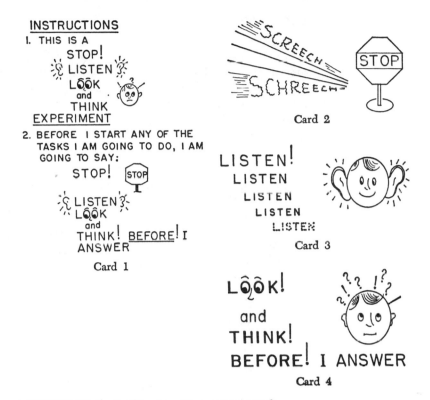

FIGURE 10–6   Self-instruction cuing cards

Note. From "Porteus maze performance of hyperactive boys after training in self-directed verbal commands" by H. Palkes, M. Stewart, & K. Kahana, *Child Development,* 1968, *39,* 817–826. Copyright 1968 by Child Development. Reprinted by permission.

(4) Making the focus of instructions specific. For example, Mishel and Patterson (1976) found that nursery school children were better able to resist talking to a puppet if they specifically instructed themselves not to talk to the clown than if they reminded themselves with general instructions to work on their assigned tasks. (O'Leary & Dubey, 1979, p. 78)

Results of self-instruction training have been somewhat inconsistent. Billings and Wasik (1985) recently failed to replicate the results of Bornstein and Quevillon's (1976) study cited earlier. Bornstein (1985) suggests that differential effects of the procedure may result from age, gender, intelligence, race, history, or attributional or cognitive style. He also suggests, "Quite simply, it appears that self-instructional programs can be effective, although obviously they are not always effective" (p. 70). As we tell our students, *"Nothing* always works."

Most self-instructional procedures are taught in combination with self-monitoring and self-reinforcement. Thus students provide for themselves both antecedents to behavior and consequences for correct performance.

SUMMARY

This chapter has described a number of techniques for transferring behavior management from teachers to students. Several advantages come with such transfer. Students become more independent. Their behavior may be maintained and generalized in settings where no intervention is in effect. The emphasis is changed from short-term intervention designed to change a single target behavior to the acquisition of strategies that may be used to change many behaviors over the long term.

Discussion of self-management procedures may seem contradictory to the emphasis on overt, observable behavior throughout this book. Some of the processes described in this chapter, such as covert self-instruction, are not directly observable. It must be pointed out, however, that the emphasis, as always, is on changes in behaviors that are observable. For example, although we cannot observe students' covert self-instruction, we can observe that they perform academic tasks more rapidly and more accurately than before they were taught to use the procedure.

REFERENCES

BARKLEY, R., COPELAND, A., & SIVAGE, C. 1980. A self-control classroom for hyperactive children. *Journal of Autism and Developmental Disorders, 10,* 75–89.

BEM, S.L. 1967. Verbal self-control: The establishment of effective self-instruction. *Journal of Experimental Psychology, 74,* 485–491.

BILLINGS, D.C., & WASIK, B.H. 1985. Self-instructional training with preschoolers: An attempt to replicate. *Journal of Applied Behavior Analysis, 18,* 61–67.

BOLSTAD, O., & JOHNSON, S. 1972. Self-regulation in the modification of disruptive classroom behavior. *Journal of Applied Behavior Analysis, 5,* 443–454.

BORNSTEIN, P.H. 1985. Self-instructional training: A commentary and state-of-the-art. *Journal of Applied Behavior Analysis, 18,* 69–72.

BORNSTEIN, P.H., & QUEVILLON, R.P. 1976. The effects of a self-instructional package on overactive preschool boys. *Journal of Applied Behavior Analysis, 9,* 179–188.

BRODEN, M., HALL, R.V., & MITTS, B. 1971. The effect of self-recording on the classroom behavior of two eighth-grade students. *Journal of Applied Behavior Analysis, 4,* 191–199.

BROWNELL, K.D., COLETTI, G., ERSNER-HERSHFIELD, R., HERSHFIELD, S.M., & WILSON, G.T. 1977. Self-control in school children: Stringency and leniency in self-determined and externally imposed performance standards. *Behavior Therapy, 8* 442–455.

BRYANT, L.E., & BUDD, K.S. 1982. Self-instructional training to increase independent work performance in preschoolers. *Journal of Applied Behavior Analysis, 15,* 259–271.

BURGIO, L.D., WHITMAN, T.L., & JOHNSON, M.R. 1980. A self-instructional package for increasing attending behavior in educable mentally retarded children. *Journal of Applied Behavior Analysis, 13,* 443–459.

BURRON, D., & BUCHER, B. 1978. Self-instructions as discriminative cues for rule-breaking or rule-following. *Journal of Experimental Child Psychology, 26,* 46–57.

CAUTELA, J.R. 1971. Covert conditioning. In A. Jacobs & L.B. Sachs (Eds.), *The psychology of private events: Perspective on covert response systems.* New York: Academic Press.

CONNIS, R.T. 1979. The effects of sequential pictorial cues, self-recording, and praise on the job task sequencing of retarded adults. *Journal of Applied Behavior Analysis, 12,* 355–361.

DEWEY, J. 1939. *Experience and education.* New York: Macmillian.

DOUGLAS, V.I., PARRY, P., MARTON, P., & GARSON, C. 1976. Assessment of a cognitive training program for hyperactive children. *Journal of Abnormal Child Psychology, 4,* 389–410.

DRABMAN, R.S., SPITALNIK, R., & O'LEARY, K.D. 1973. Teaching self-control to disruptive children. *Journal of Abnormal Psychology, 82,* 10–16.

FELIXBROD, J.J., & O'LEARY, K.D. 1973. Effects of reinforcement on children's academic behavior as a function of self-determined and externally imposed contingencies. *Journal of Applied Behavior Analysis, 6,* 241–250.

FELIXBROD, J.J., & O'LEARY, K.D. 1974. Self-determination of academic standards by children: Toward freedom from external control. *Journal of Education Psychology, 66,* 845–850.

FREDERICKSEN, L.W., & FREDERIKSEN, C.B. 1975. Teacher determined and self-determined token reinforcement in a special education classroom. *Behavior Therapy, 6,* 310–314.

GLYNN, E.L., THOMAS, J.D., & SHEE, S.M. 1973. Behavioral self-control of on-task behavior in an elementary classroom. *Journal of Applied Behavior Analysis, 6,* 105–113.

HAYES, S.C., & NELSON, R.O. 1983. Similar reactivity produced by external cues and self-monitoring. *Behavior Modification, 7,* 183–196.

HIGA, W.R., THARPE, R.G., & CALKINS, R.P. 1978. Developmental verbal control of behavior: Implications for self-instructional training. *Journal of Experimental Child Psychology, 26,* 489–497.

HUMPHREY, L.L., KAROLY, P., & KIRSCHENBAUM, D.S. 1978. Self-management in the classroom: Self-imposed response cost versus self-reward. *Behavior Therapy, 9,* 592–601.

HUNDERT, J., & BUCHER, B. 1978. Pupils' self-scored arithmetic performance: A practical procedure for maintaining accuracy. *Journal of Applied Behavior Analysis, 11,* 304.

ISRAEL, A.C. 1978. Some thoughts on correspondence between saying and doing. *Journal of Applied Behavior Analysis, 11,* 271–276.

JAMES, J.C., TRAP, J., & COOPER, J.O. 1977. Technical report: Students' self-recording of manuscript letter strokes. *Journal of Applied Behavior Analysis, 10,* 509–514.

KANFER, F. 1975. Self-management methods. In F. Kanfer & A. Goldstein (Eds.), *Helping people change: A textbook of methods.* New York: Pergamon Press.

KAUFMAN, K.F., & O'LEARY, K.D. 1972. Reward, cost, and self-evaluation procedures for disruptive adolescents in a psychiatric hospital school. *Journal of Applied Behavior Analysis, 5,* 293–309.

KAZDIN, A.E. 1975. *Behavior modification in applied settings.* Homewood, Ill.: Dorsey Press.

KNEEDLER, R.D., & HALLAHAN, D.P. 1981. Self-monitoring of on-task behavior with learning-disabled children: current studies and directions. *Exceptional Education Quarterly, 1* 73–82.

KOHLBERG, L., YAEGER, J., & HJERTHOLM, E. 1968. Private speech: Four studies and a review of theories. *Child Development, 39,* 691–736.

LEON, J.A., & PEPE, H.J. 1983. Self-instructional training: Cognitive behavior modification for remediating arithmetic deficits. *Exceptional Children, 50,* 54–60.

LOVITT, T.C. 1973. Self-management projects with children with behavioral disabilities. *Journal of Learning Disabilities, 6,* 15–28.

LOVITT, T.C., & CURTISS, K.A. 1969. Academic response rate as a function of teacher- and self-imposed contingencies. *Journal of Applied Behavior Analysis, 2,* 49–53.

MCKENZIE, T.L., & RUSHALL, B.S. 1974. Effects of self-recording on attendance and performance in a competitive swimming training environment. *Journal of Applied Behavior Analysis, 7,* 199–206.

MCLAUGHLIN, T.F. 1976. Self-control in the classroom. *Review of Educational Research, 46,* 631–663.

MEICHENBAUM, D.H. 1977. *Cognitive-behavior modification: An integrative approach.* New York: Plenum Press.

MEICHENBAUM, D.H., & GOODMAN, J. 1971. Training impulsive children to talk to themselves: A means of developing self-control. *Journal of Abnormal Psychology, 77,* 115–126.

MISHEL, W., & PATTERSON, C.J. 1976. Substantive and structural elements of effective plans for self-control. *Journal of Personality and Social Psychology, 34,* 942–950.

NELSON, R.O., LIPINSKI, D.P., & BOYKIN, R.A. 1978. The effects of self-recorders' training and the obtrusiveness of the self-recording device on the accuracy and reactivity of self-monitoring. *Behavior Therapy, 9,* 200–208.

O'BRIEN, T.P., RINER, L.S., & BUDD, K.S. 1983. The effects of a child's self-evaluation program on compliance with parental instructions in the home. *Journal of Applied Behavior Analysis, 16,* 69–79.

O'LEARY, K.D. 1968. The effects of self-instruction on immoral behavior. *Journal of Experimental Child Psychology, 6,* 297–301.

O'LEARY, S.G., & DUBEY, D.R. 1979. Applications of self-control procedures by children: A review. *Journal of Applied Behavior Analysis, 12,* 449–465.

PALKES, H., STEWART, M., & KAHANA, K. 1968. Porteus maze performance of hyperactive boys after training in self-directed verbal commands. *Child Development, 39,* 817–826.

PETERS, R., & DAVIES, K. 1981. Effects of self-instructional training on cognitive impulsivity of mentally retarded adolescents. *American Journal of Mental Deficiency, 85,* 377–382.

RACHLIN, H. 1974. Self-control. *Behaviorism, 2,* 94–107.

RHODE, G., MORGAN, D.P., & YOUNG, K.R. 1983. Generalization and maintenance of treatment gains of behaviorally handicapped students from resource rooms to regular classrooms using self-evaluation procedures. *Journal of Applied Behavior Analysis, 16,* 171–188.

ROBIN, A.L., ARMEL, S., & O'LEARY, K.D. 1975. The effects of self-instruction on writing deficiencies. *Behavior Therapy, 6,* 178–187.

ROSENBAUM, M.S., & DRABMAN, R.S. 1979. Self-control training in the classroom: A review and critique. *Journal of Applied Behavior Analysis, 12,* 467–485.

RUEDA, R., RUTHERFORD, R., & HOWELL, K. 1980. Review of self-control research with behaviorally

disordered and mentally retarded children. *Monograph in Behavioral Disorders* (Teacher Educators for Children with Behavioral Disorders), 188–197.

SAGOTSKY, G., PATTERSON, C.J., & LEPPER, M.R. 1978. Training children's self-control: A field experiment in self-monitoring and goal-setting in the classroom. *Journal of Experimental Child Psychology, 25,* 242–253.

SANTOGROSSI, D.A., O'LEARY, K.D., ROMANCZYK, R.G., & KAUFMAN, K.F. 1973. Self-evaluation by adolescents in a psychiatric hospital school token program. *Journal of Applied Behavior Analysis, 6,* 277–287.

SNOY, M.T., & VAN BENTEN, L. 1978. Self-modification technique for the control of eating behavior for the visually handicapped. *Education of the Visually Handicapped, 10,* 20–24.

STEVENSON, H.C., & FANTUZZO, J.W. 1984. Application of the "generalization map" to a self-control intervention with school-aged children. *Journal of Applied Behavior Analysis, 17,* 203–212.

SUGAI, G., & ROWE, P. 1984. The effect of self-recording on out-of-seat behavior of an EMR student. *Education and Training of the Mentally Retarded, 19,* 23–28.

TURKEWITZ, H., O'LEARY, K.D., & IRONSMITH, M. 1975. Generalization and maintenance of appropriate behavior through self-control. *Journal of Consulting and Clinical Psychology, 43,* 577–583.

WALLACE, I. 1977. (Introduction by J.J. Pear). Self-control techniques of famous novelists. *Journal of Applied Behavior Analysis, 10,* 515–525.

WOOD, R., & FLYNN, J.M. 1978. A self-evaluation token system *versus* an external evaluation token system alone in a residential setting with predelinquent youths. *Journal of Applied Behavior Analysis, 11,* 503–512.

11

Teaching Behavior-Change Procedures to Others

- While Ms. Rose was using backward chaining to teach Cynthia to put on her sweater at school, Cynthia's mother was hurriedly dressing her daughter every day without giving the girl a chance to demonstrate or practice what she knew. The mother never even knew when Cynthia had learned to dress herself at school.

- Jim, the part-time aide in Ms. Rose's classroom, spent the early days of Cynthia's dressing program trying to get Cynthia to "start at the beginning" in putting on her sweater.

- Suzie, one of Cindy's classmates, decided on her own that she would help Cindy learn to put on her sweater like Mike was helping Ronnie learn to raise his hand before speaking. Suzie had been giving Cindy some of her own tokens every time she saw Cindy put on her sweater.

Ms. Rose found that a teacher's work in changing behavior does not stop with the students. Parents, aides, and other students can be effective monitors of behavior-change programs, but they can also interfere with or obstruct those programs if they don't understand or haven't been informed. This chapter gives some guidelines for working with helpers of many kinds.

Teachers are significant agents of change in the lives of their students. They are, however, by no means the only people who influence the behavior of a particular student. Every person who comes in contact with a student affects the student's behavior in some way. The behavior of the student, in turn, changes the behavior of other people. The teacher who is determined to provide students with the best possible learning environment must be concerned with all of these interactions.

The other people whose behavior a teacher is most often concerned with include parents, paraprofessionals such as teachers' aides, and the students' peers. As you probably surmised from the opening example, these people's efforts to change behavior may either facilitate or hinder the teacher in helping students move toward academic and behavioral goals. A parent who inadvertently reinforces inappropriate behavior in his child may make such behavior virtually impervious to extinction at school. A parent's inability to manage a handicapped child at home may in some cases lead to otherwise unnecessary institutionalization of the child.

Students' learning in school is influenced by classroom personnel other than teachers. In recent years, more schools have responded to the demand for increased individualization of instruction by hiring noncertified personnel to serve as assistants or aides to teachers. Many teachers consider this a mixed blessing. Although almost always well-meaning, aides may be unaware of the needs of students and may bring to the classroom preconceptions that prevent their being helpful to students.

Peers as well are very influential in determining patterns of behavior (Patterson, Littman, & Bricker, 1967; Solomon & Wahler, 1973). This pattern is significant even with younger children but becomes especially strong when students reach adolescence. It is wasteful not to harness this influence for good. Students affect not only the behavior of their peers but that of adults as well. An understanding of the effects their behavior has on parents and teachers can enable students to be active managers of their environment.

The influence of parents, paraprofessionals, and peers can be positive as well as negative. Teachers can help these other change agents to be helpful by working with them. Sharing the principles and techniques of applied behavior analysis with others, the focus of this chapter, is one way to provide this help.

PARENTS

Few educators would disagree with the statement that parents and schools should work together. However, how often this cooperation actually occurs varies widely. Some administrators and teachers who verbally support parent involvement become hostile and defensive when parents seek information or attempt to make suggestions. Even when the involvement of parents is required, as it is for children identified as needing special education services, efforts to include parents may be perfunctory. Merely obtaining parents' signatures on various forms is hardly an ideal cooperative effort. Making parents a part of the educational process requires a good deal of effort, but it can result in much better progress for students.

Sometimes teachers' well-meaning efforts to solicit parent involvement antagonize parents. For example, many parents, even if able to do so, resent being asked to teach their children at home. Indeed, consider the very understandable reply to the following note received by a parent:

Dear Parent,
Virginia missed 10 problems on her math test. Please work on this at home.
REPLY:
Dear Teacher,
Virginia left all her clothes on the bathroom floor again last night. Please work on this at school.

If the parent and teacher, after discussing Virginia's progress, had agreed that messiness contributed to Virginia's difficulty with math, each might determine ways to help Virginia become neater, thus making her more successful both academically and socially. Some parents may be able and willing to help academically, but many feel that teaching is the responsibility of the school. How, then, can parents and schools work together? Through regular conferences, parents and teachers can cooperate in planning and in implementing behavior-change programs. Teachers may be consulted about parents' problems in managing behavior at home and may be asked to conduct training programs for parents concerned about behavior management.

Increasing concern with providing generalized behavior change is reflected in efforts to teach parents and others consistently to use ABA procedures correctly. Although these efforts have not always been successful (Forehand & Atkeson, 1977), perhaps it is time for professionals to concentrate on developing procedures that will enable parents to be effective behavior modifiers.

Planning with Parents

Parents of all young people in school should be aware of the teacher's plans for the year and, if possible, should also be included in making these plans. Parent involvement in planning is mandated by law for children in special education programs, but it can be an asset for all students. It may be difficult to reach some parents even to inform them about what is planned, but the effort should be made.

Some educators state that many parents don't seem to care about their children. This is almost invariably an unjust assumption. Parents may not respond to efforts to communicate with them for any number of reasons including a generalized avoidance reaction to schools because of aversive experiences during childhood. Many parents have demands on their time, energy, and abilities to cope that are difficult for teachers to understand. Some parents may place less value on the educational process than do teachers, but this should not be interpreted as a lack of interest in their child. After several years of listening to negative reports about their offspring, parents may develop—rather than a lack of concern abut the child—a total lack of interest in discussing her or him with teachers and other school personnel. A quick review of he discussion of punishment in Chapter 7 will make the dynamics of this situation abundantly clear.

Advantages of including parents in planning.

Including parents in the planning process has many advantages for the teacher. Parents who are familiar with educational goals, teaching procedures, and management techniques are much more likely to be supportive. For example, a special education teacher who uses a time-out procedure will receive fewer irate phone calls about shutting students in rooms by themselves if parents are familiar with the technique and its purpose. The regular classroom teacher who elects to make field trips contingent upon appropriate behavior had best make this clear to parents in advance. Any teacher who uses behavior-change procedures in the classroom, either formally or informally, should at least familiarize parents with behavioral principles.

Implementation with Parental Support

Many behaviors teachers wish to change may persist because they are, inadvertently or otherwise, reinforced outside of school. Parents may be unaware of the fact that they are reinforcing such behaviors as tantrums, whining, procrastinating, or dependence. Such behaviors are often as maladaptive at home as at school. Many parents, once made aware of the problem and provided with suggestions for changing behavior, will be of enormous support to the teacher's efforts to help students.

A word of caution is in order. Modifying the behavior of one's own child may be more difficult than modifying the behavior of other people's children. This scientifically unverified principle is illustrated in the following vignette.

"If LaDonna needs hair ribbons, let her use her own shoelaces."

The Behavior Modifier's Child Has No Shoelaces

Bas's mother teaches behavior management courses to university students and offers in-service training to teachers on the same topic. Bas attends the Campus School at his mother's university and reports to her office after school. For two years, he arrived almost every afternoon with his shoelaces untied. On several occasions, however, the shoelaces were missing entirely, having fallen out during math class while Bas was seated perfectly still at his desk. (This episode occurred in the pre-Velcro era.) Bas's mother, being an expert "behavior modifier," employed several sophisticated techniques to rectify the situation. Among the most notable of these were nagging, yelling, begging, and smacking Bas with his sneakers, which, predictably, tended to fall off when he walked fast.

Recently, however, Bas's shoelaces have been tied every afternoon. Naturally, his mother stopped mentioning shoelaces under these circumstances, until Bas asked if she wanted to know why they were tied. It seems that one of mother's colleagues, whose office Bas must pass in order to reach hers, became aware of the problem—not being profoundly deaf—and suggested that Bas might like to earn a balloon by having his shoes tied in the afternoon. Results were immediate, and the behavior has been maintained for several months on a VR10 day schedule of reinforcement.

When asking parents to use behavior-change procedures with their children, long-standing patterns of behavior may have to be broken before success will occur. Professionals must be prepared to offer support and encouragement.

Consultation with Parents

Regularly scheduled conferences with parents, supplemented by those called to deal with specific problems, provide many opportunities to share information. Parents frequently ask teachers to make suggestions about management of their children at home. Most people who regularly work with parents have found it advisable to avoid superficial advice, offering quick and easy solutions to problems. People who ask for advice frequently want only sympathy or confirmation that they are doing the most possible under the circumstances. Jong (1977) has suggested that "advice is what you ask for when you already know the answer but wish you didn't" (p. 55). A clue to the sensitive teacher is the motorboat syndrome, in which every suggestion is met with a protest: but, but, but, but, but. This should be a signal to stop making suggestions and listen for the real concern. Helpful suggestions for conducting parent conferences may be found in Price and Marsh (1985).

Clue for advice-givers.

If conferences are viewed as an opportunity to share information rather than as a vehicle merely to inform parents from a vantage of superior knowledge and wisdom, the teacher may learn a great deal about management of the child. Parents have the advantage of having known their child considerably longer than the teacher and may therefore be able to offer helpful insights into conditions precipitating inappropriate behavior and consequences that are likely to be reinforcing.

When specific suggestions about handling behavior appear to be appropriate, it is advisable to phrase such suggestions in as nontechnical a manner as possible. While the language of applied behavior analysis offers a precise means of communication among professionals, there is a fine line between precision of language and jargon. The teacher who urges parents to consider the importance of stimulus generalization in facilitating transfer across settings (when he means that the student needs a desk at home where she can do her homework) has probably crossed this line. It is possible to be precise without pedantry.

Formal Parent Training

Efforts to teach parents behavioral techniques for managing their children have been reported since 1959, when Williams documented the home-based modification of temper tantrums. A wide range of behaviors has been modified, from autistic and self-injurious behavior of severely handicapped children (Allen & Harris, 1964; Risley, 1968) to what the authors described as the brat syndrome (Bernal, Duryee, Pruett, & Burns, 1968). Parents have been trained individually in conferences and by telephone, through home visits (Johnson & Katz, 1973; Risley, 1968; Wetzel, Baker, Roney, & Martin, 1966), and through complicated laboratory procedures involving electronic cueing (Wahler, Winkel, Peterson, & Morrison, 1965). Other programs have used group training procedures (Galloway & Galloway, 1971; Lindsley, 1978; Patterson, Shaw, & Ebner, 1969). Parents have even learned to teach their children self-management strategies (O'Brien, Riner, & Budd, 1983). O'Dell (1974) reviewed the literature on parent training and concluded that such training could be successful only if attention were given to choosing optimal methods of training and to providing for generalization and maintenance of the skills learned. The parent-training program outlined in this chapter addresses both of these considerations. The following sections also discuss group organization, planning, and implementation.

ORGANIZING PARENT GROUPS

Steps in organizing a parents' group.

The first step in organizing a group of parents interested in learning behavior-change procedures is choosing a group leader or leaders. Group leaders fill several roles in parent groups. They provide basic information in the form of lecture, demonstration, discussion, and practice activities. They answer questions and provide feedback. They provide positive reinforcement for parents' efforts to learn principles and to change their children's behavior. In short, they are teachers.

Ability to teach should not, however, be the only criterion for choosing leaders. Personal characteristics may be more important than impressive credentials. A parent who has successfully completed a training program and who is perceived as warm, caring, and sympathetic may be a more effective teacher than a clinical psychologist who lectures dryly and who may be perceived as cold, superior, and unsympathetic. Of course, the same would be true of a warm, caring psychologist and a cold, unsympathetic parent. Teachers are often called upon as parent trainers and may have the advantage of familiarity with at least some of the parents and their children.

It is frequently advantageous to have two group leaders. Aside from the obvious considerations of not having to cancel sessions in case of illness or other unexpected events, two group leaders may complement one another in experience, personality, and point of view. The leaders may be a psychologist and teacher, parent and teacher, or any other combination of people who work well together. It is most important that the leaders agree on basic principles and techniques. In the event that both group leaders are professionals, their believability will be enormously enhanced if at least one of them is a parent as well. Parents are often understandably reluctant to accept suggestions about how to manage children at home from people who have never undertaken such a project.

Once group leaders have been chosen, the composition of the group can be considered. Parents should volunteer for membership; requiring parents to attend parent-training programs in order to keep their children enrolled in a given program is difficult to enforce and may result in hostile, uncooperative groups. Advertisement may encourage parents to volunteer for the initial group. If the first group of parents find the process helpful, word of mouth will encourage other parents to attend subsequent sessions. Considerable thought should be given to characteristics of the parents in a group and to the characteristics of their children. A mixed group can be successful, but we recommend "different groups for different troops" (Arnold, Rowe, & Tolbert, 1978, p. 124). There are advantages to achieving homogeneity along several criteria.

Severity of Handicap

Parents of normal children, gifted children, and mildly handicapped children often have very similar management problems. On the other hand, the needs of parents of moderately or severely handicapped children may best be met in a group of parents whose children are similarly handicapped. Parents who take turns sitting up all night with an autistic child who seems never to sleep and who leaves the house and wanders onto the expressway if left unattended may react negatively to the concern of the parents of a gifted child who won't go to bed on time because she wants to read the encyclopedia all night.

Age of Children

Management problems are very different with children of different ages. It is almost impossible to deal specifically with everything from toilet training to the hazards of drug addiction and sexual promiscuity in one group. Possibly separation into preschool, elementary, junior high or middle school, and secondary age groups would provide a more helpful setting.

Socioeconomic and Educational Level

Management of children is a problem for parents at all socioeconomic levels. Neither affluence nor an extensive educational background guarantees superior child-rearing skills. A group of parents from many sectors of the community may find surprising commonalities in their concerns about their children, but there are also significant differences. Available solutions may also be different. If the majority of the group is, for example, from an affluent suburban area, a solitary member who is less well-educated, less articulate, and noticeably less well-dressed may feel very uncomfortable. The problems faced by parents who are poor or culturally different may be very different from those of the more affluent. The best two approaches may be choosing either a group with a number of members from various socioeconomic levels or one with members from similar communities.

Single Parents

The problems encountered by single parents are in some ways unique. Divorce, separation, or death of a parent often increases children's behavior problems (Gardner, 1978; Kessler, 1966). The single parent is also faced with the necessity of managing such behavior alone. Single parents may feel that their problems are different from those of two-parent homes and that the difficulties others have are minimized by the presence of two parents to handle them. This attitude may undermine group effectiveness.

No group will ever have perfect homogeneity, whether of children or parents. We do, however, suggest considering the effects of the group's composition on its success.

The size of the group is also important. Lindsley (1978) suggested beginning with a large group, perhaps 30, anticipating some dropouts. It has been the author's experience that sometimes a smaller group, of 10 or 12, results in more commitment and fewer dropouts. It is difficult, however, to maintain a task-oriented relationship in a very small group, although 6 to 10 participants are recommended for parent counseling (Dinkmeyer & Carlson, 1973). Because the purpose of the groups we are describing is teaching rather than counseling, slightly larger groups are recommended.

PLANNING THE SESSION

In planning for parent training, the group leaders should keep in mind that the principles that affect children's learning also affect adults' learning. The following several aspects of good teaching are applicable when working with parents:

1. The arrangement of the physical environment affects behavior by providing cues for certain acts. If parents found their school years aversive, they may react to meeting in a classroom by becoming withdrawn or hostile. In contrast, if meetings are held around the fire in someone's living room, behavior such as extraneous conversation more appropriate in social occasions may occur. A businesslike, comfortable location, such as a conference room with tables and chairs, will provide an environment

most conducive to a task-oriented discussion. It is vital to remember at all times that the process underway is teaching, not counseling. Group therapy or counseling is a complex process for which teachers are not qualified. If parents need therapy or counseling, referral should be made.

2. The group leaders' verbal and nonverbal behavior at the outset will set the tone for future group interaction. A leader who is introduced as Dr. Jones, a well-known expert on child-rearing, may occasion different responses from one who is introduced as Samantha, whom some of you may remember as the hillbilly in the PTA program. This may be true even if both introductions refer to the same person. A leader dressed in jeans and sweatshirt should expect different responses from one wearing a designer suit at the extreme of fashion. The wise leader will consider the custom of the community in regard to both dress and manner of address and attempt to conform to some degree.

3. Teaching occurs only when a response is required. Telling is not teaching with adults any more than with children. Participants must have the opportunity to apply the procedures discussed and to receive feedback.

4. Positive reinforcement accelerates learning for adults as well as for children. Although changes in their children's behavior may provide some reinforcement, this may not be sufficient. Various incentives have been used—from reduced fees to trips to the hairdresser and steak dinners (Johnson & Katz, 1973). One of the authors once led a group that provided an evening's babysitting to each family that attended all sessions and completed all assignments. The sitters were high-school members of a Future Teachers of America club, who volunteered their time as part of their public service commitment. The praise and encouragement of the group leaders and other groups members is also a potential reinforcer. Positive reinforcement for group members should be directed toward increasing attendance, participation, record-keeping, and persistence in implementing difficult procedures, such as extinction, with their children.

General Format

Once leaders have been selected and the group formed, a decision must be made about how often and for how long the group will meet. There is no hard-and-fast rule about either of these decisions. The authors have found that once-a-week sessions, lasting from an hour to an hour and a half, and meeting for about 10 weeks, appear to work well. It is usually easier for parents to arrange to attend once a week than more frequently. Flexible time scheduling allows the leaders to complete the formal part of the session and allow time for individual questions that parents may not feel comfortable bringing up in front of the group. A statement such as "The group will meet from 7:30 to 8:30, but we (the leaders) will plan to be here until around 9:00 if some of you have questions" may encourage parents to bring up their individual concerns.

Each session follows the same general format:

1. Discussion of previous week's assignment
2. A teaching segment consisting of lecture, demonstration, or discussion
3. Practice activities
4. Assignment for the next week

The following sections follow this format in describing one possible 10-week course. To demonstrate the process parents may go through as they progress through the group, we shall apply each procedure to Woody, a student described in the following vignette.

Woody

Woody is an 8-year-old student who attends second grade at Parkview School. He spends about half of his school day in the regular classroom and half in a resource room for children with learning and behavior problems. Woody is making excellent progress academically in the resource room. However, he frequently fails to complete tasks, both in the resource room and in the regular class. His work is invariably messy and disorganized. A recent conference with Woody's parents revealed that they are having many problems with him at home. He often fails to come directly home from school and, on at least one occasion, was so late that his mother called the police. He dawdles and procrastinates when getting ready for school, eating meals, doing homework, and performing any task he is asked to do. He consistently leaves his clothes and toys strewn from one end of the house to another.

Woody's parents took the following 10-session parent training program.

Session 1 Pinpointing Behavior

1. *Discussion of previous week's assignment.* Because this is the first session, there will, of course, have been no assignment. Leaders should introduce themselves (after due consideration of the possible effects of such introductions) and ask parents to introduce themselves and to describe their child or children, including the children's ages and what general problems have brought the parents to the group. At this point, the advantage of the group leader's having children may be demonstrated. The leaders may model the desired behavior before asking parents to perform. A leader who has children may describe some behavior of his or her child that causes concern and then carry out behavior-change projects right along with the group members throughout the sessions.

2. *Teaching segment.* The major goal of the first session is to help participants describe behavior accurately. After introductions, one of the leaders describes the process of pinpointing in a 5- or 10-minute minilecture, using several examples.

3. *Practice activities.* Each parent is asked to refine the description of the child's problem given during the introduction. Parents should be reinforced for approximations until a behavioral description has been accomplished. It often takes considerable effort to get from

lack of respect	to	swears at mother
poor table manners	to	belches at the dinner table
irresponsibility	to	loses his lunchbox
easily frustrated	to	screams when expected to dress herself

Once participants are able to describe behavior in observable terms, they must choose a countable behavior they wish to increase. This requirement may produce a long moment of total silence; many parents, like teachers, tend to focus on behaviors they want to decrease. Behaviors appropriate to event recording should be chosen. Although some behaviors parents want to increase may require time-sampling (homework, staying in bed after bedtime) or latency recording (getting dressed, compliance), these procedures should not be introduced until later in the course.

4. *Assignment for next week.* Each parent should be provided with a chart similar to the one in Figure 11–1, which provides space for each day of the week. This simple chart may be used for behaviors that occur once a day (use a plus or minus) or for more frequently occurring behaviors (use tally marks). Each parent should write the target behavior on the chart during the session so that no confusion will occur. The assignment is to observe, count, and chart the target behavior for at least 5 consecutive days and bring the chart back for Session 2.

Session 2 Positive Reinforcement

1. *Discussion of previous week's assignment.* The beginning of this session should be devoted to discussing the charts. The trainers, as in every session, will begin by displaying their charts. Parents should be congratulated for recording the behavior and for remembering to bring charts. Parents who are present but who failed to bring materials should not receive an undue share of the trainer's attention, as is likely to happen if elaborate explanations of the reasons for failure are allowed. The majority of trainer time should be devoted to those who followed instructions. Any problems should be discussed and cleared up. Woody's chart is shown in Figure 11–2.

2. *Teaching segment.* The principle of positive reinforcement should be discussed in very simple terms. Words such as *reward* or *positive*

FIGURE 11–1
Chart for parents to record behavior

Chart 1

Name of child _____

Behavior:

Day	M	T	W	T	F	S	S

FIGURE 11–2
Woody's chart

Chart 1

Name of child _____
Behavior:

Day

M	T	W	T	F	S	S
+	−	−	+	−	◿	◿

consequence may be used if parents are likely to be resistant to the technical term. Reinforcers may be described as

a. Tangible (including money)

b. Activities

c. Praise

Although money is technically a token reinforcer, for this audience it may be grouped with tangibles. Parents should be encouraged to choose either a tangible or an activity as a reinforcer of the target behavior, and they should be urged to pair it consistently with praise. The list in Table 11–1 may be used to help parents identify a positive reinforcer. Parents should be urged to choose something that may be earned daily rather than weekly and that can be administered immediately. For example, Woody might be allowed to have a special snack, usually not available, on days when he gets home by 3:30. The major goal of this first assignment is to ensure that parents are reinforced for implementing the behavior-change project. This will occur only if the project works.

3. *Practice activities.* Demonstration and rehearsal should be used until parents are able to

a. Describe the program clearly to the child: for example, "Woody, it's important to me that you come straight home from school. On days when you do, you may have a Supercookie for a snack. The Supercookies are only for days when you are home before Robotech starts at 3:30."

b. Demonstrate positive social reinforcement: "Woody, you got home from school before 3:30. I'm proud of you. You can eat your Supercookie in the den while you watch Robotech."

c. Demonstrate a verbal procedure to use when the child fails to perform the behavior: "Woody, it's 3:45. To get a Supercookie, you have to be home before 3:30, so no Supercookie today."

4. *Assignment for next week.* The assignment for the next week is to implement the reinforcement strategy and to continue to record the behavior on the chart, which should be brought to Session 3.

TABLE 11–1 *Suggested positive reinforcers (to be used by parents)*

Preschool Children

Tangibles: any food the child likes which is limited because of concern about nutrition or dental health
small toys
activity or coloring books
pennies
Underoos® (perfect for toilet training)

Activities: television time
playing records
excursions—to library, museum, zoo, department store
helping parents with housework
games

Elementary School-aged Children

Tangibles: food, otherwise unavailable
records, tapes
toys
comic books
magazines
money

Activities: television time
late bedtime
reading in bed
sleeping on the floor with a sleeping bag
having a friend sleep over
cooking
picnics (eating peanut butter and jelly sandwich in the backyard)
excursions

Adolescents

Tangibles: money
clothing
records, tapes
something expensive and possibly unattractive (to parents) done to hair, face, or body (e.g., frizzy perm)

Activities: access to car keys (may be earned in 15-minute increments)
late curfew
getting a part-time job

Session 3 Token Systems

A "stroke" is a positive reinforcer.

1. *Discussion of previous week's assignment.* The chances of 100% success on this first project are high. Everyone will be ready to try for more changes as quickly as possible. Some parents may not have been successful, however, and will need much encouragement. Every effort should be made to determine what went wrong and to rectify the situation. Lindsley (1978) suggested the following reasons for parents' lack of success with behavior-change procedures:

a. *Projecting stroke preference*: Just because a father enjoyed playing Monopoly as a child, does not mean that the opportunity to play Monopoly will be rewarding to his son. It may even be punishing! "The child knows best."

b. *Loopholes in the welfare system*: A child would have to be stupid to change his behavior to earn a cookie if he can raid the cookiejar any time he wishes regardless. Parents need to close all leaks in the system, including indulgent grandparents. If necessary, locks can be put on cupboard or refrigerator. The behavior manager (parent) needs to control the strokes.

c. *Reaching for the unreachable star*: Idealism may be an admirable philosophy but can be frustrating to a child and his parent. Behavior goals and the intermediate steps to be rewarded need to be realistic and attainable so the child and parent can experience success.

d. *Failing to grow with the child*: An initially challenging goal can become dull and aversive as the child masters it and girds for bigger and better goals. The behavior manager needs to keep pace with the child's progress.

e. *Trying too hard or wanting too much for the child*: The parent who continually reminds the child about the reward and what must be done to earn it is not allowing the behavior mod system to work. Among other things, such a parent is contaminating the reward with an aversive stimulus—nagging. (p. 89)

2. *Teaching segment*. Once any problems have been discussed, the trainers will describe token reinforcement, relating the use of tokens to the modification of adult behavior with money.

3. *Practice activities*. Parents can now choose two or three behaviors to use on a slightly more complex chart. All should be behaviors to increase. They may or may not include the behavior from Chart 1. If that program is going well, it can be left as is. A sample chart is illustrated in Figure 11–3. The process of pinpointing may have to be described, practiced, and reinforced again.

Once target behaviors have been selected, parents should specify what tokens they wish to use. Points are the easiest to manage, but some parents may prefer to use stars, happy faces, or some other symbol. Back-up reinforcers should also be specified at the meeting. While there are advantages to including the child in this process, they are outweighed by the danger of having parents make their reinforcer menu at home. Parents sometimes have a great deal of difficulty selecting back-up reinforcers that are cheap enough to make the system work. It is important to avoid extremely long-term requirements and to make some items available at least daily. Figure 11–4 illustrates a sample reinforcer menu prepared by Woody's parents.

4. *Assignment for next week*. The assignment to be brought to Session 4 is the recording of the implemented home token system on the chart.

**Session 4
Extinction**

1. *Discussion of previous week's assignment*. Any problem with the token system should be handled at the beginning of the meeting. Woody's chart is illustrated in Figure 11–5. Changes in target behaviors, tokens received, and back-up reinforcers should be made if necessary. For the time being, parents should make such changes only after consulting with the group leaders.

2. *Teaching segment*. The extinction principle should be described in simple terms. Parents will probably agree that children do many inappropriate things "just to get attention." What it may be difficult to do is to convince them that the children will ultimately *stop* doing them if the attention is not forthcoming. The trainers must make it clear that the

Chart 2

Name _____   Dates _____
Behavior: Points:

Day	M	T	W	T	F	S.	S
Total earned							
Total spent							
Saved							
Available							

Points for bank _____

FIGURE 11–3   A chart for a home token system

**FIGURE 11–4
A sample reinforcer
menu for a home
token system**

What Woody's Points Earn

1 point	gum
5 points	make popcorn choose dinner menu
10 points	15 minutes late bedtime skip bath (no more than 20 points in one night may be spent)
25 points	pack of 3 comic books 50 ¢ for ice cream dinner at McDonald's
50 points	skating on Saturday Saturday afternoon movie
100 points	have a friend spend the night (only Friday or Saturday)

*FIGURE
11–5 Woody's point
chart*

Chart 2

Name _____   Dates _____

Behavior:	Points:
Ready to leave for school at 7:45	5
Clothes in hamper by 8:30 bedtime	5 (jeans, pants, shirt, 2 socks)
Toys in room by 8:30	5 (−1 for each toy *not* in room)

Day	M	T	W	T	F	S	S
Ready for school	5	0	5	5	5		
Clothes	4	3	4	4	5	4	5
Toys	2	3	3	2	4	2	5
Total earned	11	6	12	11	14	8	10
Total spent	5	6	0	0	0	25	5
Saved	6	1	12	11	14	0	5
Available	6	7	19	30	44	27	33

Points for bank _____

behavior will escalate at first but that persistence will pay off. Examples of behaviors frequently mentioned by parents are tantrums, crying, whining, begging, and interrupting phone conversations. Some of these behaviors are somewhat difficult to define precisely, but it can be done. Parents may be encouraged to engage in some activity incompatible to attending to the behavior, such as counting to 100, reading a magazine, running the vacuum, engaging in a conversation with someone other than the child, listening to music with headphones, or, in extreme cases, standing under the shower. Parents should also reinforce themselves when they successfully ignore inappropriate behavior.

3. *Practice activities.* It may be helpful to have parents role-play ignoring an escalating behavior. By now the members of the group will probably be

familiar enough with one another to be comfortable doing this, particularly if the session leaders demonstrate first!

4. *Assignment for next week.* The assignment is to continue the token program and to record the behavior to be decreased for 3 days before beginning extinction and for 4 days afterwards. A chart similar to that in Figure 11–1 may be used.

Session 5 Factors of Time in Changing Behavior (Latency, Duration)

1. *Discussion of previous week's assignment.* The discussion of the extinction procedure will almost certainly be lengthy. The group leaders should reinforce persistence even if no improvement has taken place. Parents should continue to record this behavior for another week.

2. *Teaching segment.* Each participant should now identify a behavior in which time is a factor, such as obeying instructions promptly, finishing some task quickly, or persisting in some activity for a specified period of time. Specific examples might include obeying within 10 seconds (slow, subvocal count—one thousand one, one thousand two, and so on), getting dressed within 20 minutes of being awakened, finishing dinner in 25 minutes, reading a library book for 15 minutes, and working on homework for 1 hour. Although we are, in reality, considering aspects of latency and duration here, parents will continue to use event recording, simply counting the instances of occurrence of the time-related behavior. This behavior will be included in the existing token economy.

Using a kitchen timer is recommended for behaviors measured in minutes rather than seconds. For example, if his mother wants Woody to finish his dinner within 25 minutes, she will set the timer for 25 minutes when Woody sits down at the table. She will record points earned if he has finished before the timer rings.

3. *Practice activities.* Rehearsal of instructions to the child and practicing pairing of praise with points would take place if there is time.

4. *Assignment for next week.* The assignment includes adding the new (time-related) behavior to the chart already in use and continuing to record the results of the extinction procedure.

Session 6 Thinning Reinforcers

1. *Discussion of previous week's assignment.* The beginning of this session should be used for a general review of progress made so far, as well as for a discussion of the previous week's progress. The trainers should provide a recapitulation of what has been taught and parents should assess the overall effectiveness of their own programs.

2. *Teaching segment.* Parents may have begun to wonder whether they will have to continue to reinforce every instance of appropriate behavior forever. This is a good time to introduce the concept of intermittent reinforcement. The leaders may suggest that parents explain to their child that one of the behaviors on the chart must be performed for 2 days in a row before points can be earned. The schedule for only one behavior should be affected this week.

3. *Practice activities.* Parents may be asked to rehearse what they will say to their child about thinning the schedule of reinforcement. Problems such as protests ("It's not fair") should be anticipated and responses to them practiced.

4. *Assignment for next week.* The assignment is to continue charting for the token economy. The recording of extinction may be dropped if success has been achieved.

**Session 7
Punishment**

1. *Discussion of previous week's assignment.* Any problems that occurred with thinning the schedule can be discussed at this time. Parents may choose to thin the reinforcement schedule for a second behavior.

2. *Teaching segment.* The function of punishment should be described. Parents may find it difficult to relate the loss of points to what they have previously attempted to use as punishers—spanking, loss of privileges, and so on. Few parents will have difficulty, however, selecting a behavior to decrease. The behavior should be written on the bottom of the chart along with the contingency: for example, Woody *loses* 5 points every time he leaves his bedroom after 8:45 P.M. It is especially important to impress upon parents that point loss is to occur only for the specific behavior described. Some parents become so taken with this procedure that they begin removing points whenever the child behaves in some way that annoys them.

3. *Practice activities.* Parents should rehearse instructions and pairing verbal punishment with removal of points: "Woody, you are not to leave your room for any reason after 8:45. That costs you 5 points." Rehearsal should include anticipated problems, such as arguing and crying, and parents should be encouraged to ignore such behavior.

4. *Assignment for next week.* The assignment is to continue charting, to thin the schedule of reinforcement for another behavior, and to implement the response cost procedure.

Session 8 Time-out

1. *Discussion of previous week's assignment.* Charts should be reviewed and any questions answered.

2. *Teaching segment.* Parents often confuse time-out with sending a child to his room. It is important to emphasize that no positive reinforcement should be available in a time-out room and that most children's rooms contain a variety of potential reinforcers. An area such as a hallway that contains nothing but a chair should be used. Time-out is also different from parents' usual practice when sending a child to her room, because it is limited to short periods, with 15 minutes after the child stops protesting suggested as a maximum.

After describing the procedure, trainers should ask parents to identify another behavior they want to decrease. Such behaviors as hitting and deliberate destructiveness are often suggested. Time-out should be reserved for *one* seriously inappropriate behavior at this point. Unfortunately, some parents who never think to apply positive reinforcement to any except specified target behaviors generalize the use of punishment procedures without the slightest hesitation.

3. *Practice activities*. Rehearsal should include pairing verbal punishment with time-out and effecting release from time-out. This practice should anticipate such problems as yelling and crying while in time-out. Refusal to go to time-out should result in physical removal; we do not suggest time-out procedures to parents whose children are so large or so aggressive as to make such physical removal impossible.

4. *Assignment for next week*. The assignment, in addition to implementing time-out, includes continuing to thin reinforcement schedules. One behavior may be removed from the chart at this time. It may be replaced by another if there are still behaviors parents wish to increase.

Session 9 Stimulus Control

1. *Discussion of previous week's assignment*. Any problems related to the time-out procedure may be discussed and suggestions offered.

2. *Teaching segment*. The description of stimulus control should focus on rules, modifications to the physical environment, and generalization. The discussion of rules should be related to the token charts. Parents have used rules when stating contingencies to their children. Parents have used rules in describing expected behavior before guests arrive, before entering a restaurant, and before making a telephone call. Parents should be told that after compliance with instructions has been consistently reinforced over time, instructions will be effective even when following them is not reinforced. Their children will have learned to "do as they are told."

Other stimulus control procedures likely to be helpful to parents include modifying the physical environment and using stimulus control to promote generalization. For example, providing plenty of places to store possessions increases neatness, as does labeling storage areas for toys and clothing. Giving the child a desk at which to do homework will increase studying, if such behavior is reinforced at school.

3. *Practice activities*. Parents should be able to think of other examples where using stimulus control would improve behavior. They may also ask their child's teacher for suggestions on arranging events at home to encourage appropriate behavioral patterns taught at school.

4. *Assignment for next week*. The assignment is to continue charting and to try at least one environmental modification.

Session 10 Evaluation

During this last session, parents have an opportunity to evaluate their progress in managing their children's behavior. The charts should provide tangible evidence of improvement. A single 10-week training program cannot be expected to change a child from a holy terror into a model person. However, dramatic changes do sometimes occur, and some parents may feel able to cope without further consultation. For those who do not feel ready, several alternatives are available.

1. Parents may re-enroll in subsequent sessions. For parents whose children present major problems, continuous enrollment may be appropriate. The feedback and reinforcement provided by weekly meetings may help such parents to persist in the effort to manage their children more effectively.

2. Individual follow-up may be provided. In any case, parents who have completed training should be encouraged to communicate with the group leaders if problems arise. Such communication may be by telephone or individual conference. Parents who continue using the token economy should share information with their child's teacher. In this way, conferences between parents and teachers can provide continuous follow-up.

3. Some parents may need referral to agencies other than the school. If family problems are so serious as to require counseling or therapy, referral may be made at this time.

4. The end of training is a good time to provide parents with a reading list. Many books are available that are suitable for or directed specifically toward parents. Bernal and North (1978) listed a number of resources. Table 11–2 includes those particularly suitable for parents, listed by level of readability.

TABLE 11–2 *Applied behavior analysis texts for parents*

Text	Reading Grade Level
Dardig, J. C., & Heward, W. L. *Sign here: A contracting book for children and their parents.* Kalamazoo, Mich.: Behaviordelia, 1976.	6
Becker, W. C. *Parents are teachers.* Champaign, Ill.: Research Press, 1971.	7
Sloane, H. N., Jr. *Behavior guide series: Stop that fighting; Dinner's ready; Not 'til your room's clean; No more whining; Because I said so.* Fountain Valley, Calif.: Telesis, 1976.	7
Smith, J. M., & Smith, D. E. P. *Child management: A program for parents and teachers.* Champaign, Ill.: Research Press, 1976.	7
Waggonseller, B. R., Burnett, M., Salzberg, B., & Burnett, J. *The art of parenting: Communication; Assertion training; Behavior management; Motivation; Behavior management: Methods; Behavior management: Discipline.* Champaign, Ill.: Research Press, 1977.	8–9
Azrin, N. H., & Foxx, R. M. *Toilet training in less than a day.* New York: Simon & Schuster, 1977.	8–9
Macht, J. *Teaching our children.* New York: Wiley & Sons, 1975.	8–9
Patterson, G. R., & Gullion, M. E. *Living with children: New methods for parents and teachers* (Rev. ed.). Champaign, Ill.: Research Press, 1976.	8–9
Alvord, J. R. *Home token economy: An incentive program for children and their parents.* Champaign, Ill.: Research Press, 1973.	10–12
Christopherson, E. R. *Little people: Guidelines for common sense rearing.* Lawrence, Kan.: H & H Enterprises, 1977.	10–12
Rettig, E. B., *ABC's for parents.* Va Nuys, Calif.: Associates for Behavior Change. 1973.	13–16

PARAPROFESSIONALS, AIDES, AND VOLUNTEERS

Many classroom teachers, faced with the necessity for providing individualized instruction for as many as 35 children having a wide range of aptitude, academic achievement, and social behavior, feel that another pair of hands in the form of a teacher's aide would provide the solution to all their problems. While the advantages of such classroom help are obvious, unless classroom helpers are carefully chosen, trained, and supervised, their presence may prove more of a hindrance than a help.

Selection of Classroom Helpers

In many cases, the selection of classroom aides is the responsibility of someone other than the teacher. The helper who is a paid paraprofessional will probably be hired through the usual personnel procedures of the school system and presented to the teacher without prior consultation. In cases where teachers have some choice in the matter, they will naturally be concerned with the prospective helper's general knowledge and specific skills. However, when selecting a person who will spend several hours a day in the room with the teacher and students, the teacher should consider several purely subjective questions:

1. Is the potential aide a pleasant person with whom the teacher will enjoy interacting?

2. Does the candidate have mannerisms the teacher finds irritating?

3. Does the candidate show obvious signs of thinking he or she knows more about teaching than the teacher?

It may seem peculiar to emphasize personal characteristics in forming what is, after all, a professional relationship. However, any teacher who has spent 6 hours a day isolated from other adult contact with someone whose voice grates on the ear is aware of the devastating effect such a situation can have on the ability to teach.

The teacher who has no paid helper may, with the approval of the school administration, explore various sources of possible volunteer aid. Wiederholt, Hammill, and Brown (1978) suggest community service organizations, such as the Junior League, Kiwanis, and Lions' Clubs. Brighton (1972) describes the use of older students as classroom volunteers. Parents may also be classroom aides, either in their own child's class—if one goal is to increase generalization from school to home—or in classes other than their child's, as is most often appropriate. Some special education programs have extensive parent volunteer programs organized and coordinated by the parents themselves. High-school students can be recruited, trained, and placed for an hour a day in elementary classrooms if at least part of a staff member's time is devoted to coordinating of such a program. Students from nearby universities may be required to spend time in classrooms as part of their training.

There are several kinds of helpers whose presence in classrooms, unless carefully managed, will do more harm than good. Unfortunately, such people seem often to be drawn to work with children, particularly handicapped children. The following examples describe three "unhelpful" helpers who are familiar to the authors (and many other teachers).

The Lover of Handicapped Children

Anyone who states that he or she just loves handicapped children should be immediately suspect. Such a person may express love by positive reinforcement of inappropriate behavior. If a woman thinks that handicapped children are cute, for example, she may encourage dependency, immature behavior, and helplessness and thus prevent the child from becoming less handicapped. Bateman (1971) has described this attitude as comparable to that of a person who, confronted by a broken toaster, states that there is nothing wrong with it, that it is just different from other toasters in that it doesn't toast bread. "But it does lend a decorative touch to the kitchen and we should accept it and cherish it for what it does do" (p. 70). Of course we value every child, handicapped or not, but the teacher's goal should always be to minimize the effects of the handicap by setting goals that will help each student become the least handicapped individual possible. The following vignette illustrates what may happen when an aide reinforces dependent behavior.

Ms. Jackson Sets Things Straight

Mark is a fourth grader identified as learning disabled who attends a special education resource room three times a week for instruction in mathematics. He is making excellent progress and has almost reached grade level in math skills. He participates enthusiastically in small-group instruction and particularly enjoys demonstrating a problem on the board and receiving his teacher's praise for doing it correctly. However, when he is expected to work independently while the teacher instructs another group, Mark sits and cries until the aide, Ms. Jackson, sits down next to him and helps him do the assignment. Mark almost never completes an assignment independently. The teacher has asked Ms. Jackson not to help Mark, explaining that her attention reinforces his crying and that Mark will never be able to function in the regular classroom unless he can work independently. Ms. Jackson's answer was that, while she might not understand all those charts and graphs, she did know that the poor little thing needed love. Ms. Jackson insisted that the teacher would understand things better if she had children of her own.

The Frustrated Therapist

The second variety of unhelpful helper is the one who really wants to be a psychiatrist or psychologist. This seems most often to be a person who has recently acquired an undergraduate degree in psychology and has discovered its dubious usefulness as vocational training. This helper, too, may reinforce inappropriate behavior because he or she feels that students should "do their own thing" and not be required to meet standards set by other people.

The Handicapped Helper

Occasionally, a teacher may be presented with an aide who is so intellectually limited or so emotionally disturbed as to make it very difficult for the person to be helpful. One teacher has described such a situation as very much like having one more student in the program.

Even these three potentially harmful types of aides can often be taught to be effective classroom helpers. Careful planning and explicit teaching are a necessity, however, even with helpers whose entering skills are much better.

Role Description

Effective utilization of classroom aides depends to a great extent upon the explicit description of expectations, clear statement of responsibilities, and precise description of the task to be undertaken. Not all components of teaching can or should be undertaken by nonprofessionals. Smith, Krouse, and Atkinson (1961) suggest that the diagnostic/prescriptive process, including selection of the content or skill to be taught, the materials to be used, and the evaluation of effectiveness must remain the responsibility of the teacher. However, a number of instructional and noninstructional roles may be filled by classroom aides. Appropriate assignments include but are not limited to

1. Direct instruction under the guidance of the teacher.
2. Observation and recording of academic and social behavior.
3. Preparation of instructional materials.
4. Supervision of children as they work independently.

The role definition of any particular classroom helper will depend on a number of factors, such as the characteristics of the students in the class and the special skills or talents of the aide. The important thing is that the definition be explicit and known to all concerned.

Training Classroom Helpers

Any person who is hired or who volunteers as a classroom aide should receive some preservice training. Such training may be limited, because of time or economic considerations, but should include at least a description of the characteristics of the students to be served, the objectives of the program, the principles upon which teaching techniques are based, and of course, a description of the expectations the helper is to fulfill. A training program following the outline for parents presented earlier may be used, if time is available. When people will be working in the classroom, however, it is important to familiarize them with the vocabulary of applied behavior analysis and with any more elaborate observational procedures that may be used. Training should include emphasis on stimulus control, particularly in terms of all adults' providing students with the same instructions, cues, prompts, and so on.

Preservice training will almost certainly be carried out by someone other than the classroom teacher. The teacher's task is to take the aide—with whatever personal characteristics, training, expectations, and goals he or she brings to the situation—and teach that person to become a productive member of the team responsible for meeting the educational needs of each student in the classroom.

Orientation

The initial interaction between teacher and teacher's aide is of paramount importance in establishing a good working relationship. In most cases, the teacher and the aide will have some time before the students arrive, whether several days at the beginning of the school year or just a half hour before the buses arrive. Such time should be used to familiarize the aide with the physical arrangement of the school and the classroom. Mundane considerations, such as expectations for signing-in, storage locations for personal items during the day, dress regulations or conventions, and areas where smoking and eating are permitted, can be handled at this time. The aide should also be introduced to other faculty members and to such vital personnel as the school secretary and the maintenance staff. The aide has the right to be treated as a colleague (Shank & McElroy, 1970), and elementary courtesy is a part of this right. The teacher's behavior during this first orientation will permanently affect future relations with the aide. A common mistake made by teachers is that, in an effort to put a new aide at ease, they fail to make clear that, while certainly a colleague, the aide is expected to defer to the teacher's decisions regarding educational planning and procedures. Such well-meant statements as "Just do things your way" or "I'm sure you know more about some of these things than I do" may well come back to haunt the teacher who makes them.

The teacher who follows the suggestions in Chapter 12 regarding structure and organization of the classroom will now reap yet another benefit of such preparation. It is much easier to orient a classroom helper to a classroom whose rules, schedules, expectations, and procedures are explicitly stated for the students. The teacher who manages a classroom using less explicitly stated procedures may have considerable difficulty teaching someone else to fit into the plan. The aide should become familiar with the overall organization of the classroom and with the plan for each student. The more information aides have, the more effective they will be.

Once the students arrive, the aide should be introduced. Few teachers would advocate making a distinction as to authority in such introductions. It is, however, almost inevitable that the students, unless severely handicapped, will know very soon who the real teacher is, just as they always seem to be aware—no matter how hard we try to fool them—which reading group is the top and which the bottom. This should cause no problem, as long as the teacher and the aide are consistent in their expectations and support one another.

The teacher may use many of the same procedures in training the aide as in teaching students. Specific techniques applicable in this situation are

1. *Modeling:* The aide should have an opportunity to observe the teacher before being expected to perform most tasks. Such observation, combined with an explanation of the specific procedure, the reasons for its use, and its effect, provide an excellent teaching mechanism.

2. *Instructions:* The instructions the teacher gives to the aide must be as clear as those given to students. The aide who is asked to "Help Harold with his math" may interpret this to mean anything from offering encouragement from across the room to doing the problems for Harold. If an aide is asked to check on Harry every 5 minutes and give him a bonus point for each problem done correctly, that aide knows exactly what to do. Providing such instructions in writing may help the aide to remember and follow them, as well as provide a mechanism for evaluation and feedback.

3. *Feedback:* Aides, like all other learners, need to know how they are doing. Feedback may take the form of positive reinforcement, when the aide has done something the teacher wants to see repeated, or correction, when a mistake has been made. Correction should always be combined with a suggestion about how to handle the situation in the future and should always occur privately.

If, in spite of best efforts, a teacher finds it impossible to teach an aide effectively, the aide's presence has become a liability. The source of the difficulty, particularly if the aide is older than the teacher, may be that the teacher finds it impossible to assert authority. If the aide is an older person, she or he may be a discriminative stimulus for obedience to the young teacher. Teachers who find themselves in such situations may profit from assertiveness training. A simple introduction to the concept of assertiveness may be found in Alberti and Emmons (1970).

We have emphasized the negative aspects of having a helper in the classroom because most teachers are unprepared for possible problems. Most helpers are an asset. They provide an opportunity for students to learn from more than one adult, thus increasing the likelihood of generalization of new patterns of behavior. They bring to the classroom new skills, abilities, and interests to be shared with teacher and students. They allow the teacher to be more flexible in planning so that, for example, a student may receive one-to-one attention if needed. Having an aide may force a teacher to analyze and possibly improve teaching procedures. The teacher who is asked why he always ignores Harold's talking out but always answers Tyrone may not even have been aware of the inconsistency. Finally, many things happen in a classroom that can be shared when two adults are present—the frustration when a student seems not to be making progress, the joy when he does, and humorous events that never seem quite so humorous when repeated to someone not involved. Having an aide means never having to say, "I guess you had to be there."

TEACHING STUDENTS TO MODIFY BEHAVIOR

The influence of students on the behavior of one another is recognized and often lamented by teachers. Indeed, much of the lamentation is justified: the behaviors most frequently reinforced by peers are those least desirable to teachers (Patterson, Littman, & Bricker, 1967; Solomon & Wahler, 1973). Some strategies for interfering with this process, such as modeling and the use of group contingencies, have been discussed in previous chapters. This section discusses the use of peers as direct change agents for one another.

That peers can observe, record, and reinforce behavior systematically has been well documented (Patterson, et al., 1969; Surratt, Ulrich, & Hawkins, 1969). Even elementary students can be effective behavior modifiers, dispensing tokens and praise contingent on appropriate academic and social behavior. Solomon and Wahler (1973) implemented a program combining social reinforcement and extinction. Sixth graders observed and recorded appropriate behavior of a peer and changed the peer's behavior using only differential attention. Axelrod, Hall, and Maxwell (1972), after failing to help a teacher reinforce appropriate academic behavior in a disruptive student, had great success by enlisting a peer. Strain, Shores, and Timm (1977) trained preschool children to modify the social behavior of their peers.

The use of peer-managed behavior-change strategies is particularly useful to the regular classroom teacher who may identify only a few students whose behavior is not controlled by the normal contingencies, such as grades, social reinforcement, or activities. The behavior of such children may be managed using a token system administered by a peer and may require investment of very little time by the teacher once training has taken place. The choice of peers is crucial, however. They must be sufficiently responsible to undertake the project and to follow through, but care should be taken to avoid choosing Goody Two-Shoes as the behavior manager. Such a student may be resented by the target students.

The process of teaching students to manage their peers' behavior should be as simple as possible. Only the specific procedures to be used need to be explained. The teacher should be certain that the peer manager can identify, count, and record the target behavior(s) and that he or she understands the contingency in effect. It is helpful if positive reinforcement for the manager can be built into the program; if not, some means of providing such reinforcement should be found. The following example illustrates a problem that might be solved using a peer-management procedure.

Tyrone

Tyrone's seventh-grade English teacher is about to refer him for special education services. Although he is clearly of at least average intelligence, he never completes an assignment, seldom brings pencil, paper, or texts to class, is frequently late, and sometimes skips class altogether. He is failing English and appears interested only in the part of the class concerned with literature. He recently astounded his teacher by confiding that he and his friend Jake were rewriting a selection from the literature book as a play and planned to produce it as a neighborhood project. Jake is an excellent student who often bails Tyrone out by providing pencils, paper, and excuses. He has been heard admonishing Tyrone to "get his act together."

Tyrone's teacher might handle this situation by appointing Jake as behavior manager. Behaviors such as attending class, being on time, bringing materials, and completing assignments could earn points toward time in class to work on the play or even to present it. The teacher's role would be only to explain the program to Jake and Tyrone and to monitor it occasionally.

Teaching Students to Modify Adults' Behavior

Bandura (1975) points out the unilateral direction of most plans for behavior change: some people are identified as deviant, and systematic attempts to alter their behavior are made by those who are putatively normal. The actual situation is a little more complicated. In any given social system, every individual's behavior is subject to modification by others (Tharp & Wetzel, 1969). While teachers are modifying their students' behavior, students—whether they or their teachers are aware of it—are modifying the teachers' behavior. The same is true of the social system in the home; everyone modifies the behavior of everyone else.

Berberich (1971) systematically investigated the effects of students' behavior on their teachers and established that a simulated student's responses changed a teacher's administration of tangible and social reinforcers and punishers, as well as the teacher's motor behavior. Sherman and Cormier (1974) found that fifth graders could be taught to change their teacher's behavior in a positive direction. Even college professors' behavior can be modified by their students (Whaley & Malott, 1968).

In perhaps the most helpful study for teachers, Graubard, Rosenberg, and Miller (1974) trained junior-high students to identify teacher-pleasing behaviors, such as asking questions, nodding wisely, and complimenting teachers on their explanations. They demonstrated that students who had been labeled as deviant could significantly increase their classroom teachers' positive responses to them by emitting higher rates of teacher-pleasing behaviors. With the current emphasis upon educating mildly handicapped students in regular classrooms, such skills can greatly increase handicapped students' acceptability. Teachers who want to give students the most powerful possible tools for success will not omit training in behavior modification. Such teachers may even find that a student trained in this way will begin consciously and systematically to reinforce the teacher's behavior, particularly when they are providing interesting instruction and high rates of positive social reinforcement. Thus, as teachers train class members to be better students, the students train the teachers to be better teachers.

SUMMARY

This chapter brings us full circle. Having already considered techniques for changing student behavior, we discussed ways to help other adults to change children's behavior and ways to teach children to change adults' behavior.

No teacher can expect to do everything for students. Each teacher must be prepared to teacher others to help students and ultimately to help the students themselves become effective managers of their environment.

REFERENCES

ALBERTI, R.E., & EMMONS, M.L. 1970. *Your perfect right: A guide to assertive behavior.* San Luis Obispo, Calif.: Impact.

ALLEN, K.E., & HARRIS, F.R. 1964. Elimination of a child's excessive scratching by training the mother in reinforcement procedures. *Behaviour Research and Therapy, 4,* 70–84.

ARNOLD, L.E., ROWE, M., & TOLBERT, H.A. 1978. Parents' groups. In L. E. Arnold (Ed.), *Helping parents help their children.* New York: Brunner/Mazel.

AXELROD, S., HALL, R.V., & MAXWELL, A. 1972. Use of peer attention to increase study behavior. *Behavior Therapy, 3,* 349–351.

BANDURA, A. 1975. The ethics and social purposes of behavior modification. In C. M. Franks and G. T. Wilson (Eds.), *Annual review of behavior therapy, theory and practice* (Vol. 3, pp. 13–20). New York: Brunner/Mazel.

BATEMAN, B.D. 1971. *The essentials of teaching.* Sioux Falls, S. Dak.: Adapt Press.

BERBERICH, J.P. 1971. Do the child's responses shape the teaching behavior of adults? *Journal of Experimental Research in Personality, 5,* 92–97.

BERNAL, M.E., DURYEE, J.S., PRUETT, H.L., & BURNS, B.J. 1968. Behavior modification and the Brat Syndrome. *Journal of Consulting and Clinical Psychology, 32,* 447–455.

BERNAL, M.E., & NORTH, J.A. 1978. A survey of parent training manuals. *Journal of Applied Behavior Analysis, 11,* 533–544.

BRIGHTON, H. 1972. *Handbook for teacher aides.* Midland, Mich.: Pendell Publishing.

DINKMEYER, D., & CARLSON, J. 1973. *Consulting: Facilitating human potential and change processes.* Columbus, Ohio: Charles E. Merrill.

FOREHAND, R., & ATKESON, B.M. 1977. Generality of treatment effects with parents as therapists: A review of assessment and implementation procedures. *Behavior Therapy, 8,* 575–593.

GALLOWAY, C., & GALLOWAY, K.C. 1971. Parent classes in precise behavior management. *Teaching Exceptional Children, 3,* 120–128.

GARDNER, R.A. 1978. Guidance for separated and divorced parents. In L.E. Arnold (Ed.), *Helping parents help their children.* New York: Brunner/Mazel.

GRAUBARD, P.S., ROSENBERG, H., & MILLER, M.B. 1974. Student applications of behavior modification to teachers and environments or ecological approaches to social deviancy. In R. Ulrich, T. Stacknick, & J. Mabry (Eds.), *Control of human behavior,* vol. 3. Glenview, Ill.: Scott, Foresman & Co.

JOHNSON, C.A., & KATZ, R.C. 1973. Using parents as change agents for their children: A review. *Journal of Child Psychology and Psychiatry, 14,* 181–200.

JONG, E. 1977. *How to save your own life.* New York: Signet.

KESSLER, J.W. 1966. *Psychopathology of childhood.* Englewood Cliffs, N.J.: Prentice-Hall.

LINDSLEY, O. 1978. Teaching parents to modify their children's behavior. In L.E. Arnold (Ed.), *Helping parents help their children.* New York: Brunner/Mazel.

O'BRIEN, T.P., RINER, L.S., & BUDD, K.S. 1983. The effects of a child's self-evaluation program on compliance with parental instructions in the home. *Journal of Applied Behavior Analysis, 16,* 69–79.

O'DELL, S. 1974. Training parents in behavior modification: A review. *Psychological Bulletin, 81,* 418–433.

PATTERSON, G.R., LITTMAN, R., & BRICKER, W. 1967. Assertive behavior in children: A preliminary outline of a theory of aggressive behavior. *Monograph of the Society for Research in Child Development, 32,* no. VI, 1–43.

PATTERSON, G.R., SHAW, D.A., & EBNER, M.J. 1969. Teachers, peers and parents as agents of change in the classroom. In A.M. Benson (Ed.), *Modifying deviant social behaviors in various classroom settings.* Eugene: University of Oregon, No. 1.

PRICE, B.J., & MARSH, G.E., II. 1985. Practical suggestions for planning and conducting parent conferences. *Teaching Exceptional Children, 17,* 274–278.

RISLEY, T.R. 1968. The effects and side effects of punishing the autistic behaviors of a deviant child. *Journal of Applied Behavior Analysis, 1,* 21–34.

SHANK, P.C., & MCELROY, W. 1970. *The paraprofessional or teacher aide.* Midland, Mich.: Pendell Publishing.

SHERMAN, T.M., & CORMIER, W.H. 1974. An investigation of the influence of student behavior on teacher behavior. *Journal of Applied Behavior Analysis, 7,* 11–21.

SMITH, E.W., KROUSE, S.W., & ATKINSON, M.M. 1961. *The educator's encyclopedia.* Englewood Cliffs, N.J.: Prentice-Hall.

SOLOMON, R.W., & WAHLER, R.G. 1973. Peer reinforcement control of classroom problem behavior. *Journal of Applied Behavior Analysis, 6,* 49–56.

STRAIN, P.S., SHORES, R.E., & TIMM, M.A. 1977. Effects of peer social initiations on the behavior of withdrawn preschool children. *Journal of Applied Behavior Analysis, 10,* 289–298.

SURRATT, P.R., ULRICH, R., & HAWKINS, R.P. 1969. An elementary student as a behavioral engineer. *Journal of Applied Behavior Analysis, 2,* 85–92.

THARP, R.G., & WETZEL, R.S. 1969. *Behavior modification in the natural environment.* New York: Academic Press.

WAHLER, R.C., WINKEL, G.H., PETERSON, R.F., & MORRISON, D.C. 1965. Mothers as behavior therapists for their own children. *Behaviour Research and Therapy, 3,* 113–134.

WETZEL, R., BAKER, J., RONEY, M., & MARTIN, M. 1966. Outpatient treatment of autistic behavior. *Behaviour Research and Therapy, 4,* 169–177.

WHALEY, D.L., & MALOTT, R.W. 1968. *Elementary principles of behavior.* Kalamazoo, Mich.: Behaviordelia.

WIEDERHOLT, J.L., HAMMILL, D.D., & BROWN, V. 1978. *The resource teacher: A guide to effective practices.* Boston: Allyn & Bacon.

WILLIAMS, C.D. 1959. The elimination of tantrum behavior by extinction procedures: Case report. *Journal of Abnormal and Social Psychology, 59,* 269.

12
Putting It All Together in the Classroom

"Oliver," said Mrs. Grundy as she was cooking breakfast, "have you noticed the improvement in your disposition? I don't know whether it's the jogging or the fact that the book is finished, but you certainly are cheerier."

"Minerva," said the Professor patiently, "I've told you many times that behaviorists don't use words like *disposition*. You can make statements about my behavior without using hypothetical constructs. I don't think you'll ever understand applied behavior analysis!"

"Perhaps not," answered Mrs. Grundy, "but it's nice to see you smiling."

As Professor Grundy drove to his office, he considered Minerva's remarks. He admitted to himself that she was right; even his colleagues had noticed a change. He arrived at the university and greeted Ms. Cadwallader cheerfully. On the way to the office, he picked up his mail and noticed that among the advertising material, catalogs, and requests from school systems for recommendations of former students, there was a single hand-addressed letter. He sat down at his desk and opened it.

Dear Professor Grundy:

I'm not sure that you will remember me, but you were the supervisor of my student teaching. I just got a job in the Quarry School System as a seventh grade remedial math teacher and I'm really worried. I'm afraid I didn't always listen very carefully to the things you said in behavior mod class, and I'm not sure exactly how to go about setting up my classroom so it will work. The suggestions you gave me in student teaching were so helpful that I'm sure you can tell me everything I need to know. Could you possibly jot down a few suggestions?

Sincerely yours,
Sandra Harper

Professor Grundy remembered Ms. Harper very well. In the past he might have thrown her letter away or, at most, written suggesting that she review her notes, but he decided "Why not?" and headed for the word processor.

My dear Ms. Harper:

Of course I remember you. I was so pleased with your performance as a student teacher—even though you got off to a rather shaky start. I'll be happy to suggest some ways to use applied behavior analysis procedures in your classroom, but you must remember that the ultimate goal must be for you to understand the principles and apply them yourself. You should review your text and notes from the class as well as reading what I have written here.

The most important part of running an efficient classroom takes place before the students arrive. You are wise to be thinking now about how you want your classroom to function. You must first plan a system for recording and documenting student progress. Particularly with remedial groups, you should plan to determine each student's level of mastery at the beginning of the year. There are probably materials available in your school to do this. If not, you must make your own tests. You can then write objectives and monitor each student's progress toward mastery. Such records can be kept in your grade book, in a notebook, or posted in the classroom. You should also plan for documenting changes in students' nonacademic behaviors— being on time to class, bringing materials, complying with instructions. Once you have accomplished this task, you can plan to arrange the antecedent stimuli in your classroom to provide cues for appropriate behavior and task completion. If you use stimulus control effectively, you will have many fewer problems to contend with. You can't prevent all problems, but you can certainly minimize them.

Students whose academic or social behavior is inappropriate or maladaptive may respond to the entire school environment as an S^D for either acting-out behavior or complete withdrawal. If you want to design a classroom environment to alter such behaviors, you must simultaneously avoid presenting old S^Ds and concentrate on establishing certain elements of the classroom as discriminative stimuli for appropriate behavior. This process may be thought of as providing appropriate classroom *structure*.

All classrooms have rules. Whether or not expectations for student behavior are made explicit, they exist. The wise teacher shares expectations for classroom behavior with students—it is difficult for students to follow rules if they don't know what those rules are. Many students follow rules; for them, rules have, by past experience, become S^Ds for compliance. The students whose behavior upsets the teacher are those who do not follow the rules, who are disruptive, or who fail to complete assigned tasks. For such students, merely specifying what is expected has little effect on behavior (Madsen, Becker, & Thomas, 1968); for them complying with rules is not under stimulus control. You must teach these students to follow rules. Adherence to the following rules about rules will facilitate this process. rules will facilitate this process.

1. Be very specific about what is expected. (Specificity)

2. Make as few rules as possible. (Economy)

3. Be explicit about the relationship between rules and consequences. (Consequences)

*"Now, don't forget to
SAVE, professor."*

Specificity

Effective rules describe behavior that is observable. An unequivocal
decision can be made about whether or not an effective rule has been
followed. Some teachers, in an effort to anticipate every situation, make
rules that are so vague as to be useless. These rules make it difficult to
decide whether a student has been a good citizen, has respected others'
rights, or has done unto others as he would have others do unto him.
Following such rules is also difficult—what does a good citizen do? What
are others' rights? Had I not better do unto him before he does unto me?
You must decide what behaviors are important in your classroom and
describe these behaviors in a set of rules. If being a good citizen means
"Complete assignments" or "Don't hit people," say that, not "Be a good
citizen."

In general, it is preferable to specify which behaviors are desired rather
than which ones are forbidden. However, some teachers, in an effort to be
completely positive, invent convoluted rules that are not at all clear. "Keep
your hands to yourself" is a poor substitute for "Don't hit people," if such a
rule is needed. A teacher once asked me for help in phrasing a rule. She
had been told that all rules must be stated positively and needed a rule
about spitting. We concluded that "Keep your saliva in your mouth" lacked
the impact and pellucid clarity of "Don't spit." Sometimes it's all right to tell
students what *not* to do.

Economy

Making too many rules is inefficient. Neither the teacher nor the students can remember a large number of rules. If you make 88 rules in an effort to anticipate every possible situation, you may find that students are challenged to find the 89th inappropriate behavior. "But there's no rule against it." This is also the primary disadvantage of allowing students to make the rules. They always seem to want to make dozens (and to invoke the death penalty for noncompliance).

Sometimes, unnecessary rules give students ideas. If on the first day of school you say to your seventh graders, "Do not bring straws back from the cafeteria. I will not have students in my classroom dampening the ends of the paper covers and blowing them upward so that they stick to the ceiling," you can expect them to do just that. For this reason, rules about hitting (especially about hitting the teacher) and other drastic forms of misbehavior should not be made unless there is incontrovertible evidence that they are necessary.

Furthermore, enforcing unnecessary rules is time-consuming. It would be interesting to collect data on the amount of time in the average classroom spent dealing with the "problem" of chewing gum. I know that you have heard (or participated in) interchanges like this:

Teacher: Are you chewing gum?

Student (swallowing rapidly): Who me?

or this:

Teacher: Are you chewing gum?

Student: No, Ms. Franklin.

Teacher: Open your mouth. Let me see! You *are* chewing gum! Spit it out this minute! I want to see it in the trashcan!

Classroom rules should concentrate on behaviors necessary for efficient instruction. Making rules into effective S^Ds is often a difficult process. Why do it unless the rule is necessary?

Consequences

For students whose behavior is not under the control of rules, systematic efforts must be made to establish a relationship between following rules and positive reinforcement or punishment. Prompts may be needed. The teacher may

1. Post the rules on a bulletin board in either written or pictorial form, thus providing a visual prompt.

2. State the rules at the beginning of each class period or ask a student to read them, thus providing a verbal prompt.

3. Draw attention to a student who is following rules, thus providing a model.

When the rules are clear and in force, the teacher must provide consistent reinforcement for following rules. Systematic efforts like these to bring rule-following under stimulus control will succeed even if students do not comply with rules in other settings, either at school or at home. This teacher and this classroom will have become S^Ds for following rules.

The physical arrangement of the classroom also provides discriminative stimuli for students. Some of the behaviors occasioned by these stimuli may be undesirable. For some students, proximity to other students (or a particular classmate) may be an S^D for off-task talking or physical abuse. Sitting in a chair may be an S^D for tilting it. The teacher may need to arrange the classroom in a way that interferes with such behaviors before more appropriate behavior can be taught.

Gallagher (1979), Haring and Phillips (1962), Hewett and Taylor (1980), and Stephens, Hartman, and Lucas (1978) have all described the relationship between classroom arrangement and student behavior. Specific suggestions from these sources include

1. Provision for easy teacher observation of all students.

2. Sufficient physical separation of students to minimize inappropriate behaviors.

3. Careful delineation of areas in which only work behaviors are reinforced from those in which more informal behavior is permitted.

4. Availability of study carrels for students so distractible that virtually any stimulus is an S^D for off-task behavior.

The physical presence of the classroom can and will, in itself, become a complex S^D for appropriate student behavior. It is important to be consistent about what behaviors are reinforced and about room arrangement. Allowing students with behavior problems to play in the work area on a rainy day may result in a weakening of the stimulus control of that area. Changing the room arrangement suddenly may also create problems for these students. In a study examining the establishment of stimulus control in the animal laboratory, Rodewald (1979) found that even cleaning pigeons' cages disrupted stimulus control. Subtle changes in the classroom may have similar effects on students.

The time of day functions as an S^D for many human behaviors. At certain times, certain behaviors are reinforced. A predictable schedule for the classroom will maximize the controlling aspects of time. Gallagher (1979) suggests prompting students by posting pictures of clock faces showing the time for scheduled activities near the classroom clock.

Other aspects of time should also be considered. There is evidence (Van Houten, Hill & Parsons, 1975; Van Houten & Thompson, 1976) that students work more quickly when their performance is timed. Apparently, being timed is an S^D for working rapidly.

A simple, relatively inexpensive kitchen timer can be of enormous help in structuring classroom time. A timer that produces a ticking noise and a single, clear bell-tone is available in department or discount stores and may be used in many ways. Individual students may be assigned a timer during individual work periods and instructed to "beat the clock." The timer may also be used to delineate segments of the class day. Students may be taught that when the timer is ticking, no talking or moving around is allowed. The ticking thus becomes an S^D for working independently and quietly. The ring becomes an S^D for stopping work and waiting for further instructions or beginning another activity. For teachers who have no aides and who work with small groups within the classroom, the students may be taught that a person who is not in the group receiving instruction may not talk to the teacher except in the case of dire emergency (the fire is out of control, or there is more than a quart of blood on the floor). It appears that the S^D, "No talking to the teacher while the timer is going," is prompted by the ticking sound. I have trained both normal and handicapped students of all ages to attend to a timer. As a matter of fact, the timer is ticking now—it is an S^D for writing, as well as an S^Δ for interruptions by colleagues. The ring is a conditioned reinforcer—coffee break time.

The teaching materials used with students also occasion responses. For many students with learning and behavior problems, traditional materials such as textbooks, workbooks, and ditto sheets may be S^Ds for a variety of inappropriate behaviors. Often, the best materials to use with such children are those that look most different from what has been used in the past. Even tearing pages out of workbooks (Gallagher, 1979) or presenting only a part of a page that has been cut into several pieces (Hewett & Taylor, 1980) may provide enough difference so that inappropriate responses will not occur. Doing academic work in more traditional forms may gradually be shaped. A great deal of creativity is required to provide some students with assignments that occasion appropriate responding. It is well worth being creative—a group of students doing long division as fast as they can are bound to be less disruptive than a group doing nothing.

The kind of reinforcers and punishers you decide to use depend on your students. I doubt that you'll want to use primary reinforcers with seventh-graders but it's very possible that students in remedial math may need more than grades to motivate them. You might want to consider a point system with points based on being in class on time, bringing materials, completing assignments, and behaving appropriately. Back-up reinforcers might include free time, contests or competitions, exemptions from homework as well as grades. You may want to consider a whole class or team contingency so that students encourage one another to work hard and behave well. You will certainly use social reinforcers as well—students of all ages respond to praise (so do professors). A word of warning—most adolescents are embarrassed by effusive praise in front of their peers. Keep it low key.

With all of your planning and systematic positive reinforcement, you should have very few students who fail to complete assignments or who become disruptive. When this happens, however, you must be prepared. You may use a response cost procedure—fines or demerits that result in loss of points or other consequences. You can use time-out by placing a chair behind a book case or screen. Some teachers assign students to write or copy essays on the subject of their misbehavior. This is certainly preferable to having students write "I must not throw spit-balls in Ms. Harper's class" 100 times. You may use overcorrection procedures—the student who throws spit-balls may be required to clean the classroom floor of all papers and other trash.

There may be a few students who do not respond to your efforts and whom you may have to refer to the administrator in charge of discipline. Try to avoid doing this regularly—it says to your students (not to mention the administrator) that you do not have control over your classroom. Solve as many problems as you can within your own classroom.

Encourage students to be responsible for their own behavior. Teach them to be self-managers and you will give them tools they can use throughout their lives.

My goodness, Ms. Harper, I got carried away—my text file is full, the timer rang an hour ago, and it's time for my run. One last word—keep your sense of humor intact. Take your job seriously, but not yourself. Remember that nothing that a 12-year-old can say or do to you diminishes your worth or dignity one scrap unless you let it.

The best of luck to you—have fun!

Sincerely yours,
Oliver Grundy

As Professor Grundy began the second mile of his daily run, he began to think about the recently completed semester. On the whole, it hadn't been too bad: his book was finished, Ms. Harper had finally completed her student teaching and got a job, DeWayne made the Dean's List, and he himself was in better shape than he had been in years, in spite of a drastically increased rate of cookie consumption. "I've learned a lot, too," he thought as he ran. "I think I finally understand what Professor Peltry was trying to tell me about applied behavior analysis when I was a student. It's not just something to think about using when someone's behavior annoys me. It's a way of looking at human behavior that helps me teach. I taught Ms. Harper to observe behavior, I taught DeWayne to write a behavioral objective, and I taught myself how to stick to a job until it was finished."

He entered his house in a state of high excitement. "Minerva," he panted, "I've finally figured out what applied behavior analysis is all about and now I think I can explain it so you'll understand! It's not just avoiding hypothetical constructs. It's not . . . behavior problems. . . . It's about *teaching*."

Mrs. Grundy looked at him quizzically. "Is that so, Oliver?" she replied. "That's nice. When you've finished showering, I'd really like for you to come and help me move all these books out of the living room. If we can get it done quickly, I'll have time to make a chocolate mousse for dessert tonight."

REFERENCES

GALLAGHER, P.A. 1979. *Teaching students with behavior disorders: Techniques for classroom instruction.* Denver, Colo.: Love Publishing.

HARING, N.G., & PHILLIPS, E.L. 1962. *Educating emotionally disturbed children.* New York: McGraw-Hill.

HEWETT, F.M., & TAYLOR, F.D. 1980. *The emotionally disturbed child in the classroom: The orchestration of success.* Boston: Allyn & Bacon.

MADSEN, C.H., BECKER, W.C., & THOMAS, D.R. 1968. Rules, praise and ignoring: Elements of elementary classroom control. *Journal of Applied Behavior Analysis, 1,* 139–150.

STEPHENS, T.M., HARTMAN, A.C., & LUCAS, V.H. 1978. *Teaching children basic skills: A curriculum handbook.* Columbus, Ohio: Charles E. Merrill.

VAN HOUTEN, R., HILL, S., & PARSONS, M. 1975. An analysis of performance feedback system: The effects of timing and feedback, public posting, and praise upon academic performance and peer interaction. *Journal of Applied Behavior Analysis, 8,* 435–448.

VAN HOUTEN, R. & THOMPSON, C. 1976. The effects of explicit timing on math performance. *Journal of Applied Behavior Analysis, 9,* 227–230.

Glossary

AB design   A single-subject experimental design. The AB design has two phases: baseline (A) and treatment (B). This design cannot demonstrate a functional relationship between dependent and independent variables.

abscissa   The horizontal or *X-axis* of a graph. The time dimension (sessions) is represented along the abscissa.

acquisition   The basic level of student response competence. It implies the student's ability to perform a newly learned response to some criterion of accuracy.

alternating conditions design   See *alternating treatments design.*

alternating treatments design   A single-subject experimental design that allows comparison of the effectiveness of two or more treatments. It differs from other single-subject designs in that treatments (sometimes including baseline) are alternated randomly rather than being presented sequentially. (Also known as *multiple schedule design, multi-element design, alternating conditions design*).

anecdotal report   A written form of continuous data recording that provides as much information about the behavior and environmental surrounding as possible. The events contained in the report are then sequenced, identifying each behavior, its antecedent, and its consequence.

antecedent stimulus   A stimulus that precedes a behavior. This stimulus may or may not serve as discriminative for a specific behavior.

aversive stimulus   A stimulus that decreases the rate or probability of a behavior when presented as a consequence; as such, it is a type of punisher. Alternatively, an aversive stimulus may increase the rate or probability of a behavior when removed as a consequence; as such, it is a negative reinforcer.

back-up reinforcer   An object or event received in exchange for a specific number of tokens, points, etc.

backward chaining   The procedure for teaching a chain of behaviors by teaching the last response in the chain first, then the next to last, and so on until the student can perform the entire chain of responses as a single complex behavior after a single instruction.

bar graph   A graph that employs vertical bars rather than horizontal lines to indicate levels of performance (also called a *histogram*).

baseline data   Data points that reflect an operant level of the target behavior. *Operant level* is the natural occurrence of the behavior before intervention. Baseline data serve a purpose similar to that from a pretest, to provide a level of behavior against which the results of an intervention procedure can be compared.

behavior   Any observable and measurable act of an individual (also called a *response*).

behavioral objective   A statement that communicates a proposed change in behavior. A behavioral objective must include statements concerning the learner, the behavior, the conditions under which the behavior will be performed, and the criteria for evaluation.

chaining   An instructional procedure that reinforces individual responses in sequence, forming a complex behavior.

changing conditions design   A single-subject experimental design that involves successively changing the conditions for response performance in order to evaluate comparative effects. This design does not demonstrate a functional relationship between variables.

changing criterion design   A single-subject experimental design that involves successively changing the criterion for reinforcement. The criterion is systematically increased or decreased in a stepwise manner.

conditioned aversive stimulus   A stimulus that has acquired secondary aversive qualities through pairing with an unconditioned aversive stimulus, such as pain or discomfort.

conditioned reinforcer   A stimulus that has acquired a reinforcing function through pairing with an unconditioned or natural reinforcer; includes most social, activity, and generalized reinforcers (also called a *secondary reinforcer*).

consequence   Any stimulus presented contingent on a particular response.

continuous behavior   A behavior with no clearly discriminable beginning and ending.

continuous schedule of reinforcement (CRF)   A schedule of reinforcer delivery that rewards each correct response. The ratio between responses and reinforcement is 1:1.

contracting   Placing contingencies for reinforcement (if. . .then. . .statements) into a written document. This creates a permanent product that can be referred to by both teacher and student.

controlled presentations   A variation of event recording. A method of predetermining the number of opportunities to respond. This method often involves presenting a specific number of trials per instructional session.

cumulative graph   A graph on which the number of occurrences of behavior observed in a given session is added to the number of occurrences of previous sessions in order to derive the data points to be plotted.

dependent variable   The behavior to be changed through intervention.

deprivation state   A condition in which the student has not had access to a potential reinforcer.

differential reinforcement of incompatible behavior (DRI)   Reinforcing a response that is topographically incompatible with a behavior targeted for reduction.

differential reinforcement of lower rates of behavior (DRL)   Delivering reinforcement when the number of responses in a specified period of time is less than or equal to a prescribed limit. This maintains a behavior at a predetermined rate, lower than at its baseline or naturally occuring frequency.

differential reinforcement of other behaviors (DRO)   Delivering reinforcement when the target behavior is not emitted for a specified period of time. Reinforcement is contingent upon the nonoccurrence of a behavior.

directionality   A distinctive ascending or descending trend of data plotted on a graph.

discrete behavior   A behavior with a clearly discriminable beginning and ending.

discrimination   The ability to differentiate among stimuli or environmental events.

duration recording   Recording the amount of time between the initiation of a response and its conclusion; an observational recording procedure.

educational goal   A statement providing the framework for planning an academic year or an entire unit of learning. It sets the estimated parameters of anticipated academic and social development for which educators are responsible (also called a *long-term objective*).

errorless learning   An instructional procedure that arranges S^Ds and prompts to occasion only correct responses.

event recording   Recording a tally or frequency count of behavior as it occurs within an observation period; an observational recording procedure.

extinction   Withholding reinforcement for a previously reinforced behavior in order to reduce the occurrence of the behavior.

fading   The gradual removal of prompts to allow the S^D to occasion a response independently.

fixed-interval schedule (FI)   See *interval schedule of reinforcement.*

fixed-ratio schedule (FR)   See *ratio schedule of reinforcement.*

fixed-response duration schedule (FRD)   See *response-duration schedule of reinforcement.*

fluency   The second level (after acquisition) of student competence. Fluency describes the rate that students accurately perform a response.

frequency   The number of times a behavior occurs during an observation period.

functional relationship   A quasicausative relationship between the dependent and independent variables. This relationship is said to exist if the dependent variable systematically changes in the desired direction as a result of the introduction and manipulation of the independent variable.

generalization   Expansion of a student's capability of performance beyond those conditions set for initial acquisition. *Stimulus generalization* refers to performance under conditions—that is, cues, materials, trainers, and environments—other than those present during acquisition. *Maintenance* refers to continued performance of learned behavior after contingencies have been withdrawn. *Response generalization* refers to changes in behaviors similar to those directly treated.

generalization conditioned reinforcer   A reinforcer associated with a variety of behaviors or with access to a variety of other primary or secondary reinforcers; may simply be called *generalized reinforcer.*

group designs   Experimental investigations that focus on data related to a number of individuals.

histogram   See *bar graph.*

independent variable   The treatment or intervention that the experimenter manipulates in order to change a behavior.

intermittent schedules of reinforcement   Schedules in which reinforcement follows some, but not all, correct or appropriate responses or follows when a period of appropriate behavior has elapsed.

These schedules include *ratio, interval,* and *response-duration schedules of reinforcement.*

interobserver agreement   See *reliability.*

interval recording   An observational recording system. An observation period is divided into a number of short intervals. The observer counts the number of intervals when the behavior occurs rather than instances of the behavior.

interval schedule of reinforcement   A schedule for the delivery of reinforcers contingent upon the occurrence of a behavior following a specified period or interval of time. In a *fixed interval (FI) schedule,* the interval of time is standard. For example, FI5 would reinforce the first occurrence of behavior following each 5-minute interval of the observation period. In a *variable interval (VI) schedule,* the interval of time varies. For example, VI5 would reinforce the first response that occurs after intervals *averaging* 5 minutes in length.

latency recording   Recording the amount of time between the presentation of the S^D and the initiation of a response.

limited hold   A procedure used with interval schedules of reinforcement that restricts the time during which the reinforcer is available.

long-term objective   See *educational goal.*

maintenance   The ability to perform a response over time, even after systematic applied behavior procedures have been withdrawn.

modeling   Demonstrating a desired behavior in order to prompt an imitative response.

multi-element baseline design   See *alternating treatments design.*

multiple baseline design   A single-subject experimental design in which a treatment is replicated across (a) two or more students, (b) two or more behaviors, or (c) two or more settings. Functional relationships may be demonstrated as changes in the dependent variables that occur with the systematic and sequenced introduction of the independent variable.

multiple schedule design   See *alternating treatments design.*

negative practice   Massed or exaggerated practice of an inappropriate behavior. Decreased occurrence results from fatigue or satiation.

negative reinforcement   The contingent removal of an aversive stimulus immediately following a response. Negative reinforcement *increases* the future rate and/or probability of the response.

nonseclusionary time-out   A time-out procedure wherein the student is not removed from the instructional setting in which reinforcers are being dispensed. The teacher denies access to reinforcement and manipulates the environment to signal a period of time during which access is denied.

observational recording systems   Methods of data collection used to record aspects of behavior while it actually occurs (*event recording, interval recording, time sampling, duration recording,* and *latency recording*).

operant conditioning   The arrangement of environmental variables to establish a functional relationship between a voluntary behavior and its consequences.

operational definition   Providing concrete examples of a target behavior. This minimizes disagreements among observers as to the behavior's occurrence.

ordinate   The vertical or *Y-axis* of a graph. The amount or level of the target behavior is represented along the ordinate.

overcorrection   A procedure used to reduce the occurrence of an inappropriate behavior. The student is taught the appropriate behavior through an exaggeration of experience. There are two forms of overcorrection. In *restitutional overcorrection* the student must restore or correct an environment he has disturbed to its condition before the disturbance. The student must then improve it beyond its original condition, thereby overcorrecting the environment. In *positive-practice overcorrection* the student, having behaved inappropriately, is required to engage in exaggerated practice of appropriate behaviors.

pairing   Simultaneous presentation of primary and secondary reinforcers in order to condition the secondary reinforcer. Once the association has been established, the secondary reinforcer takes over the reinforcing function, and the primary reinforcer is no longer necessary.

permanent product recording   Recording tangible items or environmental effects that result from a behavior; for example, written academic work (also called *outcome recording*).

pinpointing   Specifying in measurable, observable terms a behavior targeted for change.

PLACHECK   A data recording system, similar to time-sampling, in which several students are observed at the end of each interval.

positive-practice overcorrection   See *overcorrection*.

positive reinforcement   The contingent presentation of a stimulus immediately following a response, which increases the future rate and/or probability of the response.

Premack principle   A principle stating that any high-probability activity may serve as a positive reinforcer for any low-probability activity (also called *activity reinforcement*).

primary reinforcers   A stimulus (such as food) that may have biological importance to an individual; such stimuli are innately motivating (also called *natural, unlearned, unconditioned reinforcers*).

prompt   An added stimulus that increases the probability that the S^D will occasion the desired response (also known as *supplementary antecedent stimulus*).

punisher   A consequent stimulus that decreases the future rate and/or probability of a behavior.

rate   The frequency of a behavior during a defined time period:

$$\text{rate} = \frac{\text{frequency}}{\text{time}}$$

ratio graphs   Graphs on which all data are plotted at rate per minute.

ratio schedules of reinforcement   A schedule for the delivery of reinforcers contingent upon the number of correct responses. In a *fixed-ratio (FR) schedule,* the number of appropriate responses required for reinforcement is held constant. For example, FR5 would reinforce every fifth appropriate response. In a *variable-ratio (VR) schedule,* the number of appropriate responses required for reinforcement varies. For example, VR5 would reinforce on the average of every fifth appropriate response.

ratio strain   A disruption of response performance that follows when the schedule of reinforcement has been thinned so quickly that the ratio between correct responding and reinforcement is too great to maintain an appropriate rate of responding.

reinforcer   A consequent stimulus that increases or maintains the future rate and/or probability of occurrence of a behavior.

reinforcer sampling   Allowing students is to come in contact with potential reinforcers. Reinforcer sampling allows teachers to determine which reinforcers are likely to be effective with individual students. It also allows students to become familiar with previously unknown potential reinforcers.

reliability   The consistency of data-collection reports among independent observers. The coefficient of reliability is determined by the formula

$$\frac{\text{agreements}}{\text{agreements} + \text{disagreements}} \times 100$$

respondent conditioning   The process of pairing stimuli so that an unconditioned stimulus elicits a response. Most such responses are reflexive; they are not under voluntary control.

response   See *behavior*.

response cost   Reducing inappropriate behavior through withdrawal of specific amounts of reinforcer contingent upon the behavior's occurrence.

response-duration schedules of reinforcement   Schedule for the delivery of reinforcers contingent upon how long a student engages in a continuous behavior. In a *fixed-response-duration (FRD) schedule,* the duration of the behavior required for reinforcement is held constant. For example, FRD10 minute would deliver reinforcement following each 10 minutes of appropriate behavior. In a *variable-response-duration (VRD) schedule,* the amount of time required for reinforcement varies. For example, VRD10 minute would deliver reinforcement following an average of 10 minutes of appropriate behavior.

response generalization   Unprogrammed changes in similar behaviors when a target behavior is modified.

response prompt   A teaching procedure in which a student is assisted to respond until the response is under stimulus control.

restitutional overcorrection   See *overcorrection*.

reversal design   A single-subject experimental design that removes a treatment condition after intervention in order to verify the existence of a functional relationship. This design has four phases: baseline, imposition of treatment, removal of treatment (also known as *return to baseline*), and reimposition of treatment (also called *ABAB design*).

S^D See *stimulus control*.

S-delta (S^Δ)   See *stimulus control*.

satiation   A condition that occurs when there no longer is a state of deprivation.

schedules of reinforcement   The patterns of timing for delivery of reinforcers (see *interval* *schedules, ratio schedules,* and *response-duration schedules).*

seclusionary time-out   A time-out procedure that removes the student from the instructional setting as the means of denying access to reinforcement.

secondary reinforcer   A stimulus that is initially neutral but acquires reinforcing qualities through pairing with a primary reinforcer (also called a *conditioned reinforcer*).

self-instruction   The process by which a student provides verbal prompts to himself or herself in order to direct or maintain a particular behavior.

self-recording   Data collection on one's own behavior (also called *self-observation, self-evaluation,* or *self-monitoring*).

self-reinforcement (self-punishment)   Administering consequences to oneself. Students may be taught to select reinforcers (or punishers), determine criteria for their delivery, and deliver the consequences to themselves.

shaping   Teaching new behaviors through differential reinforcement of successive approximations to a specified target behavior.

single-subject designs   Experimental investigations in which each individual serves as his own control (see *AB designs, alternating treatments design, changing conditions design, changing criterion design, multiple baseline design,* and *reversal design*).

social reinforcers   A category of secondary reinforcers that includes facial expressions, proximity, contact, privileges, words, and phrases.

stimulus control   The relationship in which an antecedent occasions behavior or serves as a cue for the behavior to occur. Repeated occurrences of the behavior depend upon its being reinforced. An antecedent that serves as an appropriate cue for occasioning a response and therefore results in reinforcement is known as a *discriminative stimulus* (S^Δ). An antecedent that does not serve as an appropriate cue for occasioning a response and therefore does not result in reforcement is known as an *S-delta* (S^Δ).

stimulus generalization   See *generalization.*

stimulus overselectivity   A tendency to attend to only one or a few aspects of a stimulus rather than the stimulus as a whole.

stimulus prompt   An alteration of a stimulus to increase the probability of correct responding, often used in errorless learning procedures.

task analysis   The process of breaking down a complex behavior into its component parts.

thinning   Making reinforcement gradually available less often or contingent upon greater amounts of appropriate behavior.

time-out   Reducing inappropriate behavior by denying the student access, for a fixed period of time, to the opportunity to receive reinforcement.

time sampling   An observational recording system in which an observation period is divided into equal intervals; the target behavior is observed at the end of each interval.

topography   The physical form or description of a motor behavior.

trend   A description of data represented on a graph. An ascending or descending trend is defined as three data points in a single direction.

trial   A discrete opportunity for occurrence of a behavior. A trial is operationally defined by its three behavioral components: an antecedent stimulus, a response, and a consequating stimulus. The delivery of the antecedent stimulus marks the beginning of the trial, and the delivery of the consequating stimulus signifies the termination of the trial.

unconditioned aversive stimulus   A stimulus that results in physical pain or discomfort to an individual (also called *universal, natural,* or *unlearned aversive stimulus*).

variable   Attributes unique to the individual involved in the study or to conditions associated with the environment of the study.

variable-interval schedule (VI)   See *interval schedule of reinforcement.*

variable-ratio schedule (VR)   See *ratio schedule of reinforcement.*

variable-response-duration schedule (VRD)   See *response-duration schedule of reinforcement.*

Author Index

Subject Index